Methods in Molecular Biology

Series Editor
John M. Walker
School of Life and Medical Sciences
University of Hertfordshire
Hatfield, Hertfordshire, AL10 9AB, UK

For further volumes:
http://www.springer.com/series/7651

Chromothripsis

Methods and Protocols

Edited by

Franck Pellestor

Unit of Chromosomal Genetics, Department of Medical Genetics, CHU Montpellier, Montpellier, Hérault, France

Humana Press

Editor
Franck Pellestor
Unit of Chromosomal Genetics
Department of Medical Genetics
CHU Montpellier
Montpellier, Hérault, France

ISSN 1064-3745 ISSN 1940-6029 (electronic)
Methods in Molecular Biology
ISBN 978-1-4939-7779-6 ISBN 978-1-4939-7780-2 (eBook)
https://doi.org/10.1007/978-1-4939-7780-2

Library of Congress Control Number: 2018935180

Printed on acid-free paper

This Humana Press imprint is published by the registered company Springer Science+Business Media, LLC part of Springer Nature.
The registered company address is: 233 Spring Street, New York, NY 10013, U.S.A.

Preface

Initially described in cancer genomes, and subsequently observed in constitutional disorders, chromothripsis constitutes a new class of massive genomic alterations characterized by the simultaneous shattering of chromosome segments followed by random reassembly of the DNA fragments during a single cellular event, to form complex derivative chromosomes.

The identification of this unanticipated catastrophic phenomenon has deeply modified our perception of the genesis and the etiology of complex genomic rearrangements and has provided new insights on cellular and molecular mechanisms for genomic instability and the role of genome maintenance pathways.

Several nonexclusive causative mechanisms have been proposed to explain the diversity and the high complexity of chromothripsis events. In addition to the important question of how chromothripsis arises and how cells can manage such a genomic crisis and survive, chromothripsis is also of great interest for clinicians who must now consider how it might be exploited in the clinical perspective.

Since its discovery, advances in molecular cytogenetic technologies, in microarrays, and in next-generation DNA sequencing, as well as in bioinformatics and computer modelling, have led to the introduction of efficient methods to characterize and quantify chromothripsis. In parallel, in vitro experimental models have been developed to simulate the chaotic process of chromothripsis and validate its existence.

The aim of this first book focused on chromothripsis in the *Methods in Molecular Biology* series is to present the chromothripsis phenomenon and to provide an up-to-date view on methods and experimental approaches developed to identify and explore chromothripsis.

The book is divided into five parts: (I) Genesis and Impact of Chromothripsis, (II) The Identification of Chromothripsis, (III) The Causal Mechanisms of Chromothripsis, (IV) The Bioinformatics Tools for Chromothripsis Analysis, and (V) The Experimental Models.

The first part brings together chapters that review the genesis of chromothripsis and its biological impact.

Part II presents methods ranging from classical FISH techniques to RNA-seq analysis and topologically associating domain (TAD) mapping for the identification of chromothripsis events, not only in humans but also in plants.

Part III focuses on the potential causes of chromothripsis. This section includes the description of a variety of techniques used for studying mechanisms driving the chromothripsis process.

Part IV is devoted to bioinformatics tools and web services, which allow the automatic annotation of chromothripsis-like patterns and the integrative analysis of large-scale genomic data.

Finally, Part V addresses experimental systems recently developed for the in vitro investigation of chromothripsis. The methodologies described involve high-resolution imaging in live cells, micronucleus formation, and induction of genome chaos.

Most of the chapters provide practical guidance, step-by-step instructions, and helpful notes, based on the experience of well-versed researchers. Beyond the detailed description of methods, this book also provides a synthesis on the emergence of chromothripsis and its clinical and biological consequences in the form of updated and relevant reviews.

This book is intended to be useful not only for cell biologists, molecular biologists, cytogeneticists, and geneticists investigating chromothripsis but also for students and researchers new to the field of chromothripsis and genomic instability.

The creation of this book would not have been possible without the dedicated involvement of all the authors and coauthors who have graciously contributed to these chapters. I deeply express my gratitude to each one of them for their commitment to this project and their great contributions.

I am also grateful to Professor John M. Walker, the series editor, for his support, suggestions, and encouragement during the whole process of editing this book.

Montpellier, Hérault, France ***Franck Pellestor***

Contents

Contributors

JÉRÔME AUDOUX · *Institut de Médecine Régénératrice et de Biothérapie, INSERM U1183, CHU Montpellier, Montpellier, France; Institut de Biologie Computationnelle, Université de Montpellier, Montpellier, France*

NATHALIE BASTIEN · *Laboratoire d'Anatomopathologie et de Cytologie, Laboratoires médicaux de la Capitale Nationale et des Îles, site IUCPQ-UL, Québec, QC, Canada*

ANNE-LAURE BOUGÉ · *Institut de Médecine Régénératrice et de Biothérapie, INSERM U1183, CHU Montpellier, Montpellier, France; Institut de Biologie Computationnelle, Université de Montpellier, Montpellier, France*

HAOYANG CAI · *Center of Growth, Metabolism, and Aging, Key Laboratory of Bio-Resources and Eco-Environment, College of Life Sciences, Sichuan University, Chengdu, Sichuan, China*

SHAWN L. CHAVEZ · *Division of Reproductive and Developmental Sciences, Oregon National Primate Research Center, Beaverton, OR, USA; Department and Physiology and Pharmacology, Oregon Health and Science University School of Medicine, Portland, OR, USA; Department of Obstetrics and Gynecology, Oregon Health and Science University School of Medicine, Portland, OR, USA*

SIXUN CHEN · *Lee Kong Chian School of Medicine, Nanyang Technological University, Singapore, Singapore; A*STAR Institute of Molecular and Cell Biology, Singapore, Singapore*

LUCA COMAI · *Genome Center & Department of Plant Biology, University of California, Davis, CA, USA*

THÉRÈSE COMMES · *Institut de Médecine Régénératrice et de Biothérapie, INSERM U1183, CHU Montpellier, Montpellier, France; Institut de Biologie Computationnelle, Université de Montpellier, Montpellier, France*

KAREN C. CRASTA · *Lee Kong Chian School of Medicine, Nanyang Technological University, Singapore, Singapore; A*STAR Institute of Molecular and Cell Biology, Singapore, Singapore; School of Biological Sciences, Nanyang Technological University, Singapore, Singapore; Department of Medicine, Imperial College London, London, UK*

BRITTANY L. DAUGHTRY · *Department of Cell, Developmental and Cancer Biology, Oregon Health and Science University School of Medicine, Portland, OR, USA; Division of Reproductive and Developmental Sciences, Oregon National Primate Research Center, Beaverton, OR, USA*

RÉGEN DROUIN · *Division of Medical Genetics, Department of Pediatrics, Laval University and Centre Hospitalier Universitaire de Québec, Quebec City, QC, Canada*

GREG FITZHARRIS · *Centre de Recherche du Centre Hospitalier de l'Université de Montréal, Montréal, QC, Canada; Département d'Obstétrique-Gynécologie, Faculté de Médecine, Université de Montréal, Montréal, QC, Canada*

MAKI FUKAMI · *Department of Molecular Endocrinology, National Research Institute for Child Health and Development, Tokyo, Japan*

Jean Baptiste Gaillard · *Unit of Chromosomal Genetics, Department of Medical Genetics, Arnaud de Villeneuve Hospital, CHU Montpellier, Montpellier Cedex 5, France*
Vincent Gatinois · *Unit of Chromosomal Genetics, Department of Medical Genetics, Arnaud de Villeneuve Hospital, CHU Montpellier, Montpellier Cedex 5, France; INSERM U1183 Unit "Genome and Stem Cell Plasticity in Development and Ageing", Institute for Regenerative Medicine and Biotherapy, St Eloi Hospital, Montpellier, France*
Anna Genescà · *Cell Biology Unit, Department of Cell Biology, Physiology and Immunology, Biosciences School, Universitat Autònoma de Barcelona, Bellaterra, Spain*
Roy M. Golsteyn · *Cancer Cell Laboratory, Department of Biological Sciences, University of Lethbridge, Lethbridge, AB, Canada*
Benoit Guibert · *Institut de Médecine Régénératrice et de Biothérapie, INSERM U1183, CHU Montpellier, Montpellier, France; Institut de Biologie Computationnelle, Université de Montpellier, Montpellier, France*
Thomas Guignard · *Unit of Chromosomal Genetics, Department of Medical Genetics, Arnaud de Villeneuve Hospital, CHU Montpellier, Montpellier Cedex 5, France*
Dustin C. Hancks · *Eccles Institute of Human Genetics, School of Medicine, University of Utah, Salt Lake City, UT, USA; Department of Immunology, UT Southwestern Medical Center, Dallas, TX, USA*
Henry H. Heng · *Center for Molecular Medicine and Genomics, Wayne State University School of Medicine, Detroit, MI, USA; Department of Pathology, Wayne State University School of Medicine, Detroit, MI, USA*
Isabelle M. Henry · *Genome Center & Department of Plant Biology, University of California, Davis, CA, USA*
Wigard P. Kloosterman · *Division of Biomedical Genetics, Department of Genetics, Center for Molecular Medicine, University Medical Center Utrecht, Utrecht, The Netherlands*
Hiroki Kurahashi · *Division of Molecular Genetics, Institute for Comprehensive Medical Science, Fujita Health University, Toyoake, Aichi, Japan*
Jeannie X. T. Lee · *Lee Kong Chian School of Medicine, Nanyang Technological University, Singapore, Singapore*
Thomas Liehr · *Institute of Human Genetics, Jena University Hospital, Friedrich Schiller University, Jena, Germany*
Bo Liu · *State Key Laboratory of Biotherapy, Collaborative Innovation Center of Biotherapy, West China Hospital, Sichuan University, Chengdu, Sichuan, China*
Guo Liu · *Center for Molecular Medicine and Genomics, Wayne State University School of Medicine, Detroit, MI, USA*
Monique N. H. Luijten · *Lee Kong Chian School of Medicine, Nanyang Technological University, Singapore, Singapore*
Ruth N. MacKinnon · *Department of Medicine (St. Vincent's), University of Melbourne, Melbourne, VIC, Australia; Victorian Cancer Cytogenetics Service, St Vincent's Hospital, Melbourne, VIC, Australia*
Sabine Mai · *Department of Physiology and Pathophysiology, Cell Biology, University of Manitoba, Research Institute of Oncology and Hematology, CancerCare Manitoba, Winnipeg, MB, Canada*
Alessio Marcozzi · *Division of Biomedical Genetics, Department of Genetics, Center for Molecular Medicine, University Medical Center Utrecht, Utrecht, The Netherlands*
Marta Martín · *Cell Biology Unit, Department of Cell Biology, Physiology and Immunology, Biosciences School, Universitat Autònoma de Barcelona, Bellaterra, Spain*

CHRISTINA MENDIOLA · *Department of Pathology and Laboratory Medicine, University of Texas Health Science Center, San Antonio, TX, USA*

VERONICA ORTEGA · *Department of Pathology and Laboratory Medicine, University of Texas Health Science Center, San Antonio, TX, USA*

FRANCK PELLESTOR · *Laboratory of Chromosomal Genetics, Department of Medical Genetics, Arnaud de Villeneuve Hospital, Montpellier CHRU, Montpellier, France; INSERM Unit Plasticity of the Genome and Aging, Institute of Functional Genomics, Montpellier, France; Unit of Chromosomal Genetics, Department of Medical Genetics, Arnaud de Villeneuve Hospital, CHU Montpellier, Montpellier Cedex 5, France; INSERM U1183 Unit "Genome and Stem Cell Plasticity in Development and Ageing", Institute for Regenerative Medicine and Biotherapy, St Eloi Hospital, Montpellier, France*

MARTIN POOT · *Department of Human Genetics, University of Würzburg, Würzburg, Germany*

JACQUES PUECHBERTY · *Unit of Chromosomal Genetics, Department of Medical Genetics, Arnaud de Villeneuve Hospital, CHU Montpellier, Montpellier Cedex 5, France*

SÉBASTIEN RIQUIER · *Institut de Médecine Régénératrice et de Biothérapie, INSERM U1183, CHU Montpellier, Montpellier, France; Institut de Biologie Computationnelle, Université de Montpellier, Montpellier, France*

FLORENCE RUFFLÉ · *Institut de Médecine Régénératrice et de Biothérapie, INSERM U1183, CHU Montpellier, Montpellier, France; Institut de Biologie Computationnelle, Université de Montpellier, Montpellier, France*

OUMAR SAMASSEKOU · *3D Signatures Holdings Inc. MaRS Centre, South Tower, Toronto, ON, Canada*

SATISH SATI · *Chromatin and Cell Biology Group, CNRS—Institute of Human Genetics, Montpellier, France*

ANOUCK SCHNEIDER · *Unit of Chromosomal Genetics, Department of Medical Genetics, Arnaud de Villeneuve Hospital, CHU Montpellier, Montpellier Cedex 5, France*

LUCY H. SWIFT · *POETIC Laboratory for Preclinical and Drug Discovery Studies, Departments of Pediatrics and Oncology, Alberta Children's Hospital Research Institute, Arnie Charbonneau Cancer Institute, University of Calgary, Calgary, AB, Canada*

EK HAN TAN · *School of Biology and Ecology, University of Maine, Orono, ME, USA*

MARIONA TERRADAS · *Cell Biology Unit, Department of Cell Biology, Physiology and Immunology, Biosciences School, Universitat Autònoma de Barcelona, Bellaterra, Spain; Hereditary Cancer Program, Catalan Institute of Oncology IDIBELL, Hospitalet de Llobregat, Barcelona, Spain*

CAYETANA VÁZQUEZ-DIEZ · *Centre de Recherche du Centre Hospitalier de l'Université de Montréal, Montréal, QC, Canada*

GOPALRAO V. N. VELAGALETI · *Department of Pathology and Laboratory Medicine, University of Texas Health Science Center, San Antonio, TX, USA*

JU YAN · *Cytogenetics and Molecular Cytogenetics laboratory, Beijing Boren Hospital, Beijing, China*

JIAN YANG · *Center of Growth, Metabolism, and Aging, Key Laboratory of Bio-Resources and Eco-Environment, College of Life Sciences, Sichuan University, Chengdu, Sichuan, China*

KEVIN YAUY · *Unit of Chromosomal Genetics, Department of Medical Genetics, Arnaud de Villeneuve Hospital, CHU Montpellier, Montpellier Cedex 5, France*

CHRISTINE J. YE · *The Division of Hematology/Oncology, Department of Internal Medicine, University of Michigan, Ann Arbor, MI, USA*

Part I

Genesis and Impact of Chromothripsis

Chapter 1

The Genomic Characteristics and Origin of Chromothripsis

Alessio Marcozzi, Franck Pellestor, and Wigard P. Kloosterman

Abstract

In 2011 a phenomenon involving complex chromosomal rearrangements was discovered in cancer genomes. This phenomenon has been termed chromothripsis, on the basis of its chromosomal hallmarks, which point to an underlying process involving chromosome (*chromo*) shattering (*thripsis*). The prevailing hypothesis of cancer genome evolution as a gradual process of mutation and selection was challenged by the discovery of chromothripsis, because its patterns of chromosome rearrangement rather indicated an one-off catastrophic burst of genome rearrangement. The initial discovery of chromothripsis has led to many more examples of chromothripsis both in cancer genomes as well as in patients with congenital diseases and in the genomes of healthy individuals. Since then, a burning topic has been the study of the molecular mechanism that leads to chromothripsis. Cumulating evidence has shown that chromothripsis may result from lagging chromosomes encapsulated in micronuclei, as well as from telomere fusions followed by chromosome bridge formation. In this chapter, we will outline the genomic characteristics of chromothripsis, and we present genomic methodologies that enable the detection of chromothripsis. Furthermore, we will give an overview of recent insights into the mechanisms underlying chromothripsis.

Key words Chromothripsis, Chromoanasynthesis, Chromoplexy, Cancer, Congenital disease, Telomeres, Micronuclei, Next-generation sequencing

1 Introduction

Chromosomes have been discovered in the nineteenth century [1–4], and the Boveri-Sutton chromosome theory postulated in the early years of the twentieth century that chromosomes are vectors of heredity [5, 6].

Since their initial discovery, chromosomes have continuously challenged the certainty of our knowledge of them. For example, in 1923 the American zoologist Theophilus Painter established that humans had 24 pairs (48) of chromosomes [7]. It took more than 20 years before it was recognized that the correct number of chromosome pairs is 23 [8]. An even greater misconception was surrounding the structure and organization of chromosomes. The main view throughout the last century was following the original

Franck Pellestor (ed.), *Chromothripsis: Methods and Protocols*, Methods in Molecular Biology, vol. 1769,
https://doi.org/10.1007/978-1-4939-7780-2_1,

Mendelian concept that individuals inherit exactly two copies of each gene, one from the mother and one from the father. Consequently, it was believed that any gene (except the one located on the sex chromosomes) must be biallelic. In 1991, Lupski et al. [9] challenged this vision by publishing a study that linked the Charcot-Marie-Tooth disease type 1A to a duplication in chromosome 17, leading to a triallelic gene.

Nowadays genome researchers are accustomed to the concept of structural variations (SVs), a term that encompasses a variety of genetic variations that occur at the chromosome level. The most common representatives of such events are deletions and duplications that together constitute copy-number variations in the human genome. As opposed to "unbalanced" copy-number variations, SVs that do not alter copy number are referred to as "balanced" and include inversions, insertions, or reciprocal translocations.

Besides forming a prominent class of genetic variations, SVs have been associated with human congenital disease and cancer [10, 11]. The formation of SVs underlying human genetic disease may be driven by genomic context, such as repeated sequences. In some cases, complex rearrangement of chromosomal segments occurs in patient genomes. Alternatively, SVs may be formed through random chromosome breakage and nonhomologous repair, which is a typical mechanism for the generation of chromosomal translocations [12, 13].

Over the last decade, the power of dissecting genome sequences with next-generation sequencing technologies has led to considerable insights into the complexity of genomic structural variations. These efforts culminated in the discovery of chromothripsis, a phenomenon leading to a complex form of structural genomic rearrangement. Chromothripsis is characterized by dozens of clustered chromosomal rearrangements that appear to arise as a result of a single catastrophic event (Fig. 1). A simple model that fits the observed rearranged chromosomes involves a shattering of chromosomes into pieces, followed by reassembly of the chromosomal segments into newly formed, mosaic, chromosomes (Fig. 1). Several forms of complex chromosome rearrangements related to chromothripsis have since been discovered. In this chapter, we will introduce the characteristics and mechanisms of the origin of chromothripsis and other complex chromosome rearrangements. Furthermore, we will discuss genomic methods for discovery of chromothripsis.

2 Genomic Characteristics of Chromothripsis

The term chromothripsis, derived from the Greek words *chromos* (chromosome) and *thripsis* (shattering into pieces), was first coined by Stephens et al. [14] in 2011 to describe the massive genomic rearrangements they found while screening patients with chronic lymphocytic leukemia (CLL). One of the patients had 42 somatically

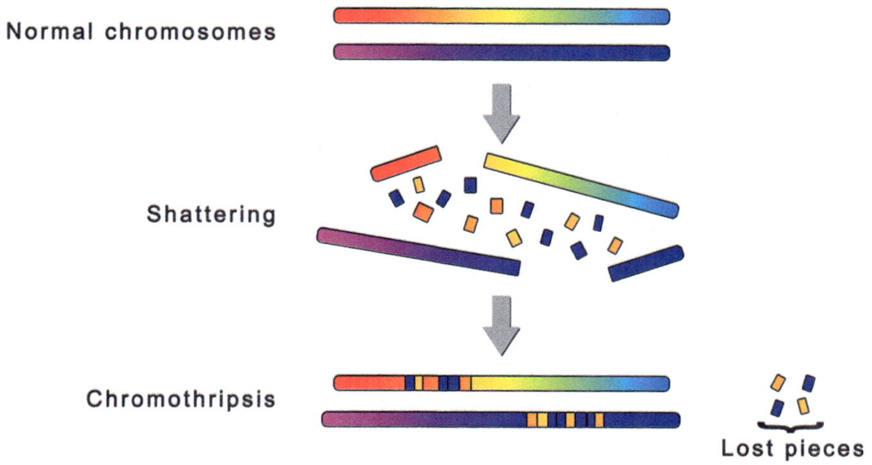

Fig. 1 Model of chromothripsis formation. In a first step, chromosomes are shattered into small chromosomal fragments. The chromosomal fragments are subsequently reassembled into a new chromosome, but the order and orientation of the fragments are changed compared to the structure of the original chromosome. Some of the chromosomal fragments may not be included in the reassembly process and are lost

acquired genomic rearrangements localized on the long arm of chromosome 4. Both the vast number and the clustering of those rearrangements were unexpected and contradicting textbook definitions on the evolvement of cancer [15]. As opposed to a gradual increase in mutational burden over time, with mutations distributed across all chromosomes, the CLL patient harbored massive rearrangements affecting just one chromosomal arm.

Further analysis of similarly complex cancer genomes shed light on common genomic features that mark chromosomes rearranged by chromothripsis [14, 16]. Besides the "geographic" localization of rearrangement breakpoints within a confined chromosomal region, a striking characteristic of chromothripsis is that such complex rearrangements occur simultaneously with copy-number changes that involve just two (occasionally three) states. Genomic segments of copy number one are the result of deletions of chromosomal segments, while regions with copy number two represent sequences that were retained in the final rearranged chromosomes. Such an alternation between two copy-number states cannot be explained by a stepwise accumulation of rearrangements during consecutive cell divisions. Furthermore, copy-neutral rearranged regions retained heterozygosity, while this was lost for monosomic regions. Another prediction that directly arises from the model depicted in Fig. 1 concerns an expected equal amount of inverted and non-inverted breakpoint-junctions in a chromosome rearranged by chromothripsis. Indeed, all four possible combinations of joining of chromosomal segments (head-to-tail, tail-to-head, head-to-head, and tail-to-tail) were equally present in chromosomes that underwent chromothripsis.

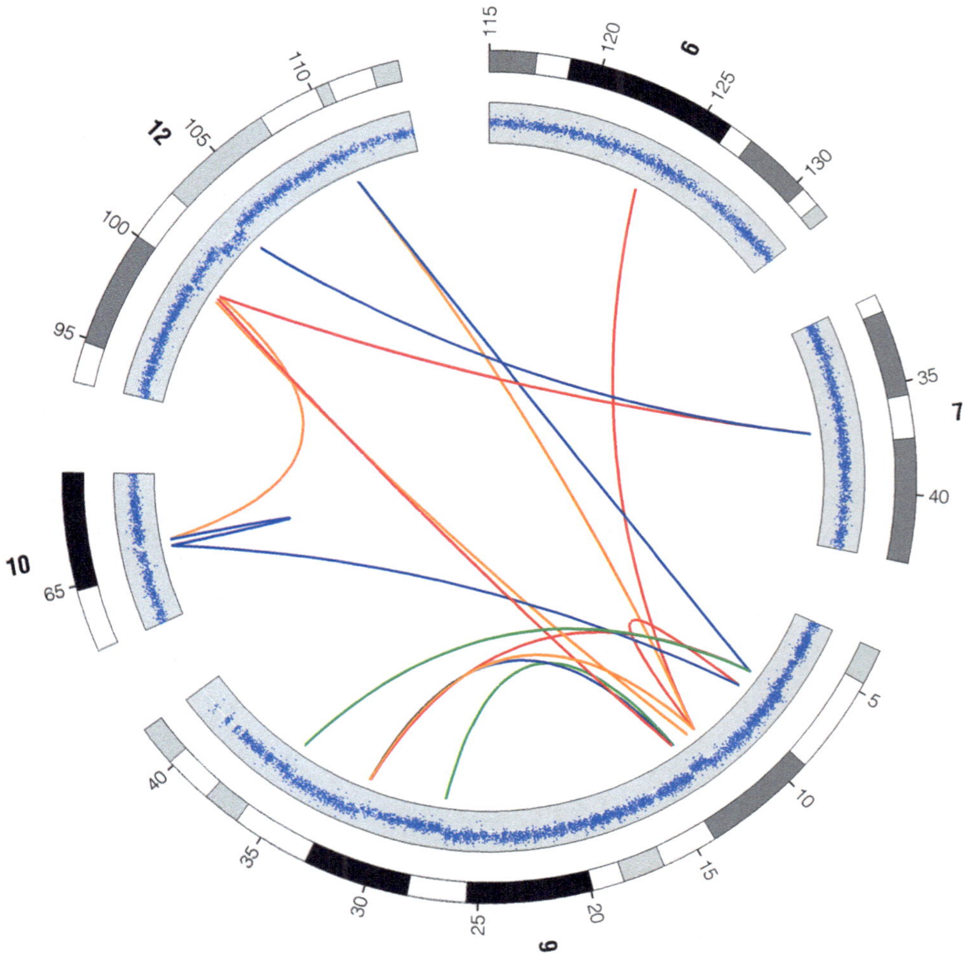

Fig. 2 Circos plot showing a constitutional chromothripsis event involving five chromosomes. The example involves a patient reported in an earlier study [17]

Although many cases of chromothripsis have been documented where only a single chromosome or chromosomal arm is affected, several instances of chromothripsis involving multiple chromosomes have also been reported. For example, the first notion of chromothripsis came from a CLL cancer, which displayed chromothripsis involving four chromosomes [14]. Apparently, multiple chromosomes and chromosomal regions can be subject to catastrophic rearrangement at the same time, a finding that is supported by patterns of linked breakpoint-junctions connecting the rearranged chromosomal segments from different chromosomes (Fig. 2).

These abovementioned features, consolidated by the observation that chromothripsis is normally evident in all tumor cells and not only in subclones, support the hypothesis that this phenomenon is the result of a single catastrophic event during cancer development.

3 Prevalence of Chromothripsis in Human Genomes

3.1 Cancer Chromothripsis

Chromothripsis was initially observed in at least 2–3% of all human cancers, on the basis of SNP array-based copy-number profiling (*see* Box 1) among 746 cancers, and this finding was confirmed in 2792 cancer samples [14]. Chromothripsis has a

Box 1 Technologies enabling detection of chromothripsis in genomes. **Array comparative genomic hybridization** (arrayCGH) is a technique used to detect copy-number changes in a genome by direct comparison of two DNA samples [24, 25]. This technique is able to resolve deletions and duplications larger than 10 kb, depending on the density of the probes on the array [25, 26]. The schematic workflow of the arrayCGH assay is the following: first, a reference DNA and a sample DNA are sheared into small fragments, and then the fragments are labeled with two different dyes, one for the reference and one for the sample. Next, the labeled DNA mixture is hybridized to an array of pre-spotted single-stranded DNA probes on a chip. The reference and sample DNA will compete for the binding on the probes. Once the labeled DNA is bound to the probes, the fluorescence emitted from each spot is measured. Finally, the ratio between the signal intensities of the two dyes (reference and sample) is used to determine the copy-number state, thus the number and the distribution of SVs is estimated. There are several types of arrayCGH chips commercially available, and they differ in the number and variety of DNA probes. A frequently used arrayCGH platform is the Agilent SurePrint series that spans from fifty thousand up to one million probes printed on glass chips. **Single nucleotide polymorphism arrays** (SNP arrays) are used to detect common biallelic polymorphisms within a genome [27]. The most commonly used SNP arrays are the Illumina BeadChips. These chips are based on SNP-specific probes that are only forming an amplification product if a specific SNP allele is present. Each of the two possible alleles at each SNP locus is amplified using a specific color. The labeled amplification products are bound to an array with probes for each SNP. The array is subsequently scanned for fluorescence intensity, which reveals the genotype for each SNP as well as the overall intensity of the signal for each SNP. These measurements provide an estimate of DNA copy number and of heterozygosity. For example, on chromothripsis chromosomes, regions with the lowest copy-number state typically show loss of heterozygosity (LOH) [27, 28].The most powerful method for detection of (complex) structural variations at genome-wide scale is **next-generation sequencing** (NGS) [29]. Next-generation paired-end sequencing involves shotgun sequencing of short DNA fragments from both ends. Subsequently, the resulting paired sequencing reads are mapped to reference genomes, using specific read mapping software [30]. The mapping characteristics, mapping position, and mapping orientation of each pair of sequencing reads are used to infer the presence of structural variations. Coverage depth analysis—that is, the number of sequencing reads at a given genomic locus—is informative for detection of copy-number variations. However, coverage analysis does not directly detect the breakpoint-junctions of structural variations nor can it identify copy-neutral variants. Coverage analysis can be augmented with information on the mapping orientation of reads. Concordantly mapping read pairs have an expected distance between the mapping distance of the two reads, and the mapping is oriented normally (opposite direction of the forward and reverse read for Illumina paired-end sequencing). In contrast, if read pairs bridge a structural variation breakpoint-junction, the reads are discordantly mapped. Discordant read pairs are a strong indication of, for the presence of a structural variation, both copy-neutral and copy-number variable. A discordant read-pair signal can be augmented by split mapping of single reads, given that the single reads are of sufficient length to be split mapped. The exact breakpoint-junction sequence can be inferred from split-mapped reads, thus providing information on possible homologous or microhomologous sequences at the breakpoints.

(continued)

Box 1 (continued)

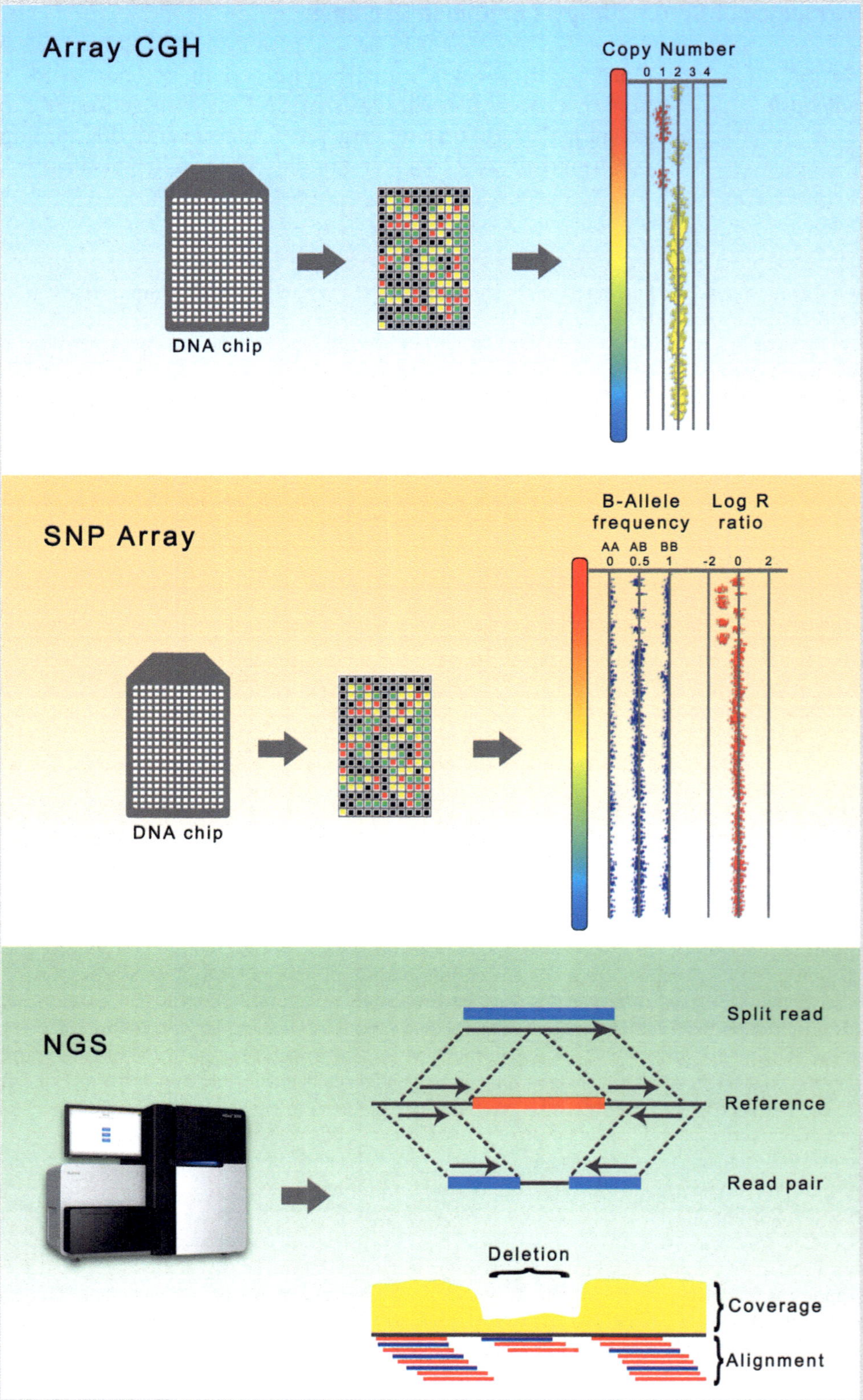

particularly high prevalence in bone cancers with frequencies up to 33% for osteosarcoma [14, 18, 19] and similar frequencies in glioblastoma [19–21]. It should be noted that such estimates are dependent on the technology that is used to identify chromothripsis (*see* Box 1) and the criteria that are used to discern chromothripsis from other complex chromosome rearrangements. In this context, tools have been developed that use systematic criteria to detect chromothripsis in copy-number data of cancer genomes [22]. CTLP scanner is an example of such a tool, accessible through a web browser, and allows users to define customized settings for chromothripsis detection [23].

A major question, that was immediately raised following the discovery of chromothripsis in a broad range of cancer, concerns the effect of chromothripsis on cell viability. Usually, somatic cells undergo apoptosis when affected by severe DNA damage. In fact, this is what will happen in most cases when chromothripsis occurs in a cell. Exceptionally, however, chromothripsis can cause disruption of tumor suppressor genes and activation of oncogenes [14, 19, 31], leading to cell survival probably determined by a selective advantage caused by the massive chromosome rearrangement. Furthermore, pieces of the shattered chromosomes can be reassembled into extrachromosomal structures, termed double minutes [14, 31–33], that like normal chromosomes are composed of chromatin and can be normally replicated during the S phase of the cell cycle. However, unlike canonical chromosomes, double minutes are circular and do not contain a centromere and telomeres. Asymmetrical segregation of double minutes during cell division may lead to daughter cells with more than one copy of the double minute. If double-minute chromosomes contain oncogenes, they may provide a selective advantage leading to high copy number in a single-cell and elevated oncogene expression.

3.2 Constitutional Chromothripsis

Shortly after the discovery of chromothripsis in cancer, Kloosterman and coworkers reported the first case of chromothripsis that was present constitutionally in an individual [34]. Analyzing the genome of a patient affected by a complex congenital phenotype characterized by psychomotor retardation, speech delay, hypertelorism, and kyphoscoliosis, the researchers mapped a complex rearrangement involving three chromosomes (1, 4, and 10). The breakpoints of the rearrangement were localized in distinct genomic regions, and the breakpoint-junctions were supporting a random shattering and repair process, just like chromothripsis observed in cancer (Fig. 2). Subsequent work consolidated these findings and revealed that chromothripsis occurs in approximately 5% of cases with karyotypically visible chromosome rearrangements [17, 35, 36], with some cases of germline chromothripsis displaying extreme complexity with up to 65 breakpoint-junctions [37].

A remarkable recent finding pertains the discovery of chromothripsis in healthy individuals [38]. When examining the genomes of patients with severe congenital abnormalities, it was noted that each of the patients inherited chromothripsis from their mothers. Strikingly, the mothers were healthy despite the largely balanced chromothripsis events in their genomes and the disruption of multiple genes by the chromothripsis breakpoints. Part of the affected chromosomes from the mothers were transmitted to their children, leading to imbalances that are most likely causal to their disease. Although chromothripsis is regularly observed among carriers of karyotypically visible balanced chromosome rearrangements, the presence of chromothripsis in genomes of healthy people is likely very low. For example, among 250 families from the Genome of the Netherlands project [39], chromothripsis was not observed.

There are some notable differences between germline chromothripsis and cancer chromothripsis [40]. In particular, copy-number states oscillating between two states—as prevalent in cancer chromothripsis—were not observed for germline chromothripsis. Instead, a more balanced form of chromothripsis is frequently observed in patients with congenital disease [35]. Additionally, the amount of breakpoints for germline chromothripsis is typically lower than for cancer chromothripsis, with a maximum reported number of breakpoints of 65 [37]. One possible explanation for these facts is that large numbers of breakpoints and imbalances would be strongly counterselected during fetal development.

4 Multiple Classes of Complex Chromosome Rearrangements: Chromothripsis, Chromoplexy, and Chromoanasynthesis

The use of next-generation sequencing and copy-number profiling to map complex chromosome rearrangements provided a detailed view of breakpoint characteristics and led to their classification into different categories (Table 1) [41, 44].

4.1 Chromoanasynthesis

The study of complex copy-number changes in the genomes of patients with congenital disease led to the discovery of chromoanasynthesis as a distinct entity among complex chromosome rearrangements [42]. A key difference between chromoanasynthesis and chromothripsis is the presence of copy gains (duplication, triplication) in addition to deletions and copy-neutral chromosomal regions. These multiple copy-number changes could not be explained by a simple process of shattering and reassembly. Instead, the rearrangements were proposed to be caused by a replication-based mechanism—termed microhomology-mediated break-induced replication (MMBIR). MMBIR explained both the loss

Table 1
Overview of different classes of complex chromosomal rearrangements

	Chromothripsis	Chromoanasynthesis	Chromoplexy
Number of events	Single	Single or multiple	Multiple
Structural variation	Balanced or deletions	Balanced, deletions, duplications/triplications	Balanced with occasional deletions
Involved chromosomes	Few (1–4)	Few (often 1)	Multiple (>5)
Number of breakpoints	Many (more in cancer than germline rearrangements)	Fewer (usually 5–25)	Fewer (usually 5–25)
Breakpoint signature	Blunt ends (small insertions)	Microhomology	Blunt ends
Proposed mechanism	NHEJ	MMBIR/FoSTes	NHEJ
References	[19, 34, 35]	[35, 41, 42]	[41, 43]

and gain of chromosomal material and was described before as a mechanism explaining complex chromosome rearrangements [45, 46]. MMBIR occurs when a replication fork collapses as a result of a nick in the template strand. The resulting free DNA end is resected to leave a 3′ single-stranded DNA end which can invade other areas with microhomology, leading to a new replication fork and formation of complex rearrangements [46].

4.2 Chromoplexy

The sequencing of prostate cancer led to the discovery of chromoplexy, a distinct type of complex chromosome rearrangement, characterized by the involvement of multiple chromosomes [43]. The contributing chromosomes often show fewer rearrangements than the ones observed in chromothripsis. Another stark distinction between chromothripsis and chromoplexy lies in the patterns of breakpoints-junctions. In chromoplexy, the rearrangements often occur as closed chains with (nearly) precise junctions and almost no deletions [43, 47]. As a result, the characteristic copy-number oscillation observed on chromothripsis chromosomes is not found for chromoplexy. Statistical analysis indicated that the chromoplexy events may occur during multiple cell cycles [48]. Chromoplexy breakpoints seem to cluster in actively transcribed DNA with high GC content and are associated with active transcription and open chromatin configurations [48]. It was proposed that nonhomologous end joining is the primary repair mechanism of chromoplexy breaks as the breakpoint-junctions are blunt ended without any sequence homology or insertions.

Besides chromothripsis, chromoanasynthesis, and chromoplexy that are most often mentioned as the group of massive chromosome rearrangements, other rare chromothripsis-like events have been reported [22, 49–51]. To clarify relations between all the described events, Holland and Cleveland [41] proposed to use the umbrella term *chromoanagenesis*.

5 Mechanisms of Chromothripsis

The genomic characteristics of chromothripsis suggested that a single catastrophic event should be responsible for its formation [14]. Several possible triggers have been proposed that could account for a massive rearrangement of one or just a few chromosomes. For example, ionizing irradiation has been shown to cause chromosome shattering resembling chromothripsis [52]. Also, various types of cellular stresses may cause chromosomal fragmentation rather than apoptosis [53]. However, the shattering generated by such triggers affects the whole genome and not just a single chromosome or chromosomal regions, indicating that other mechanisms should be responsible for chromothripsis.

5.1 Micronuclei

A cellular mechanism that can explain the localization of breakpoints to one chromosome involves small cellular structures called micronuclei (Fig. 3a). Micronuclei are frequently found in cancer cells, and they can be formed around single chromosomes following errors in mitosis [54, 55]. Observations made by the Pellman Lab showed that micronuclei have deficiencies in the process of DNA replication and replication in the micronucleus is asynchronous compared to the primary nucleus [55, 56]. Therefore, the majority of the chromosomes entrapped in micronuclei are still replicating DNA in the G2 phase, while the main nucleus has already replicated its chromosomal content. When nuclear envelope breakdown follows, the still replicating chromatin in the micronucleus is prematurely condensed [55, 56]. When chromosomes from S-phase nuclei are forced to undergo chromosome condensation, an event triggered by signals from chromosomes in mitotic cells, the incompletely replicated chromosomes shatter into pieces [57].

During the successive cell division after micronucleation, the micronucleus can be degraded within the cytoplasm, which results in loss of the DNA material in the micronucleus [58]. Another possibility is that the micronucleus and its DNA content are reincorporated into the primary nucleus of one of the daughter cells. In an elegant series of experiments, coupling of live-cell imaging and single-cell genome sequencing demonstrated that the physical isolation of one or a few chromosomes into micronuclei can indeed generate a spectrum of complex genomic rearrangements, with

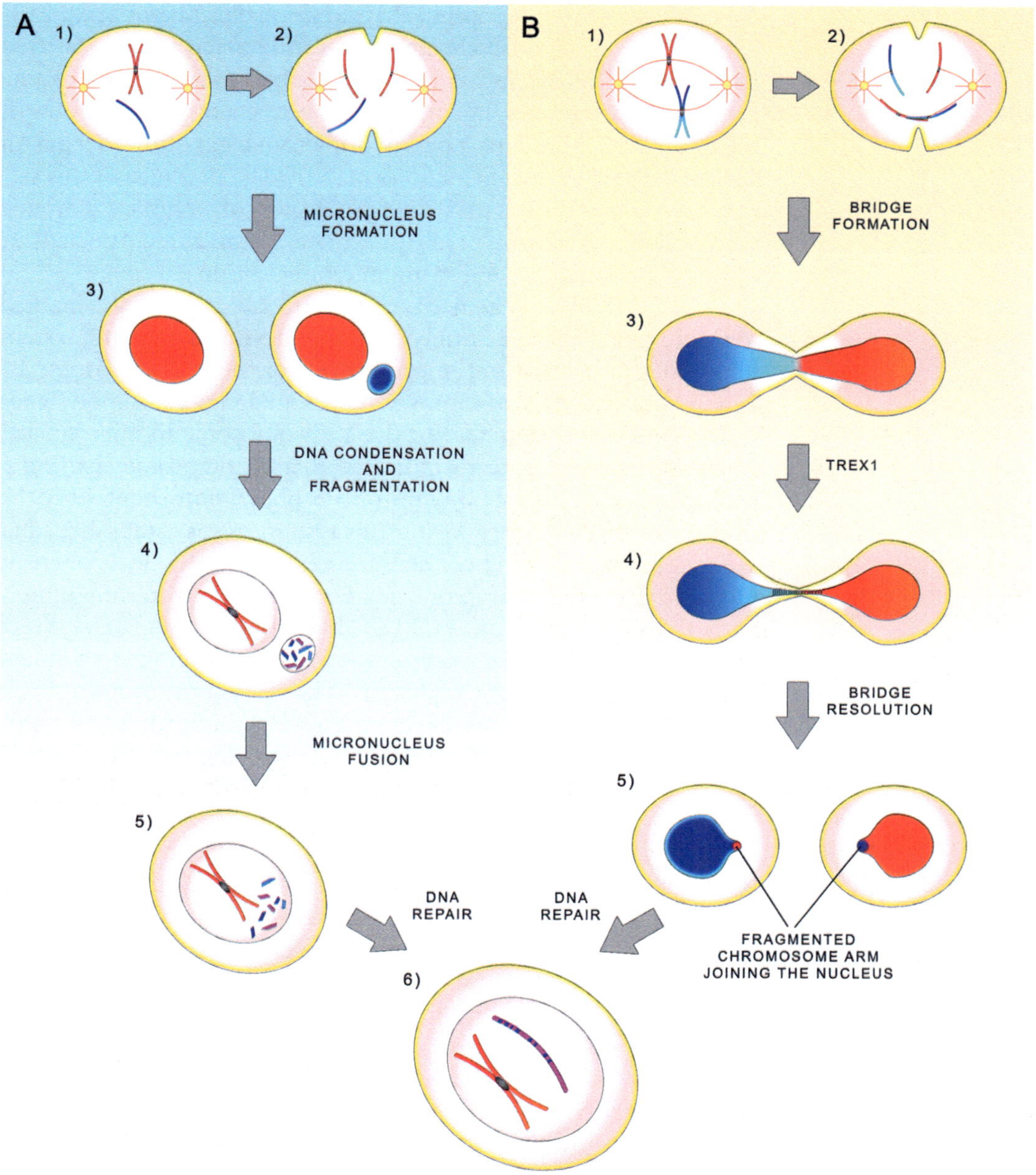

Fig. 3 Two mechanisms for chromothripsis. (**a**) Micronucleus. The missegregation of a chromosome pair during cell division (*1* and *2*) causes the entrapment of the lagging chromosome in the micronucleus (*3*). Chromosomes in the micronucleus undergo premature condensation followed by shattering (*4*). A subsequent round of mitosis incorporates the shattered chromosome in the main nucleus (*5*) where it will be repaired into the chromothriptic chromosome (*6*). (**b**) Chromosome fusion. The formation of a dicentric chromosome (*1*) will cause the formation of a DNA bridge during mitosis (*2*) causing the cell to stall during cytokinesis (*3*). The nuclease TREX1 cuts the DNA causing the resolution of the bridge and the shattering of the DNA (*4*). Once cytokinesis is complete, the fragmented chromosome fragments possibly join the nucleus (*5*) where they will be repaired into the chromothriptic chromosome (*6*)

hallmarks of chromothripsis [28]. Additionally, Zhang et al. [28] observed small circular chromosomes after micronucleation that could represent the initial steps of double-minute chromosome formation. Some rearrangements that were characterized on chromosomes that were entrapped in a micronucleus showed signs of MMBIR-based rearrangement formation. The presence of this signature could indicate that multiple forms of complex genomic rearrangements are potentially generated within a micronucleus or during repair of DNA following micronucleation.

A new cellular system developed by Ly et al. [59] enabled missegregation of specifically the Y chromosome. The system involved replacement of the centromere of the Y chromosome by a dysfunctional one that cannot initiate kinetochore assembly, leading to enriched entrapment of the Y chromosome in micronuclei. Analysis of mitotic spreads indicated that chromosome shattering took place after passage of cells into S phase, consistent with the hypothesis of PCC [59]. Using this system it was established that premature chromosome condensation is indeed the cause of chromosome shattering in micronuclei. Furthermore, by testing inhibitors of DNA repair pathways, Ly et al. [59] provided the first direct molecular evidence that nonhomologous end joining is the main pathway involved in the repair of shattered chromosomes. Such repair pathways are not active in micronuclei, indicating that chromosome reassembly can only occur when fragmented chromosomes are reincorporated in the primary nucleus.

5.2 Telomere Crisis

Another process that has been suggested to trigger chromothripsis is telomere shortening (Fig. 3b). In healthy cells, telomeres are protected by a protein complex called shelterin that binds telomeric repeats [60]. When cells undergo continued proliferation, telomeres shorten and lose the terminal repeats that in turn cause the loss of shelterin [61, 62]. Usually, unprotected chromosome ends will trigger senescence or apoptosis, but in case of mutations in key regulatory genes, like—for example—*TP53* or *RB*, cells with shortened chromosomes continue to grow. The accumulation of unprotected chromosomes may result in the formation of end-to-end fused dicentric chromosomes [63–65].

During mitosis, dicentric chromosomes can be attached to opposite spindle poles eventually leading chromosome breakage. When during the next cell cycle broken chromosomes again fuse together, a phenomenon called breakage-fusion-bridge (BFB) cycling is occurring, resulting in massive chromosomal rearrangements over multiple cell cycles [66, 67]. Alternatively, dicentric chromosomes could form lagging chromosomes during mitosis and entrap in micronuclei [68], possibly leading to chromothripsis.

Indeed, several studies have suggested an association between BFB cycles and chromothripsis [69, 70].

When telomere crisis is experimentally introduced by inhibiting shelterin in *RB*- and *TP53*-deficient cells, a mutagenic process that resembles chromothripsis can be observed [71]. Using live-cell imaging to determine the fate of dicentric chromosomes, formed during telomere crisis, it was shown that the induced dicentric chromosomes do not break during mitosis but form long stretching bridges between the two daughter cells [71]. The chromatin bridges contain a continuous nuclear envelope, which connects the two nuclei. Upon rupture of the nuclear envelope the DNA in the chromosome bridge can be digested by the cytoplasmic exonuclease TREX1, resulting in extensive single-stranded DNA and eventually resolution of the chromatin bridge by fragmentation. Sequencing of postcrisis cell clones revealed that ~50% of the clones harbor genomic rearrangements, exhibiting features of chromothripsis [71]. Noticeably, all rearranged chromosomes were derivatives of two chromosomes and thus probably derived from the end-to-end fusion of chromosomes. These results provide evidence that telomere crisis can induce chromothripsis, through the formation of dicentric chromosomes.

6 Conclusion

As a mechanism for the localized and chaotic restructuring of chromosomes, chromothripsis constitutes a unique and fascinating event, hitherto unsuspected in the field of chromosomal mechanics. His discovery has greatly renewed interest in the study of the genesis of chromosomal abnormalities and genomic instability. Increasingly reported chromothriptic events in cancers, in patients with congenital diseases, in embryos, but also in healthy individuals [18–20, 38, 40, 72–79] suggest that chromothripsis is much more frequent than what we were able to imagine.

Today, different models of experimental study of chromothripsis have been developed [28, 52, 56, 59, 71, 80–82], proving the existence of this cataclysmic mechanism, in front of doubts of some researchers [83–85]. The study of chromothripsis allows us to extend our knowledge about the underlying mechanisms of complex rearrangement formation and to better understand how the cell can manage such chaotic events [86, 87], now grouped under the term of chromoanagenesis.

Finally, the discovery of chromothripsis and the rapid alterations of the genome that it causes have also emphasized an interesting, still little-invested question, namely, the role of chromosomal abnormalities in the biological evolution.

References

1. Schwann T, Schleiden MJ, Smith H (1847) Microscopical researches into the accordance in the structure and growth of animals and plants. Translated from the German of Theodor Schwann, by Henry Smith. The Sydenham Society, London
2. Schleiden MJ (1847) Contributions to phytogenesis. In: Schwann T, Schleiden MJ, Smith H (eds) Microscopical researches into the accordance in the structure and growth of animals and plants. Translated from the German of Theodor Schwann, by Henry Smith. The Sydenham Society, London
3. Virchow RLK (1860) Cellular pathology. De Witt RM, New York
4. Butschli O (1876) Studien über Die Ersten Entwicklungsvorgänge Der Eizelle, Die Zelltheilung Und Die Conjugation Der Infusorien. Winter C, Frankfurt a. M
5. Sutton WS (1902) On the morphology of the chromosome group in Brachystola Magna. Biol Bull 4(1):24–39
6. Boveri T (1904) Ergebnisse über Die Konstitution Der Chromatischen Substanz Des Zellkerns. Fischer G, Jena
7. Painter TS (1923) Further observation on the sex chromosomes of mammals. Science 58(1500):247–248
8. Tjio JH, Levan A (1956) The chromosome number of man. Hereditas 42:1–6
9. Lupski JR, de Oca-Luna RM, Slaugenhaupt S et al (1991) DNA duplication associated with Charcot-Marie-Tooth disease Type 1A. Cell 66(2):219–232
10. Carvalho CMB, Ramocki MB, Pehlivan D et al (2011) Inverted genomic segments and complex triplication rearrangements are mediated by inverted repeats in the human genome. Nat Genet 43(11):1074–1081
11. Carvalho CMB, Lupski JR (2016) Mechanisms underlying structural variant formation in genomic disorders. Nat Rev Genet 17(4):224–238
12. Nambiar M, Raghavan SC (2011) How does DNA break during chromosomal translocations? Nucleic Acids Res 39(14):5813–5825
13. Küppers R, Dalla-Favera R (2001) Mechanisms of chromosomal translocations in B cell lymphomas. Oncogene 20(40):5580–5594
14. Stephens PJ, Greenman CD, Fu B et al (2011) Massive genomic rearrangement acquired in a single catastrophic event during cancer development. Cell 144(1):27–40
15. Stratton MR, Campbell PJ, Futreal PA (2009) The cancer genome. Nature 458(7239):719–724
16. Korbel JO, Campbell PJ (2013) Criteria for inference of chromothripsis in cancer genomes. Cell 152(6):1226–1236
17. Kloosterman WP, Tavakoli-Yaraki M, van Roosmalen M et al (2012) Constitutional chromothripsis rearrangements involve clustered double-stranded DNA breaks and nonhomologous repair mechanisms. Cell Rep 1(6):648–655
18. Behjati S, Tarpey PS, Haase K et al (2017) Recurrent mutation of IGF signalling genes and distinct patterns of genomic rearrangement in osteosarcoma. Nat Commun 8(June):15936
19. Kloosterman WP, Koster J, Molenaar JJ (2014) Prevalence and clinical implications of chromothripsis in cancer genomes. Curr Opin Oncol 26(1):64–72
20. Malhotra A, Lindberg M, Faust GG et al (2013) Breakpoint profiling of 64 cancer genomes reveals numerous complex rearrangements spawned by homology-independent mechanisms. Genome Res 23(5):762–776
21. Erson-Omay EZ, Henegariu O, Bülent Omay S et al (2017) Longitudinal analysis of treatment-induced genomic alterations in gliomas. Genome Med 9(1):12
22. Cai H, Kumar N, Bagheri HC et al (2014) Chromothripsis-like patterns are recurring but heterogeneously distributed features in a survey of 22,347 cancer genome screens. BMC Genomics 15(January):82
23. Yang J, Liu J, Ouyang L et al (2016) CTLP scanner: a web server for chromothripsis-like pattern detection. Nucleic Acids Res 44(W1):W252–W258
24. Costa JL, Meijer G, Ylstra B et al (2008) Array comparative genomic hybridization copy number profiling: a new tool for translational research in solid malignancies. Semin Radiat Oncol 18(2):98–104
25. Lassmann S, Weis R, Makowiec F et al (2007) Array CGH identifies distinct DNA copy number profiles of oncogenes and tumor suppressor genes in chromosomal- and microsatellite-unstable sporadic colorectal carcinomas. J Mol Med 85(3):293–304
26. Buffart TE, Tijssen M, Krugers T et al (2007) DNA quality assessment for array CGH by isothermal whole genome amplification. Cell Oncol 29(4):351–359
27. LaFramboise T (2009) Single nucleotide polymorphism arrays: a decade of biological, computational and technological advances. Nucleic Acids Res 37(13):4181–4193

28. Zhang CZ, Spektor A, Cornils H et al (2015) Chromothripsis from DNA damage in micronuclei. Nature 522(7555):179–184
29. Alkan C, Coe BP, Eichler EE (2011) Genome structural variation discovery and genotyping. Nat Rev Genet 12(5):363–376
30. Li H, Durbin R (2009) Fast and accurate short read alignment with burrows-wheeler transform. Bioinformatics 25(14):1754–1760
31. Rausch T, Jones DTW, Zapatka M et al (2012) Genome sequencing of pediatric medulloblastoma links catastrophic DNA rearrangements with TP53 mutations. Cell 148(1–2):59–71
32. Nones K, Waddell N, Wayte N et al (2014) Genomic catastrophes frequently arise in esophageal adenocarcinoma and drive tumorigenesis. Nat Commun 5(October):5224
33. Cheng C, Zhou Y, Li H et al (2016) Whole-genome sequencing reveals diverse models of structural variations in esophageal squamous cell carcinoma. Am J Hum Genet 98(2):256–274
34. Kloosterman WP, Guryev V, van Roosmalen M et al (2011) Chromothripsis as a mechanism driving complex de novo structural rearrangements in the germline. Hum Mol Genet 20(10):1916–1924
35. Chiang C, Jacobsen JC, Ernst C et al (2012) Complex reorganization and predominant non-homologous repair following chromosomal breakage in karyotypically balanced germline rearrangements and transgenic integration. Nat Genet 44(4):390–397
36. Redin C, Brand H, Collins RL et al (2017) The genomic landscape of balanced cytogenetic abnormalities associated with human congenital anomalies. Nat Genet 49(1):36–45
37. Collins RL, Brand H, Redin C et al (2017) Defining the diverse spectrum of inversions, complex structural variation, and chromothripsis in the morbid human genome. Genome Biol 18(1):36
38. de Pagter M, van Roosmalen MJ, Baas A et al (2015) Chromothripsis in healthy individuals affects multiple protein-coding genes and can result in severe congenital abnormalities in offspring. Am J Hum Genet 96(4):651–656
39. Hehir-Kwa J, Marschall T, Kloosterman WP et al (2016) A high-quality reference panel reveals the complexity and distribution of structural genome changes in a human population. Nat Commun 7:12989. https://doi.org/10.1038/ncomms12989
40. Kloosterman WP, Cuppen E (2013) Chromothripsis in congenital disorders and cancer: similarities and differences. Curr Opin Cell Biol 25(3):341–348
41. Holland AJ, Cleveland DW (2012) Chromoanagenesis and cancer: mechanisms and consequences of localized, complex chromosomal rearrangements. Nat Med 18(11):1630–1638
42. Liu P, Erez A, Sreenath Nagamani SC et al (2011) Chromosome catastrophes involve replication mechanisms generating complex genomic rearrangements. Cell 146(6):889–903
43. Shen MM (2013) Chromoplexy: a new category of complex rearrangements in the cancer genome. Cancer Cell 23(5):567–569
44. Zhang CZ, Leibowitz ML, Pellman D (2013) Chromothripsis and beyond: rapid genome evolution from complex chromosomal rearrangements. Genes Dev 27(23):2513–2530
45. Zhang F, Khajavi M, Connolly A et al (2009) The DNA replication FoSTeS/MMBIR mechanism can generate genomic, genic and exonic complex rearrangements in humans. Nat Genet 41(7):849–853
46. Hastings PJ, Ira G, Lupski JR (2009) A microhomology-mediated break-induced replication model for the origin of human copy number variation. PLoS Genet 5(1):e1000327
47. Wang K, Wang Y, Collins CC (2013) Chromoplexy: a new paradigm in genome remodeling and evolution. Asian J Androl 15(6):711–712
48. Baca SC, Prandi D, Lawrence MS et al (2013) Punctuated evolution of prostate cancer genomes. Cell 153(3):666–677
49. Kim TM, Xi R, Luquette LJ et al (2012) Functional genomic analysis of chromosomal aberrations in a compendium of 8000 cancer genomes. Genome Res 23(2):217–227
50. Mandahl N, Magnusson L, Nilsson J et al (2017) Scattered genomic amplification in dedifferentiated liposarcoma. Mol Cytogenet 10:25. https://doi.org/10.1186/s13039-017-0325-5
51. Gu H, Jiang JH, Li JY et al (2013) A familial cri-du-chat/5p deletion syndrome resulted from rare maternal complex chromosomal rearrangements (CCRs) and/or possible chromosome 5p chromothripsis. PLoS One 8(10):e76985
52. Morishita M, Muramatsu T, Suto Y et al (2016) Chromothripsis-like chromosomal rearrangements induced by ionizing radiation using proton microbeam irradiation system. Oncotarget 7(9):10182–10192
53. Stevens JB, Abdallah BY, Liu G et al (2011) Diverse system stresses: common mechanisms of chromosome fragmentation. Cell Death Dis 2(June):e178

54. Hatch EM, Fischer AH, Deerinck TJ et al (2013) Catastrophic nuclear envelope collapse in cancer cell micronuclei. Cell 154(1):47–60
55. Crasta K, Ganem NJ, Dagher R et al (2012) DNA breaks and chromosome pulverization from errors in mitosis. Nature 482(7383):53–58
56. Terzoudi G, Karakosta M, Pantelias A et al (2015) Stress induced by premature chromatin condensation triggers chromosome shattering and chromothripsis at dna sites still replicating in micronuclei or multinucleate cells when primary nuclei enter mitosis. Mutat Res Genet Toxicol Environ Mutagen 793:185–198
57. Rao PN, Johnson RT (1970) Mammalian cell fusion: studies on the regulation of DNA synthesis and mitosis. Nature 225(5228):159–164
58. Terradas M, Martín M, Laura Tusell L et al (2010) Genetic activities in micronuclei: is the DNA entrapped in micronuclei lost for the cell? Mutat Res 705(1):60–67
59. Ly P, Teitz LS, Kim DH et al (2017) Selective Y centromere inactivation triggers chromosome shattering in micronuclei and repair by non-homologous end joining. Nat Cell Biol 19(1):68–75
60. de Lange T (2005) Shelterin: the protein complex that shapes and safeguards human telomeres. Genes Dev 19(18):2100–2110
61. Rodriguez R, Müller S, Yeoman JA et al (2008) A novel small molecule that alters shelterin integrity and triggers a DNA-damage response at telomeres. J Am Chem Soc 130(47):15758–15759
62. Hockemeyer D, Collins K (2015) Control of telomerase action at human telomeres. Nat Struct Mol Biol 22(11):848–852
63. Titen SWA, Golic KG (2008) Telomere loss provokes multiple pathways to apoptosis and produces genomic instability in drosophila melanogaster. Genetics 180(4):1821–1832
64. Lowden MR, Meier B, Lee T et al (2008) End joining at caenorhabditis elegans telomeres. Genetics 180(2):741–754
65. Boukamp P, Popp S, Krunic D (2005) Telomere-dependent chromosomal instability. J Investig Dermatol Symp Proc 10(2):89–94
66. Gisselsson DL, Pettersson M, Höglund M et al (2000) Chromosomal breakage-fusion-bridge events cause genetic intratumor heterogeneity. Proc Natl Acad Sci U S A 97(10):5357–5362
67. Bignell GR, Santarius T, Pole JCM et al (2007) Architectures of somatic genomic rearrangement in human cancer amplicons at sequence-level resolution. Genome Res 17(9):1296–1303
68. Pampalona J, Roscioli E, Silkworth WT et al (2016) Chromosome bridges maintain kinetochore-microtubule attachment throughout mitosis and rarely break during anaphase. PLoS One 11(1):e0147420
69. Li Y, Schwab C, Ryan SL et al (2014) Constitutional and somatic rearrangement of chromosome 21 in acute lymphoblastic leukaemia. Nature 508(7494):98–102
70. Garsed DW, Marshall OJ, Corbin VDA et al (2014) The architecture and evolution of cancer neochromosomes. Cancer Cell 26(5):653–667
71. Maciejowski J, Li Y, Bosco N et al (2015) Chromothripsis and kataegis induced by telomere crisis. Cell 163(7):1641–1654
72. Ratnaparkhe M, Hlevnjak M, Kolb T et al (2017) Genomic profiling of acute lymphoblastic leukemia in ataxia telangiectasia patients reveals tight link between ATM mutations and chromothripsis. Leukemia 31(10):2048–2056. https://doi.org/10.1038/leu.2017.55
73. Notta F, Chan-Seng-Yue M, Lemire M et al (2016) A renewed model of pancreatic cancer evolution based on genomic rearrangement patterns. Nature 538(7625):378–382
74. Abaigar M, Robledo C, Benito R et al (2016) Chromothripsis is a recurrent genomic abnormality in high-risk myelodysplastic syndromes. PLoS One 11(10):e0164370. https://doi.org/10.1371/journal.pone.0164370
75. Wang JC, Fisker T, Sahoo T (2015) Constitutional chromothripsis involving chromosome 19 in a child with subtle dysmorphic features. Am J Med Genet A 167A(4):910–913
76. Weckselblatt B, Hermetz KE, Rudd MK (2015) Unbalanced translocations arise from diverse mutational mechanisms including chromothripsis. Genome Res 25(7):937–947
77. Bertelsen B, Nazaryan-Petersen L, Sun W et al (2016) A germline chromothripsis event stably segregating in 11 individuals through three generations. Genet Med 18(5):494–500
78. Macera MJ, Sobrino A, Levy B et al (2015) Prenatal diagnosis of chromothripsis, with nine breaks characterized by karyotyping, FISH, microarray and whole-genome sequencing. Prenat Diagn 35(5):299–301
79. Daughtry BL, Chavez SL (2016) Chromosomal instability in mammalian pre-implantation embryos: potential causes, detection methods, and clinical consequences. Cell Tissue Res 363(1):201–225
80. Mardin BR, Drainas AP, Waszak SM et al (2015) A cell-based model system links chromothripsis with hyperploidy. Mol Syst Biol 11(9):828. https://doi.org/10.15252/msb.20156505

81. Chen JM, Cooper D (2016) A mechanistic link between retrotransposition and chromothripsis. Hum Mutat 37(4):329. https://doi.org/10.1002/humu.22870
82. Middelkamps S, van Heesch S, Braat AK et al (2017) Molecular dissection of germline chromothripsis in a developmental context using patient-derived iPS cells. Genome Med 9(1):9. https://doi.org/10.1186/s13073-017-0399-z
83. Righolt C, Mai S (2012) Shattered and stitched chromosomes-chromothripsis and chromoanasynthesis-manifestations of a new chromosome crisis ? Genes Chromosom Cancer 51(11):975–981
84. Sozano CO, PAscual-Montano A, Sanchez de Diego A et al (2013) Chromothripsis: breakage-fusion-bridge over and over again. Cell Cycle 12(13):2916–2023
85. Kinsella M, Patel A, Bafna V (2014) The elusive evidence for chromothripsis. Nucleic Acids Res 42(13):8231–8242
86. Ivkov R, Bunz F (2015) Pathways to chromothripsis. Cell Cycle 14(18):2886–2890
87. Fukami M, Shima H, Suzuki E et al (2017) Catastrophic cellular events leading to complex chromosomal rearrangements in the germline. Clin Genet 91(5):653–660

Chapter 2

Clinical Consequences of Chromothripsis and Other Catastrophic Cellular Events

Maki Fukami and Hiroki Kurahashi

Abstract

Chromothripsis was initially described as a novel cause of chromosomal rearrangements in cancer cells and was subsequently implicated in the development of gross chromosomal rearrangements in the germline. Other catastrophic cellular events such as chromoanasynthesis and chromoplexy have also been observed in human cells. Such events have been associated with various phenotypes including mental retardation and congenital malformations. Here, we introduce representative cases of human disorders arising from somatic or germline chromothripsis or similar catastrophic events. In this chapter, we use the term "chromoanagenesis" to indicate all catastrophic events including chromothripsis.

Key words Germline, Rearrangement, Gene, Deletion, Duplication, Congenital anomaly, Tumor

1 Germline Chromothripsis/Chromoanagenesis as Genetic Causes of Chromosomal Rearrangements

1.1 How Gross Chromosomal Rearrangement Occurs

Development of gross chromosomal rearrangements (GCR) is fundamentally dependent on two distinct factors: the generation of double-strand breaks (DSBs) and the error of their repair. DSB can result either from exogenous agents such as ionizing radiation and certain chemotherapeutic drugs or from endogenously generated reactive oxygen species and from mechanical stress on the chromosomes [1]. Such DSBs are usually repaired appropriately by an error-free repair system called homologous recombination (HR), which requires homologous DNA-like sister chromatid or chromosome homologue as a template for accurate repair. Without such homologous DNA, DSBs are occasionally repaired by an illegitimate error-prone repair system, nonhomologous end joining (NHEJ), which often leads to GCRs. Actually, experimentally induced two DSBs are sufficient to produce GCRs in mammalian cells [2]. Two DSBs that occur on the same chromosome produce deletions, inversions, or ring chromosome, while those on different chromosomes yield translocations. Recently accumulating

Franck Pellestor (ed.), *Chromothripsis: Methods and Protocols*, Methods in Molecular Biology, vol. 1769,
https://doi.org/10.1007/978-1-4939-7780-2_2, © Springer Science+Business Media, LLC, part of Springer Nature 2018

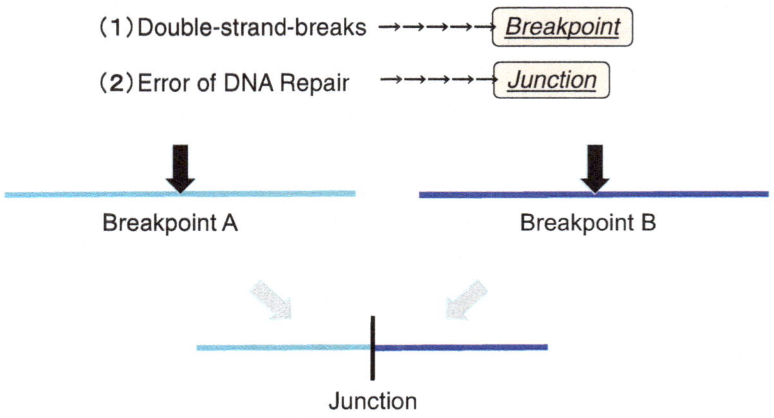

Fig. 1 Two factors that affect the development of GCR

evidence suggests that stall of DNA replication fork forms DNA ends that need to be repaired and it is involved in formation of GCR more frequently than others [3]. Such DNA ends are repaired by template switching via a limited homology, and most of nonrecurrent structural variants have microhomology at the junction.

Genome analysis is a powerful strategy that can elucidate the mechanism that is involved in the development of a certain GCR. The information of two essential factors for GCR, DSB and its repair, can be obtained by comparative genomic analysis of the nascent sequences in cells harboring a GCR with the putative original sequences and identified as "breakpoint" and "junction," respectively (Fig. 1). Genetic information of the breakpoints provides us how the DSB arose at the specific region particularly in the case of recurrent GCR. In contrast, information of the junction sequence would provide the information about how the illegitimate DNA repair arose. Historically, breakpoints had been mapped cytogenetically by multiple rounds of fluorescence in situ hybridization using probes each encompassing the breakpoint region. Cytogenetic microarray is now the standard method to obtain the breakpoint information of the GCR that affects copy-number status. Once the breakpoint could be localized in a small region using microarray, conventional PCR/Sanger sequence allows us to obtain the junction sequence and deduced breakpoint sequence. Recently, emergence of next-generation sequencing (NGS) has enabled us to analyze the information of the breakpoints and junction of the GCR much easier than we previously did. NGS provides us the information even in the highly complex rearrangements.

In earlier times, studies for mechanism of GCR were focused on recurrent nature of a subset of GCRs. One of the best-studied examples is 22q11 deletion syndrome, the most frequent congenital anomaly syndrome caused by GCR (OMIM #188400). In spite of the large spectrum of phenotypic variation, the deleted region of 90% of cases is almost identical regardless of their phenotypic difference [4]. Human genome project identified locus-specific segmental duplications (SDs) located at the 22q11 region, which were predicted to contribute to generation of the recurrent deletion. Later, patients with 22q11 duplication that had the breakpoints at these SDs have been reported and recognized as reciprocal to the 22q11 deletion [5]. During meiotic prophase, generation of DSBs is programed at the hundreds of sites throughout the human genome, which is followed by homologous chromosome pairing and subsequent error-free DNA repair by homologous recombination. A subset of recombination results in crossover that contributes to exchange in genetic materials between homologous chromosomes. However, highly homologous SDs nearby might induce mispairing of homologous chromosomes, resulting in interchromosomal nonallelic homologous recombination (NAHR). The possible outcome is a deletion on one allele and a duplication on the other allele. Another pathway that can produce SD-mediated deletion is an intrachromosomal nonallelic recombination between the SDs on the same chromosome. Further, such NAHR might occur even in the case of non-programmed DSB repair in the mitotic cells. Thus, GCRs of this type are likely to result from initial random DSBs followed by illegitimate NAHR. Most of the recurrent genomic disorders caused by deletion/duplication as well as Robertsonian translocations are generated by such homology-based mechanism [6].

The constitutional t(11;22)(q23;q11) is another example for recurrent GCR. The t(11;22) is the most frequent non-Robertsonian translocation in humans. Although balanced carriers of the t(11;22) usually manifest no clinical symptoms, they often have problems with reproduction, such as male infertility, recurrent spontaneous abortions, and the birth of offspring with chromosomal imbalance, Emanuel syndrome (MIM# 609029) [7]. The genomic analysis revealed that breakpoints on 11q23 and 22q11 cluster within palindromic AT-rich repeats (PATRRs) of approximately 500 bp in length. In spite of their similarity with regard to AT richness and palindromic structure, no substantial homology was observed between both PATRRs on 11q23 and 22q11 (58% identity). It is suggested that DSBs occur at the center of the two PATRRs followed by repair via the NHEJ pathway. In this way, studies for recurrent GCRs opened an avenue for elucidation of mechanism for general GCRs.

1.2 Chromothripsis in the Germline

Recent trends in utilizing cytogenetic microarrays or NGS to characterize disease-causing copy-number abnormalities have enabled the fine localization of GCR breakpoints. A substantial number of disease-related GCRs and benign copy-number variations have now been identified in this way, and the breakpoints have been analyzed. As a consequence, a subset of GCRs has been found to have more complex breakpoints than was previously thought. A novel concept for complex GCR was first introduced by Lupski et al. [6], who named it fork stalling and template switching (FoSTeS). Pelizaeus-Merzbacher disease is one of the hereditary neurodegenerative disorders resulting from the deletion/duplication of the chromosomal region that includes the dosage-sensitive *PLP1* gene on Xq22. In contrast to other deletion/duplication syndromes caused by NAHR, analysis of the breakpoints in this disorder has revealed no SD-like genomic structures at either end of these deletions or duplications. Instead, the junctions are characterized by complex structures harboring smaller deletions and duplications in either orientation that are joined via 2–5 nucleotides of microhomology. Based on the finding that sequence refractory fragments that preclude the progression of DNA polymerases have been consistently identified at these rearrangement junctions, a unique mechanism related to DNA replication stalling was proposed [8]. Later, since most of the cases do not show the replication barrier sequence, the name FoSTeS was revised by the same group as another term, microhomology-mediated break-induced replication (MMBIR) [3]. Genome-wide analysis of structural variations indicates that nearly 50% of breakpoints are generated via microhomology-mediated mechanisms [9]. Indeed, FoSTeS/MMBIR might be a more general mechanism that leads to the GCR and that having junction with complex structure might be exceptions for this mechanism.

In this context, a new paradigm referred to as chromothripsis, which refers to chromosome shattering followed by random reassembly, was somewhat of a sensation in this field [10]. Chromothripsis is defined on the basis of three main features: large numbers of breakpoints in fine localized region, a low number of copy-number states, and alternation of heterozygosity and loss of heterozygosity [11]. Features of junction sequences are reminiscent of repair via NHEJ. Chromothripsis was first discovered in cancer genome, and such a structure can be seen in 2–3% of cancer genomes. In contrast to the conventional multistep carcinogenesis theory, it was surprising that only one pathogenic episode might produce such complex genomic structure. Soon afterward, similar genomic structure was discovered in the germline of the patients with congenital anomaly [12]. Although these cases meet the requirement for definition of chromothripsis, the complexity is slightly simpler than those in cancer since selection might be stricter in the germline. A considerable number of case studies were subsequently published describing the involvement of chromothripsis in the mechanism of complex GCR in the germline [14, 15]. It was also reported that chromothripsis is generated via a replication-based mechanism like FoSTeS/MMBIR [13]. However, this conclusion may be somewhat premature at present as several features that were observed in replication-based mechanisms have not been observed in chromothripsis and vice versa. Although the other term, chromoanasynthesis, is proposed for such replication-based mechanisms for complex GCR, the terminology is becoming somewhat confusing and even chaotic [16, 17].

In case of the GCR that has been thought to have a simple structure based on findings observed by conventional cytogenetic analysis like G-staining, actual structure of the junction at the genomic resolution is more complicated than we previously thought. For example, chromosomal insertion, also called insertional translocation, involves translocation of a segment from one chromosome and its insertion as an interstitial region into another nonhomologous chromosome. The rearrangement requires at least three breaks followed by aberrant repair, but information on the breakpoints and junctions is scarce. A recent large-scale study using NGS of cases with an insertion identified the complex genomic structures in all cases and revealed the underlying mechanism leading to the insertion as a chromothripsis-like pathway [18, 19]. In this view, three-way translocation is the similar condition, in which three breakpoints are observed in G-staining. This might be a good candidate for genome analysis of the breakpoint regions.

1.3 How Does Such Complex GCRs Occur in the Germline

Chromosomal abnormalities in humans often manifest a gender bias with respect to parental origin. Aneuploidy, a numerical abnormality, is more likely to arise in maternal gametogenesis, whereas 80% of the GCRs are of paternal origin [1]. This is not surprising since it is formally possible that most of the GCRs arise via replication-based mechanism. Replication number is much higher in spermatogenesis than in oogenesis (Fig. 2). Likewise, a sizeable body of data has accumulated to show that most of the chromothripsis arise in male germline [20]. It is suggested that higher number of DNA replication or mitotic division reflects the predominance in the occurrence in the male germline. Post-meiosis hypothesis might be also possible since the mature sperm has no efficient DNA repair system and the sperm DNA occasionally becomes fragmented. Such DNA damage might be repaired after fertilization using maternal repair system, but how DNA is repaired is poorly understood [21].

On the other hand, recent experimental evidences suggest that chromosome segregation error such as anaphase lag during cell division leads to micronucleus containing the lagging chromosome [22, 23]. In the micronucleus, DNA replication might be unsynchronized with the nuclear DNA, leading to pulverization of the DNA in the micronucleus during the next mitotic cell division. Chromosome segregation error often arises during the first several mitotic divisions post-fertilization.

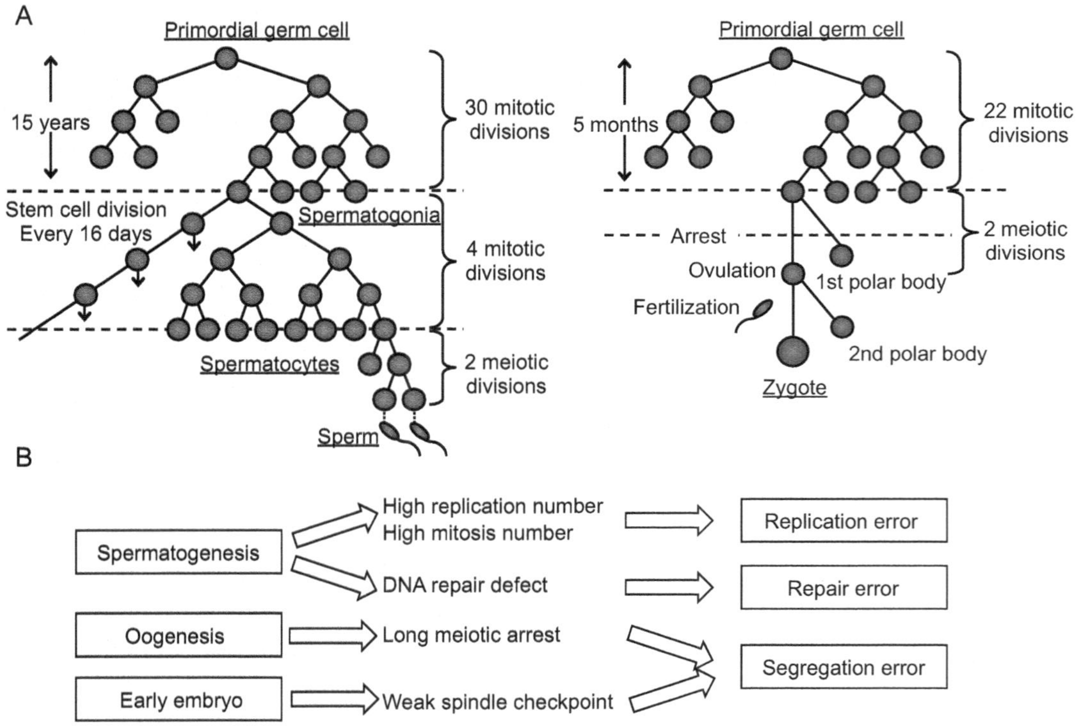

Fig. 2 Factors affecting the development of germline chromothripsis (Adapted from [1]). (**a**) Schematic representation of the differences between male and female gametogenesis. (**b**) Factors affecting the development of germline chromothripsis in gametogenesis

2 Clinical Cases with Germline Chromothripsis/Chromoanagenesis

2.1 Cases of Germline Chromoanagenesis

In 2011, Kloosterman et al. first reported patients with chromothripsis-compatible complex rearrangements in the germline [12]. These patients presented with congenital malformation syndromes. Subsequent studies provided further evidence that chromoanagenesis contributes to the development of de novo GCRs in the germline. To date, more than 30 individuals carrying chromoanagenesis-compatible germline genomic rearrangements have been reported (summarized in [24, 25]). In most of these cases, chromoanagenesis involved only one chromosome, but in a few cases, two to five chromosomes were affected. The affected chromosomes in these cases were variable and included almost all autosomes and the X chromosome. To our knowledge, Y chromosomal chromoanagenesis has not been identified in humans. Chromoanagenesis-mediated genomic rearrangements in the germline tend to have fewer number of breakpoints than those in cancer cells. These germline rearrangements usually arise as de novo abnormalities during meiosis, although in rare cases, they are inherited from phenotypically normal parents. Chromoanagenesis seems to account for a certain percentage of the etiology of complex rearrangements in the germline. Indeed, Kloosterman et al. analyzed genomes of ten individuals with complex germline rearrangements and identified chromothripsis in eight cases [15].

Importantly, de novo germline chromothripsis in previously reported cases invariably affected paternally derived chromosomes [20]. Thus far, there is no report of chromoanagenesis occurring during oogenesis. This suggests that chromoanagenesis takes place predominantly during male meiosis. It has been suggested that chromothripsis occurs primarily during sperm maturation or post-fertilization [20]. However, chromoanagenesis may also occur during premeiotic cell divisions [26]. The cause of paternal bias in the occurrence of de novo chromoanagenesis remains to be clarified. This phenomenon may reflect the relatively low DNA repair activity and/or high susceptibility to DNA damages of spermatids. However, considering the small number of reported cases, further studies are necessary to clarify whether germline chromoanagenesis can also occur during maternal meiosis. Moreover, it remains to be clarified whether germline chromoanagenesis is induced by environmental factors. In particular, increased parental age may play a role in the development of chromoanagenesis, because aging impairs DNA repair activity in germ cells [20].

2.2 Phenotypes of Patients with Germline Chromoanagenesis

2.2.1 Germline Chromoanagenesis of Autosomes

Complex autosomal rearrangements indicative of germline chromoanagenesis have been identified primarily in patients with mental retardation and dysmorphic features, with or without additional clinical features [24–28]. These rearrangements were frequently associated with copy-number gains and/or losses and occasionally with interchromosomal translocations. The phenotypes of these patients are likely to be caused by dysfunction, disruption, and/or dysregulation of several developmental genes on the affected chromosomes. Furthermore, such complex rearrangements may create aberrant chimeric genes at the fusion junction. Complex germline rearrangements indicative of chromoanagenesis have also been detected in a few individuals without clinically discernible abnormalities [27]. This suggests that chromoanagenesis is not necessarily pathogenic, particularly in the absence of copy-number alterations.

Although the frequency of chromoanagenesis-mediated rearrangements among fetuses has not been examined, such rearrangements may be relatively common in spontaneously aborted fetuses. Indeed, gross chromosomal alterations are known as the major cause of pregnancy loss [29]. Complex genomic rearrangements often exert deleterious effects on fetal growth and development by altering the expression of a number of functionally important genes. It is possible that, of all chromoanagenesis-mediated rearrangements, only those with relatively minor effects permit live birth. In this regard, it is noteworthy that chromoanagenesis-compatible rearrangements in previously reported cases were more

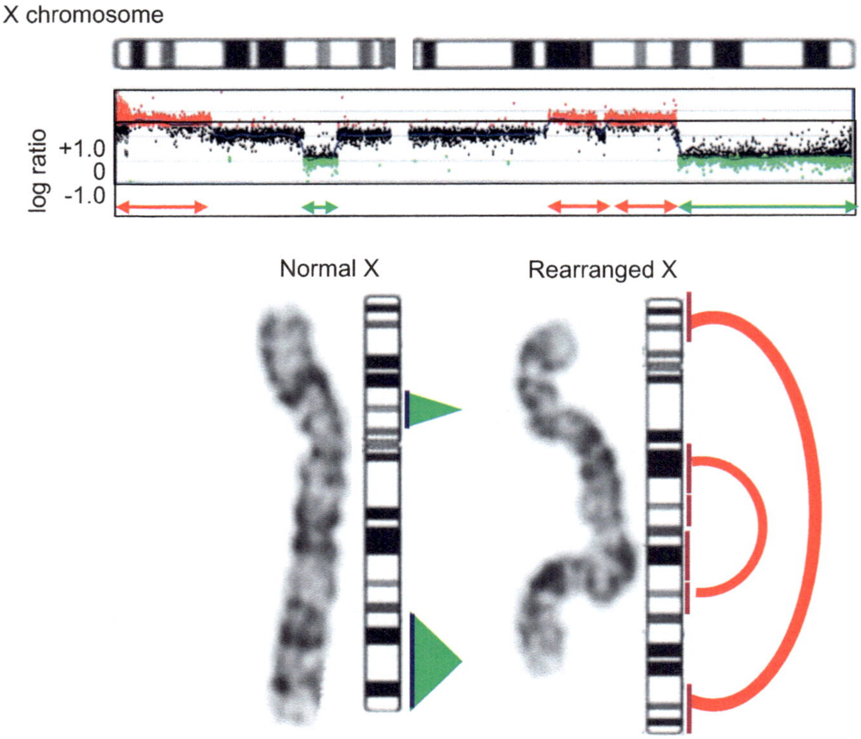

Fig. 3 X chromosomal chromothripsis identified in our previous study [30]. The results of array-based comparative genomic hybridization (the upper panel) and high-resolution karyotyping (the lower panel) are shown. This female patient carried a gross X chromosomal rearrangement with multiple copy-number gains (shown in red) and losses (green)

frequently associated with copy-number gains than copy-number losses [24–28]. Since gene deletions are more likely to exert damaging effects on human health than gene overdosage, the rearrangements in these cases may have relatively minor effects. It remains to be clarified whether chromoanagenesis-compatible GCRs, particularly those associated with copy-number losses, are common genetic causes of spontaneous abortion.

2.2.2 Germline Chromothripsis/Chromoanagenesis of the X Chromosome

Previous studies have identified chromoanagenesis-compatible X chromosomal rearrangements in a small number of female individuals [30–32]. Most of these individuals showed relatively mild manifestations compared to patients with chromoanagenesis-mediated rearrangements involving autosomes. A girl with an X chromosomal rearrangement, reported by Auger et al. [32], showed short stature and pulmonary alveolar proteinosis due to an unmasked recessive mutation in *CSF2RA*, but had no apparent developmental delay or dysmorphism. Likewise, we have identified a woman with a chromoanagenesis-compatible X chromosomal rearrangement, who had ovarian dysfunction and a thyroid disorder as sole clinically recognizable features (Fig. 3) [30]. The relatively mild phenotypic effects of this X chromosomal chromoanagenesis can be explained by the selective inactivation of the rearranged X chromosome. It is known that rearranged X chromosomes, except for those with balanced X-autosomal translocations, usually undergo selective inactivation [33]. Since 70–80% of genes on a normal X chromosome are subjected to X inactivation, skewed X inactivation serves to rescue the phenotypic severity of X chromosomal chromoanagenesis to some extent. Indeed, we confirmed completely skewed X chromosomal inactivation in the blood sample of our patient with chromoanagenesis-compatible X chromosomal rearrangement [30].

The phenotypes of women with chromoanagenesis-compatible X chromosomal rearrangements can be ascribed to gross chromosomal imbalance and dysfunction or dysregulation of genes that escape X inactivation. Moreover, gross X chromosomal rearrangements may cause clinical features by altering the DNA methylation status of some genes that usually escape inactivation. Indeed, we observed abnormal methylation of the *SHOX* gene promoter region in a woman with highly complex rearrangements on the X chromosome [30]. This phenomenon may be due to a perturbed physical interaction between inactivation-escaping genes and the X inactivation center. In addition, heterozygous deletions in the genome may unmask cryptic mutations of some genes responsible for recessive disorders, as seen in the case reported by Auger et al. [32].

2.3 Chromothriptic Cure of an Autosomal Dominant Disorder

In an exceptional case, chromothripsis has been shown to exert beneficial effects on the human health [34]. The case was a patient with WHIM (warts, hypogammaglobulinemia, infections, and myelokathexis) syndrome, an autosomal dominant immunodeficiency syndrome caused by gain-of-function mutations in *CXCR4*. In this case, a chromothripsis-medicated somatic deletion of chromosome 2 harboring a mutated *CXCR4* gene significantly lessened the clinical severity of the disease. This case represents a model for the somatic rescue of congenital disorders by chromothripsis.

2.4 Molecular Methods to Detect and Characterize Chromoanagenesis-Mediated GCR in the Germline

Chromoanagenesis-compatible rearrangements in the germline can be detected by karyotyping, NGS, array comparative genomic hybridization (CGH), single-nucleotide polymorphism (SNP) array analysis, and fluorescence in situ hybridization (FISH). Of these techniques, cytogenetic analyses using karyotyping and/or array CGH are commonly used for molecular diagnosis of patients with mental retardation and dysmorphic features and are, therefore, particularly useful to screen complex rearrangements. In recent studies, NGS was employed to detect and characterize chromosomal rearrangements in individuals with developmental delay [15, 24–28, 35, 36]. Data from these studies indicate that NHEJ and MMBIR are likely to be involved in the development of chromoanagenesis in the germline.

3 Chromothripsis/Chromoanagenesis in Cancer Cells

3.1 Cases of Somatic Chromoanagenesis

Chromothripsis was initially identified as a cause of somatic GCR in cancer cells. In 2011, Stephens et al. [10] identified 42 somatically acquired genomic rearrangements in a 62-year-old female individual with chronic lymphocytic leukemia. Most breakpoints were clustered on the short arm of chromosome 4, while a few were located on other chromosomes. Stephens et al. [10] detected additional complex genomic rearrangements in osteosarcoma, chordoma, and other cancer cell lines, which were also indicative of all-at-once catastrophic events.

To date, chromoanagenesis-compatible complex genomic rearrangements have been identified in several tumor cells. The overall frequency of chromoanagenesis in cancers remains unknown, although Stephens et al. [10] proposed that chromothripsis-compatible complex genomic rearrangements account for at least 2–3% of all cancer cells. It appears that the contribution of chromothripsis is variable depending on cancer types. Chromothripsis-compatible rearrangements are seen in more than 30% of esophageal adenocarcinomas, glioblastomas, and osteosarcomas, but they are relatively rare in ovarian cancers, glioneuronal tumors, and pheochromocytomas [37–40].

Previous studies have shown that chromoanagenesis can affect various autosomes and sex chromosomes in cancer cells. Typically, chromoanagenesis-mediated genomic rearrangements involve only one or a few chromosomes. However, Stephens et al. [10] identified a complex rearrangement in a chordoma sample that affected five chromosomes.

3.2 Pathogenic Effects of Somatic Chromoanagenesis

GCRs due to chromoanagenesis are likely to promote tumor development through diverse mechanisms. First, such rearrangements may alter the function or expression of tumor suppressor genes on the affected chromosomes. For example, chromosomal deletions involving *CDKN2A*, a tumor suppressor gene on chromosome 9, were observed in cancer cell lines [16]. Furthermore, Stephens et al. [10] reported a chordoma case in which chromoanagenesis resulted in disruption of three tumor suppressor genes. Second, some genomic rearrangements increase the expression of oncogenes on the affected chromosomes by increasing gene copy number or by disrupting the regulatory machinery. It is known that DNA fragments generated by chromoanagenesis can form small circles referred to as double minutes [41], which may contain oncogenes. Indeed, chromoanagenesis of chromosome 8 has been shown to facilitate the creation of a double minute chromosome containing the MYC oncogene, whose amplification induces tumor progression. Third, complex genomic rearrangements may create aberrant fusion genes with oncogenic activity. Since chromoanagenesis-mediated genomic rearrangements in cancer cells often have more than 100 breakpoints [39], some of the fusion junctions may serve as the source of chimeric genes.

The phenotypic effects of somatic chromothripsis appear to be variable, possibly depending on the affected chromosomes and the origin of the cancers. Chromoanagenesis has initially been associated with poor prognosis, although chromoanagenesis-compatible genomic rearrangements have also been observed in tumor cells with relatively low malignancy.

3.3 Molecular Methods to Detect and Characterize Chromoanagenesis-Mediated GCR in Tumor Cells

Various molecular technologies can be utilized to investigate somatic genomic rearrangements created by chromoanagenesis. Of these, paired-end whole-genome sequencing (WGS) using NGS is particularly powerful for characterizing complex genomic rearrangements. WGS can provide information about the underlying mechanisms of the rearrangements by identifying the sequences at the breakpoint junctions. It has been shown that NHEJ and MMBIR play major roles during chromoanagenesis in cancer cells [10, 13].

In addition to WGS, genome-wide copy-number analyses, such as array CGH and SNP array assays, are frequently used to identify deletions and amplifications in the genome. Karyotyping and FISH are used to determine the structure of rearranged chromosomes. Furthermore, advanced FISH technologies, such as multicolor FISH, whole chromosome painting, and spectral karyotyping (SKY) are highly beneficial for the detailed characterization of complex rearrangements involving multiple chromosomes [29]. FISH and its derivative techniques often uncovered additional complexity in previously identified chromosomal rearrangements.

References

1. Kurahashi H, Bolor H, Kato T et al (2009) Recent advance in our understanding of the molecular nature of chromosomal abnormalities. J Hum Genet 54(5):253–360. https://doi.org/10.1038/jhg.2009.35
2. Richardson C, Jasin M (2000) Frequent chromosomal translocations induced by DNA double-strand breaks. Nature 405:697–700
3. Hastings PJ, Lupski JR, Rosenberg SM et al (2009) Mechanisms of change in gene copy number. Nat Rev Genet 10(8):551–564. https://doi.org/10.1038/nrg2593
4. Shaikh TH, Kurahashi H, Saitta SC et al (2000) Chromosome 22-specific low copy repeats and the 22q11.2 deletion syndrome: genomic organization and deletion endpoint analysis. Hum Mol Genet 9(4):489–501
5. Edelmann L, Pandita RK, Spiteri E et al (1999) A common molecular basis for rearrangement disorders on chromosome 22q11. Hum Mol Genet 8(7):1157–1167
6. Lupski JR (1998) Genomic disorders: structural features of the genome can lead to DNA rearrangements and human disease traits. Trends Genet 14(10):417–422
7. Kurahashi H, Inagaki H, Ohye T et al (2010) The constitutional t(11;22): implications for a novel mechanism responsible for gross chromosomal rearrangements. Clin Genet 78(4):299–309. https://doi.org/10.1111/j.1399-0004.2010.01445.x
8. Lee JA, Carvalho CM, Lupski JR (2007) A DNA replication mechanism for generating nonrecurrent rearrangements associated with genomic disorders. Cell 131(7):1235–1247
9. Conrad DF, Bird C, Blackburne B et al (2010) Mutation spectrum revealed by breakpoint sequencing of human germline CNVs. Nat

Genet 42(5):385–391. https://doi.org/10.1038/ng.564

10. Stephens PJ, Greenman CD, Fu B et al (2011) Massive genomic rearrangement acquired in a single catastrophic event during cancer development. Cell 144(1):27–40. https://doi.org/10.1016/j.cell.2010.11.055
11. Forment JV, Kaidi A, Jackson SP (2012) Chromothripsis and cancer: causes and consequences of chromosome shattering. Nat Rev Cancer 12(10):663–670. https://doi.org/10.1038/nrc3352
12. Kloosterman WP, Guryev V, van Roosmalen M et al (2011) Chromothripsis as a mechanism driving complex de novo structural rearrangements in the germline. Hum Mol Genet 20(10):1916–1924. https://doi.org/10.1093/hmg/ddr073
13. Liu P, Erez A, Nagamani SC et al (2011) Chromosome catastrophes involve replication mechanisms generating complex genomic rearrangements. Cell 146(6):889–903. https://doi.org/10.1016/j.cell.2011.07.042
14. Chiang C, Jacobsen JC, Ernst C et al (2012) Complex reorganization and predominant non-homologous repair following chromosomal breakage in karyotypically balanced germline rearrangements and transgenic integration. Nat Genet 44(4):390–397, S1. https://doi.org/10.1038/ng.2202
15. Kloosterman WP, Tavakoli-Yaraki M, van Roosmalen MJ et al (2012) Constitutional chromothripsis rearrangements involve clustered double-stranded DNA breaks and nonhomologous repair mechanisms. Cell Rep 1(6):648–655. https://doi.org/10.1016/j.celrep.2012.05.009
16. Holland AJ, Cleveland DW (2012) Chromoanagenesis and cancer: mechanisms and consequences of localized, complex chromosomal rearrangements. Nat Med 18(11):1630–1638. https://doi.org/10.1038/nm.2988
17. Kurahashi H, Ohye T, Inagaki H et al (2012) Mechanism of complex gross chromosomal rearrangements: a commentary on concomitant microduplications of MECP2 and ATRX in male patients with severe mental retardation. J Hum Genet 57(2):81–83. https://doi.org/10.1038/jhg.2011.143
18. Weckselblatt B, Hermetz KE, Rudd MK (2015) Unbalanced translocations arise from diverse mutational mechanisms including chromothripsis. Genome Res 25(7):937–947. https://doi.org/10.1101/gr.191247.115
19. Kato T, Ouchi Y, Inagaki H et al (2017) Genomic characterization of chromosomal insertions: implication for mechanism leading to the chromothripsis. Cytogenet Genome Res. https://doi.org/10.1159/000481586
20. Pellestor F, Gatinois V, Puechberty J et al (2014) Chromothripsis: potential origin in gametogenesis and preimplantation cell divisions. A review. Fertil Steril 102(6):1785–1796. https://doi.org/10.1016/j.fertnstert.2014.09.006
21. Marchetti F, Essers J, Kanaar R et al (2007) Disruption of maternal DNA repair increases sperm-derived chromosomal aberrations. Proc Natl Acad Sci U S A 104(45):17725–17729
22. Zhang CZ, Spektor A, Cornils H et al (2015) Chromothripsis from DNA damage in micronuclei. Nature 522(7555):179–184. https://doi.org/10.1038/nature14493
23. Ly P, Teitz LS, Kim DH et al (2017) Selective Y centromere inactivation triggers chromosome shattering in micronuclei and repair by non-homologous end joining. Nat Cell Biol 19(1):68–75. https://doi.org/10.1038/ncb3450
24. Fukami M, Shima H, Suzuki E et al (2017) Catastrophic cellular events leading to complex chromosomal rearrangements in the germline. Clin Genet 91(5):653–660. https://doi.org/10.1111/cge.12928
25. Poot M, Haaf T (2015) Mechanisms of origin, phenotypic effects and diagnostic implications of complex chromosome rearrangements. Mol Syndromol 6(3):110–134. https://doi.org/10.1159/000438812
26. Kloosterman WP, Cuppen E (2013) Chromothripsis in congenital disorders and cancer: similarities and differences. Curr Opin Cell Biol 25(3):341–348. https://doi.org/10.1016/j.ceb.2013.02.008
27. de Pagter MS, van Roosmalen MJ, Baas AF et al (2015) Chromothripsis in healthy individuals affects multiple protein-coding genes and can result in severe congenital abnormalities in offspring. Am J Hum Genet 96:651–656
28. Masset H, Hestand MS, Van Esch H et al (2016) A distinct class of chromoanagenesis events characterized by focal copy number gains. Hum Mutat 37:661–668
29. Pellestor F, Anahory T, Lefort G et al (2011) Complex chromosomal rearrangements: origin and meiotic behavior. Hum Reprod Update 17(4):476–494. https://doi.org/10.1093/humupd/dmr010
30. Suzuki E, Shima H, Toki M et al (2017) Complex X-chromosomal rearrangements in two women with ovarian dysfunction: implications for chromothripsis/chromoanasynthesis-dependent and independent origins of complex genomic alterations. Cytogenet Genome Res

150(2):86–92. https://doi.org/10.1159/000455026

31. Ochalski ME, Engle N, Wakim A et al (2011) Complex X chromosome rearrangement delineated by array comparative genome hybridization in a woman with premature ovarian insufficiency. Fertil Steril 95:2433.e9–2433.15
32. Auger J, Bonnet C, Valduga M et al (2013) De novo complex X chromosome rearrangement unmasking maternally inherited CSF2RA deletion in a girl with pulmonary alveolar proteinosis. Am J Med Genet A 161A:2594–2599
33. Zhong Q, Layman LC (2012) Genetic considerations in the patient with turner syndrome—45,X with or without mosaicism. Fertil Steril 98:775–779
34. McDermott DH, Gao JL, Liu Q et al (2015) Chromothriptic cure of WHIM syndrome. Cell 160:686–699
35. Zanardo ÉA, Piazzon FB, Dutra RL et al (2014) Complex structural rearrangement features suggesting chromoanagenesis mechanism in a case of 1p36 deletion syndrome. Mol Gen Genomics 289:1037–1043
36. Plaisancié J, Kleinfinger P, Cances C et al (2014) Constitutional chromoanasynthesis: description of a rare chromosomal event in a patient. Eur J Med Genet 57:567–570
37. Kloosterman WP, Koster J, Molenaar JJ (2014) Prevalence and clinical implications of chromothripsis in cancer genomes. Curr Opin Oncol 26:64–67
38. Mehine M, Kaasinen E, Mäkinen N et al (2013) Characterization of uterine leiomyomas by whole-genome sequencing. N Engl J Med 369:43–53
39. Storchová Z, Kloosterman WP (2016) The genomic characteristics and cellular origin of chromothripsis. Curr Opin Cell Biol 40:106–113
40. Rode A, Maass KK, Willmund KV et al (2016) Chromothripsis in cancer cells: an update. Int J Cancer 138(10):2322–2333. https://doi.org/10.1002/ijc.29888
41. Hatch EM, Hetzer MW (2015) Linking micronuclei to chromosome fragmentation. Cell 161:1502–1504

Check for updates

Chapter 3

Potential Role of Chromothripsis in the Genesis of Complex Chromosomal Rearrangements in Human Gametes and Preimplantation Embryo

Franck Pellestor and Vincent Gatinois

Abstract

The discovery of a new class of massive chromosomal rearrangement, baptized chromothripsis, in different cancers and congenital disorders has deeply modified our understanding on the genesis of complex genomic rearrangements. Several mechanisms, involving abortive apoptosis, telomere erosion, mitotic errors, micronuclei formation, and p53 inactivation, might cause chromothripsis. The remarkable point is that all these plausible mechanisms have been identified in the field of human reproduction as causal factors for reproductive failures and chromosomal abnormality genesis. Specific features of gametogenesis and early embryonic development may contribute to the emergence of chromothripsis. Multiple lines of evidence support the assumption that chromothripsis may arise more frequently than previously thought in both gametogenesis and early human embryogenesis.

Key words Chromothripsis, Massive chromosomal rearrangement, Spermatogenesis, Oogenesis, Fertilization, Embryo

1 Introduction

The emergence of a new type of massive and chaotic chromosomal rearrangement, coined chromothripsis, has radically modified the view of how complex rearrangements are generated.

Initially described in cancers [1–3], and then in congenital disorders [4–6], chromothripsis is characterized by the shattering of one or a few chromosomal segment(s) followed by a chaotic reassembly of the chromosomal fragments, both events occurring during one unique cellular event. Several causative mechanisms, involving abortive apoptosis, telomere erosion, mitotic errors, micronuclei formation, and p53 inactivation, have been proposed [7–11]. The remarkable point is that all these potential mechanisms have been identified in the field of human reproduction as causal factors for reproductive failures and chromosomal abnormality formation [12]. Recent reports validate the occurrence of

Franck Pellestor (ed.), *Chromothripsis: Methods and Protocols*, Methods in Molecular Biology, vol. 1769, https://doi.org/10.1007/978-1-4939-7780-2_3, © Springer Science+Business Media, LLC, part of Springer Nature 2018

chromothripsis in the human germline or in early embryonic development but also confirm its compatibility with viability and transmission to subsequent generations [13–15].

2 Hallmarks of Chromothripsis

Chromothripsis was defined on the basis of a few features common to all chromothripsis rearrangements: the generation of remarkable numbers of rearrangements locally on one single chromosome or a few chromosomes, the maintenance of a low number of copy number states oscillating between 1 and 2, and the alternation of preserved heterozygosity and segments with loss of heterozygosity (LOH) in the chromothriptic areas.

The occurrence of chromothripsis has also been strongly associated with dysregulation or loss of p53 tumor suppressor genes. Known as the guardian of the genome, p53 plays a major role in maintaining genome stability by mediating cell-cycle arrest, apoptosis, and cell senescence in response to DNA damages [16, 17].

Congenital chromothripsis is characterized by their relatively balanced status. This does not signify that unbalanced chromothripsis events don't occur in germline or during post-zygotic divisions, but it strongly emphasizes the effect of selective pressure as a bias factor in the assessment of congenital chromothripsis, since only balanced chromothripsis outcomes compatible with life have been found in individuals to date. Altogether, these key features define a molecular signature of chromothripsis [18–20].

3 Chromothripsis and Reproduction

All the mechanisms initiating chromothripsis can theoretically occur in gametes as well as in zygotes and preimplantation embryo.

3.1 Chromothripsis in Spermatogenesis

Existing data show that most of chromothripsis arises in the male germline [21–24]. This finding underlines the great vulnerability of spermatogenesis to DNA damage and clearly points out spermatogenesis as a critical stage in the genesis of congenital chromothripsis.

The male germline displays a limited DNA repair capacity as compared to somatic tissue cells, and the mutation rate is greater in spermatogenesis than in oogenesis. In the course of spermatogenesis, chromothripsis can arise either during the premeiotic division of diploid spermatogonia, through the first meiotic division, or during the differentiation of round spermatids into spermatozoa (i.e., spermiogenesis). During each of the three steps, environmental stimuli such as ionizing radiations or free radicals can act as triggers for DNA damage.

In spermatogonia, the DNA repair appears to be significantly different than that of somatic cells, and the DNA quality control pathway operates in combination with the apoptosis process. About 50% of cells that enter meiosis I are eliminated by apoptosis [25]. During spermatogenesis, apoptosis is a mechanism controlled by Sertoli cells for the elimination of defective germ cells. Tubio and Estivill [7] proposed that chromothripsis might be caused by abortive apoptosis. Whereas apoptosis was considered as an irreversible cascade of extensive chromatin fragmentation leading to cell death, several lines of evidence have suggested that abortion of apoptosis could contribute to the occurrence of chromosome rearrangements [26]. Thus, a small subset of cells could undergo a restricted form of apoptosis and survive. The subsequent DNA repair might be accomplished through a fast and incorrect repair process, promoting the emergence of chaotic chromosomal rearrangement. To support this hypothesis, a recent report has shown that apoptosis process could be reversed [27]. Among the cells undergoing apoptosis, a few could undergo a restricted form of apoptosis or escape the process and subsequently survive after performing rapid DNA repair to maintain their genome integrity [28]. This could generate multiple and chaotic chromosome rearrangements in some spermatogonia.

In addition, spermatogonia must undergo hundreds of mitotic divisions before entering into meiotic prophase, which implies that these cells are particularly prone to DNA damage. Consistent with the role of DNA replication defects in the chromothripsis process, replicative stress can then serve as an endogenous stimulus. Replication stress and mitotic error could synergize to induce chromosomal instability and chromothripsis occurrence [29]. One could speculate that mitotic errors during chromosome segregation may be a key event in the initiation of chromothripsis, especially in human gametes which are highly sensitive to aneuploidy [30].

During meiosis, the chromatin of spermatocytes sustains programmed structural changes which require the formation and the repair of DSBs (i.e., meiotic recombination). This implies the expression of numerous genes [31] and the recruitment of DSB repair proteins on the synaptonemal complexes. The DSB occurrence and repair during meiosis require homologous pairing between non-sister chromatids, whereas normal DSB repair in somatic cells involves sister chromatids. There is increasing evidence that in the case of incomplete paring of homologous chromosomes, the formation of nonhomologous synapsis (heterosynapsis) before the activation of apoptosis is a strategy for spermatocytes to escape the pachytene checkpoint [32]. It is well established that this process promotes the formation of chromosomal structural rearrangements [33]. It could also be a way to generate chromothripsis-related rearrangements.

Finally, during spermiogenesis, spermatids undergo remarkable genomic reorganization involving the histone-to-protamine transition which leads the condensed packaging of the haploid genome in spermatozoa. Alterations in these processes can lead to DNA fragmentation, and incomplete protamination has been correlated with decreased DNA integrity [34]. This DNA damage will make spermatozoa more susceptible to post-testicular assaults, especially during transport through the epididymis [35].

At this stage, DSB cannot be repaired by homologous recombination since spermatids lack sister chromatids for homologous repair. Effective checkpoints trigger the elimination of damaged cells by apoptosis. However, this process might not always operate efficiently since a variable proportion of defective spermatozoa are observed in the ejaculate. This could be potentially associated with the occurrence of abortive apoptosis stimulating the occurrence of chromothripsis events.

3.2 Chromothripsis in Oogenesis

Oocytes are particularly vulnerable to environmental attacks and DNA alterations and can cumulate genomic and chromosomal damages for decades [12]. During oogenesis, the primordial germ cells undergo only a few mitotic divisions before undergoing complex meiotic events of synapsis and recombination and entering the protracted arrest phase in late prophase [36]. Consequently, oocytes are sensitive to various exogenous agents which can generate DSB during the meiotic prophase, the follicle formation, the oocyte growth, and meiotic completion [37]. In contrast to male germ cells, oocytes are able to repair DNA damage throughout the complete oogenesis process, using either HR or NHEJ [38]. DNA repair genes are expressed at high levels in human oocytes [39]. However, this capacity is highly variable from oocyte to oocyte.

Also, oocytes display permissive cell-cycle checkpoints [40]. The molecular basis for the lack of DNA repair pathway appears to be the oocyte failure to activate ATM kinase, an essential regulator of the DNA damage response [41]. Failures in these DNA repair pathways may have similar consequence as those previously described in male germ cells, with the potential occurrence of abortive apoptose, or replication stress as a trigger of chromothripsis. The synergistic combination of replicative stress and mitotic error might be particularly prevalent in oogenesis because of the frequent occurrence of chromosome malsegregation in female gametes. Chromosome malsegregation or premature separation of sister chromatids occurring during the two female meiotic divisions can induce genome instability, micronuclei formation, and chromothripsis events [42].

3.3 Chromothripsis in Zygotes and Preimplantation Embryos

During the fertilization process, the paternal nucleus undergoes chromatin remodeling and DNA replication corresponding to male pronucleus formation [43]. DNA breaks generated during spermiogenesis and sperm transport as well as DNA damages generated in oocytes need to be repaired before the first round of DNA replication in the zygote, in order to create a safe zygotic diploid nuclear compatible with embryonic development. This crucial phase of DNA repair is entirely dependent on the repair capacities of the oocyte since maternal gene products persist until complete activation of the embryo nucleus, i.e., at the four- to eight-cell stage. However, the ability of human oocytes to repair DNA damages from the fertilizing sperm nucleus does not exceed 8–10% of the haploid genome [44]. When confronted with extensive DNA damages, the classical error-free HR repair pathway could be saturated and inefficient, and the zygote could die. In this context, good-quality zygotes might undergo abortive apoptosis and/or initiate fast and error-prone repair machinery in order to maintain a relatively complete and balanced genome despite the generation of massive rearrangements.

Another mechanism that may initiate chromothripsis during fertilization is premature chromosome condensation (PCC) of sperm nuclei, due to asynchrony between the female nucleus blocked in the metaphase II stage and the male nucleus in the S phase [45]. Thus, incompletely replicated male chromosomes can be partially pulverized and compacted. In the vast majority of cases, zygotes with such extensive damages will die through an apoptotic process. However, the zygote with partially under-replicated and fragmented DNA might then prematurely enter mitosis and generate the formation of micronuclei in which fragmented chromosomes can undergo random repair before being reincorporated into the nucleus of a daughter embryonic cell [46].

4 Conclusion

According to the mechanisms described above, there is growing evidence that chromothripsis occurs mainly through spermatogenesis and preimplantation stages. In the biological context of fertilization and early embryonic development, an important characteristic of chromothripsis is its occurrence through a unique cellular event because of the strong impact of each early division on the genetic status of the future embryo and its viability.

Most of chromothripsis events probably lead to nonviable embryos because of the extreme complexity of generated rearrangements and their unbalanced status. Alternatively, chromothripsis resulting in balanced and stable chromosomal rearrangements would be viable and transmissible, despite a chaotic and complex chromosomal reorganization. According to the embryonic stage where the catastrophic event occurs, chromothripsis will be confined to extraembryonic tissues or be present as full chromothripsis or mosaic chromothripsis in the embryonic lineage and have pathogenic consequences in live borns [47].

Acknowledgments

Work in the unit of Chromosomal Genetics is supported by the CHU research platform ChromoStem (http://www.chu-montpellier.fr/fr/chercheurs/plateformes/les-plateformes-recherche/chromostem/).

References

1. Stephens PJ, Greenman CD, Fu B et al (2011) Massive genomic rearrangement acquired in a single catastrophic event during cancer development. Cell 144:27–40
2. Magrangeas F, Avet-Loiseau H, Munshi NC et al (2011) Chromothripsis identifies a rare and aggressive entity among newly diagnosed multiple myeloma patients. Blood 118:675–678
3. Rode A, Maass KK, Willmund KV et al (2016) Chromothripsis in cancer cells: an update. Int J Cancer 138:2322–2333
4. Kloosterman WP, Gurvey V, van Roosmalen M et al (2011) Chromothripsis as a mechanism driving complex de novo structural rearrangements in the germline. Hum Mol Genet 20:1916–1924
5. Chiang C, Jacobsen JC, Ernst C et al (2012) Complex reorganization and predominant non-homologous repair following chromosomal breakage in karyotypically balanced germline rearrangements and transgenic integration. Nat Genet 44:390–998
6. De Pagter MS, van Roosmalen MJ, Baas AF et al (2015) Chromothripsis in healthy individuals affects multiple protein-coding genes and can result in severe congenital abnormalities in offspring. Am J Hum Genet 96:651–656
7. Tubio JMC, Estivill X (2011) When catastrophe strikes a cell. Nature 470:476–477
8. Crasta K, Ganem NJ, Dagher R et al (2012) DNA breaks and chromosome pulverization from errors in mitosis. Nature 482:53–58
9. Zhang CZ, Spektor A, Cornils H et al (2015) Chromothripsis from DNA damage in micronuclei. Nature 522:179–184
10. Mardin BR, Drainas AP, Waszak SM et al (2015) A cell-based model system links chromothripsis with hyperploidy. Mol Syst Biol 11:828. https://doi.org/10.15252/msb.20156505
11. Maciejowski j LY, Bosco N et al (2015) Chromothripsis and Kataegis induced by telomere crisis. Cell 163:1641–1654
12. Pellestor F, Gatinois V, Puechberty J et al (2014) Chromothripsis: potential origin in gametogenesis and preimplantation cell divisions. A review. Fertil Steril 102:1785–1796
13. Daughtry BL, Chavez SL (2016) Chromosomal instability in mammalian pre-implantation embryos: potential causes, detection methods, and clinical consequences. Cell Tissue Res 363:201–225
14. Pellestor F (2014) Chromothripsis: how does such a catastrophic event impact human reproduction ? Hum Reprod 29:388–393
15. Fukami M, Shima H, Suzuki E et al (2017) Catastrophic cellular events leading to complex chromosomal rearrangements in the germline. Clin Genet 91:653–660
16. Rausch T, Jones DTW, Zapatka M et al (2012) Genome sequencing of pediatric medulloblastoma links catastrophic DNA rearrangements with TP53 mutations. Cell 148:59–71
17. Ivkov R, Bunz F (2015) Pathways to chromothripsis. Cell Cycle 14:2886–2890
18. Korbel JO, Campbell PJ (2013) Criteria for inference of chromothripsis in cancer genomes. Cell 152:1226–1236
19. Kloosterman WP, Cuppen E (2013) Chromothripsis in congenital disorders and cancer: similarities and differences. Curr Opin Cell Biol 25:341–348
20. Yang J, Liu J, Ouyang L et al (2016) CTLP scanner: a web server for chromothripsis-like pattern detection. Nucleic Acids Res 44(W1):W252–W258. https://doi.org/10.1093/nar/gkw434
21. Kloosterman WP, Tavakoli-Yaraki M, van Roosmalen M et al (2012) Constitutional chromothripsis rearrangements involve clustered double-stranded DNA breaks and nonhomologous repair mechanisms. Cell Rep 1:648–655
22. Macera MJ, Sobrino A, Levy B et al (2015) Prenatal diagnosis of chromothripsis, with nine breaks characterized by karyotyping, FISH, microarray and whole-genome sequencing. Prenat Diagn 35:299–301

23. Bertelsen B, Nazaryan-Petersen L, Sun W et al (2016) A germline chromothripsis event stably segregating in 11 individuals through three generations. Genet Med 15:494–500
24. Anderson SE, Kamath A, Pilz DT et al (2016) A rare example of germ-line chromothripsis resulting in large genomic imbalance. Clin Dysmorphol 25:58–62
25. Sakkas D, Alvarez JG (2010) Sperm DNA fragmentation: mechanisms of origin, impact on reproductive outcome, and analysis. Fertil Steril 93:1027–1036
26. Taylor RC, Cullen SP, Martin SJ (2008) Apoptosis: controlled demolition at the cellular level. Nat Rev Mol Cell Biol 9:231–241
27. Tang HL, Tang HM, Mak KH et al (2012) Cell survival, DNA damage, and oncogenic transformation after a transient and reversible apoptotic response. Mol Biol Cell 23:2240–2252
28. Sakkas D, Seli E, Bizzaro D et al (2003) Abnormal spermatozoa in the ejaculate: abortive apoptosis and faulty nuclear remodeling during spermatogenesis. Reprod Biomed Online 7:428–432
29. Jones MJK, Jallepalli PV (2012) Chromothripsis: chromosomes in crisis. Dev Cell 23:908–917
30. Delhanty JDA, Pellestor F (eds) (2011) Aneuploidy. Cytogenet Genome Res 133:2–4
31. Baarends WM, van der Laan R, Grootegoed A (2001) DNA repair mechanisms and gametogenesis. Reproduction 121:31–39
32. Oliver-Bonnet M, Ko E, Martin RH (2005) Male infertility in reciprocal translocation carriers: the sex body affair. Cytogenet Genome Res 111:434–346
33. Martin RH (2008) Cytogenetic determinants of male fertility. Hum Reprod Update 14:379–390
34. Aoki VW, Liu L, Carrell DT (2005) Identification and evaluation of a novel sperm protamine abnormality in a population of infertile males. Hum Reprod 20:1298–1306
35. Oliva R (2006) Protamines and male infertility. Hum Reprod Update 12:417–435
36. Carroll J, Marangos P (2013) The DNA damage response in mammalian oocytes. Front Genet 4:1–9
37. Hunt PA, Hassold TJ (2008) Human female meiosis: what makes a good egg go bad? Trends Genet 24:86–93
38. Marchetti F, Essers J, Kanaar R et al (2007) Disruption of maternal DNA repair increases sperm-derived chromosomal aberrations. Proc Natl Acad Sci U S A 104:17725–17729
39. Jaroudi S, Kakourou G, Cawood S et al (2009) Expression profiling of DNA repair genes in human oocytes and blastocysts using microarrays. Hum Reprod 24:2649–2655
40. Steuerwald N (2005) Meiotic spindle checkpoints for assessment of aneuploid oocytes. Cytogenet Genome Res 111:256–259
41. Kastan MB, Lim DS (2000) The many substrates and functions of ATM. Nat Rev Mol Cell Biol 1:179–186
42. Pellestor F, Anahory T, Hamamah S (2005) The chromosomal analysis of human oocytes. An overview of established procedures. Hum Reprod Update 11:15–32
43. McLay DW, Clarke HJ (2003) Remodeling the paternal chromatin at fertilization in mammals. Reproduction 125:625–633
44. Ahmadi A, Ng SC (1999) Developmental capacity of damaged spermatozoa. Hum Reprod 14:2279–2285
45. Eichenlaub-Ritter U, Schmiady H, Kentenich H et al (1995) Recurrent failure in polar body formation and premature chromosome condensation in oocytes from a human patient: indicators of asynchrony in nuclear and cytoplasmic maturation. Hum Reprod 10:2343–2349
46. Meyerson M, Pellman D (2011) Cancer genomes evolve by pulverizing single chromosome. Cell 144:9–10
47. Lebedev I (2011) Mosaic aneuploidy in early fetal losses. Cytogenet Genome Res 133:169–183

Chapter 4

Chromothripsis and the Macroevolution Theory

Franck Pellestor

Abstract

The recent discovery of a new class of massive chromosomal rearrangements, occurring during one unique cellular event and baptized "chromothripsis," deeply modifies our perception on the genesis of complex genomic rearrangements, but also it raises the question of the potential driving role of chromothripsis in species evolution. The occurrence of chromothripsis appears to be in good agreement with macroevolution models proposed as a complement to phyletic gradualism. The emergence of this unexpected phenomenon may help to demonstrate the contribution of chromosome rearrangements to speciation process.

Key words Chromothripsis, Chromosome, Macroevolution, Speciation

1 Introduction

Chromothripsis appears to be an extraordinary phenomenon of chaotic genomic rearrangement, affecting one or a small set of chromosomes in a one-step cellular event of chromosomal shattering and reassembly. First described in cancers through the progress of in-depth DNA sequence analysis, chromothripsis patterns have also been observed in patients with congenital disorders [1–4].

The occurrence of chromothripsis into a cell addresses the conceptual question about how chromosomes get shattered and how the pulverized fragments are stitched together through a single cellular event. Various mechanisms, involving abortive apoptosis, telomere erosion, mitotic errors, micronuclei formation, and p53 inactivation, can cause chromothripsis [5–8]. Also, specific features of gametogenesis and early embryonic development, such as the weakness of cell cycle and mitosis checkpoints and the rapid kinetics of division in germ cells and early cleavage embryos, may contribute to the emergence of chromothripsis [9, 10].

Several key features common to all chromothripsis events, such as the clustering of chromosomal breakpoints and oscillating copy number states, have been described, allowing to distinguish

Franck Pellestor (ed.), *Chromothripsis: Methods and Protocols*, Methods in Molecular Biology, vol. 1769,
https://doi.org/10.1007/978-1-4939-7780-2_4,

chromothripsis from other types of complex chromosomal rearrangements. Altogether, these key features define a molecular signature of chromothripsis [11] and facilitate its ascertainment by using bioinformatic tools for exploring chromothripsis-like genome patterns [12, 13].

2 Chromothripsis and Models of Macroevolution

The emergence of chromothripsis in both cancers and congenital disorders provides a new perception of how genomes can be rapidly altered. Beyond the impact of chromothripsis as a significant pathogenic phenomenon, the unexpected discovery of this mechanism raises the interesting question of the potential driving role of chromothripsis in species evolution.

Because chromothripsis features oppose the model of progressive acquisition of driver mutations or rearrangements, this phenomenon conceptually seems to give support to theories of the "hopeful monsters" and the "punctuated equilibrium" and thus sheds light on the links between the occurrence of chromosomal rearrangements and the speciation process. The implication of large and profound genetic alterations in speciation was originally suggested by the geneticist R. Goldschmidt in a model of macroevolution known as the "hopeful monster" model [14]. Goldschmidt postulated that macromutations in "rate" genes or "controlling" genes might modify early development and cause large effects in the adult phenotypes. According to this view, evolution proceeds by the rare success of such hopeful monsters and not by the gradual accumulation of small changes within populations. Since this hypothesis argued against the strict gradualism of neo-Darwinism, it was subjected to derision and scorn. In 1972, Eldredge and Gould [15] reformulated the macroevolution model and proposed the punctuated equilibrium theory as a complement to phyletic gradualism. This new model postulated that species undergo little variation during most of their evolutionary history (corresponding to stasis of established species), interrupted by rare but abrupt bursts of change (evolutionary shifts) leading to the splitting of lineages and the subsequent emergence of new species [16]. The punctuated changes might involve sporadic genetic mutations with large effects or chromosome rearrangements affecting gene expression. During the last decade, there has been a renewed interest for these theories in the field of evolutionary biology since some experimental data have demonstrated how single genetic alterations (mutations or recombinations) can have huge impacts in developmental evolution [17, 18]. To date, a growing number of reports have documented punctuated equilibrium and hopeful monsters in various species, introducing the notion of "transgressive segregation" for the generation and the rapid fixation in population of new genotypes that express phenotypes outside the normal range of variations observed in either parental gene pool [19, 20].

3 New Insights into the Consequences of Chromothripsis

Research in recent years has also revealed that de novo genes from noncoding DNA could emerge in various eukaryote lineages and contribute to adaptive evolution. The fact that random DNA sequence could be translated and encoded for functional protein has been recently demonstrated in *E. coli* [21], thus providing evidence for the potential bioactivity of random DNA sequence. The birth of new genes may result from a gradual process involving the formation of protogene with adaptive potential [22]. An alternative model, consistent with chromothripsis, could be the punctuated occurrence of genomic rearrangements, deeply affecting gene expression or leading to the emergence of new genes subsequently fixed by natural selection and contributing to diversity. The presence of LINE 1 in human genome might facilitate this process by mediating intra- and interchromosomal rearrangements [23, 24] and by mobilizing genomic DNA sequences flanking their 3′ ends for the creation of chimeric proteins by a process called L1-mediated transduction [25]. The chromosome configuration capture experiments and the concomitant analysis of topologically associating domains (TADs) precisely evidence how rearrangements in the layout of genes and their regulatory sequences within the same chromosomal domain may lead to altered gene expression or ectopic genomic interactions [26, 27]. Accumulating data from whole genome sequencing and 3C and Hi-C experiments yields new valuable insights in the role of TAD disruption in human diseases. The occurrence of massive chromosome breakage can mostly lead to deleterious genomic configurations but also to genetic novelty, with the formation of stable and inheritable massive rearrangements, disrupting multiple protein-coding genes without phenotypic consequences [28, 29], similar to extreme genome restructuring occurring in plants during early mitotic divisions [30].

The occurrence of chromothripsis in cancers appears to be in good agreement with these macroevolution models, since a one-step catastrophic genomic event may generate multiple chromosomal rearrangements and accelerate the evolution of a significant fraction of cancers [31]. Thus, recent data suggest that pancreatic cancer progression is not sequential. The observation of a high prevalence of chromothripsis (>65%) in pancreatic tumors supports the punctuated equilibrium model as the principal process for tumorigenesis of pancreatic cancer [32]. Concerning germline chromothripsis, recently reported studies have evidenced that chromothripsis could generate massive but balanced genomic rearrangements, compatible with life and contributing or not to developmental diseases [33–35] but also able to restore a normal function [36].

4 Chromothripsis and Selective Advantage

In light of these findings, an interesting question is whether chromothripsis could promote the emergence of subjects with selective advantages, able to stably transmit their novel genetic composition to subsequent generations. This may have essential implications for organismal evolution. The hypothesis of such abrupt changes doesn't fundamentally discredit Darwinism and its general principle of microevolution. Indeed, macromutations could result from selection-driven responses to sudden environmental changes. They may serve as "key" adaptive measures for its carriers. Sheldon [37] suggested that punctuated equilibrium could prevail in unstable environments and gradualism in stable regimes. Accumulation of macromutations such as chromothripsis events in gametogenesis or during early embryonic development could lead to profound differences among adults. However, even if a chaotic rearrangement may confer an advantage to its carrier, its fixation into a population remains uncertain. The transmission of large alterations through generations is problematic and frequently leads to hypofertility for the heterozygous carrier until a homozygous state for the rearrangement is reached. Several models of chromosomal speciation are thus based on the existence of a gametic barrier resulting from the progressive fixation of one rearrangement or a combination of various rearrangements in a population [38, 39]. An alternative hypothesis proposes that complex chromosome configurations may reduce gene flow between species by suppressing recombination [40].

Such rearrangements, which allow the genetic distinction between species, constitute evolutionary rearrangements. Genome sequence investigations have demonstrated that evolutionary rearrangements are much more numerous than initially estimated. In 1980, it was considered that human and chimpanzee genomes only differed by nine chromosomal inversions and one fusion [41]. In 2005, Newman et al. [42] identified 93 supplementary evolutionary rearrangements, ranging from 12 kb to 1 Mb. Chromosomal rearrangements assuredly play an essential role in genomic evolution and the speciation process. More than 245 large rearrangements including translocations, inversions, fusions, and deletions have been identified in the discriminative path between the mouse and humans [43], and numerous segmental duplications involved in human-specific adaptive traits have been recently characterized by sequencing human and nonhuman primate genome [44]. In gibbon genome, the insertion of a retrotransposon (LAVA) in chromosome segregation genes could be at the origin of a high rate of chromosomal rearrangements leading to the accelerated evolution of the gibbon karyotype and the emergence of different gibbon lineages [45].

5 Conclusion

Chromothripsis as a mechanism for rapid and profound genome modification in germlines and early development could be regarded as a credible molecular mechanism for evolutionary changes. Like other genomic mutational events, chromothripsis can be deleterious, neutral, or beneficial. All these new data contribute to the substitution of notions of chaos and discontinuity for the previous convictions on gradualism and predictable determinism in the field of evolutionary biology. Experimental evolutionary modeling is now required to test and integrate chromothripsis into the process of speciation. This could constitute new element of the evolutionary predictions [46].

Acknowledgments

Work in the unit of Chromosomal Genetics is supported by the CHU research platform ChromoStem (http://www.chu-montpellier.fr/fr/chercheurs/plateformes/les-plateformes-recherche/chromostem/).

References

1. Stephens PJ, Greenman CD, Fu B et al (2011) Massive genomic rearrangement acquired in a single catastrophic event during cancer development. Cell 144:27–40
2. Kloosterman WP, Gurvey V, van Roosmalen M et al (2011) Chromothripsis as a mechanism driving complex de novo structural rearrangements in the germline. Hum Mol Genet 20:1916–1924
3. Molenaar JJ, Koster J, Zwijnenburg DA et al (2012) Sequencing of neuroblastoma identifies chromothripsis and defects in neuritogenesis genes. Nature 483:589–593
4. Chiang C, Jacobsen JC, Ernst C et al (2012) Complex reorganization and predominant non-homologous repair following chromosomal breakage in karyotypically balanced germline rearrangements and transgenic integration. Nat Genet 44:390–998
5. Tubio JMC, Estivill X (2011) When catastrophe strikes a cell. Nature 470:476–477
6. Rausch T, Jones DTW, Zapatka M et al (2012) Genome sequencing of pediatric medulloblastoma links catastrophic DNA rearrangements with TP53 mutations. Cell 148:59–71
7. Jones MJK, Jallepalli PV (2010) Chromothripsis: chromosomes in crisis. Dev Cell 23:908–917
8. Crasta K, Ganem NJ, Dagher R et al (2012) DNA breaks and chromosome pulverization from errors in mitosis. Nature 482:53–58
9. Ledbetter DH (2008) Chaos in the embryo. Nat Med 5:490–491
10. Pellestor (2014) Chromothripsis: how does such a catastrophic event impact human reproduction ? Hum Reprod 29(3):388–393
11. Korbel JO, Campbell PJ (2013) Criteria for inference of chromothripsis in cancer genomes. Cell 152:1226–1236

12. Govind SK, Zia A, Hennings-Yeomans PH et al (2014) ShatterProof: operational detection and quantification of chromothripsis. BMC Bioinformatics 15:78. https://doi.org/10.1186/1471-2105-15-78
13. Yang J, Liu J, Ouyang L et al (2016) CTLPScanner: a web server for chromothripsis-like pattern detection. Nucleic Acids Res 8 44(W1):W252–W258. https://doi.org/10.1093/nar/gkw434
14. Goldschmidt R (1940) The material basis of evolution. Yale University Press, New Haven, London
15. Eldredge N, Gould SJ (1972) Punctuated equilibria: an alternative to phyletic gradualism. In: Schopf TJM (ed) Models in Paleobiology. Freeman Cooper, San Francisco, pp 82–115
16. Gould SJ, Eldredge N (1993) Punctuated equilibrium comes of age. Nature 366:223–227
17. Mallet J (2007) Hybrid speciation. Nature 446:279–283
18. Chouard T (2010) Revenge of the hopeful monster. Nature 463:864–867
19. Reiseberg LH, Archer MA, Wayne RK (1999) Transgressive segregation, adaptation and speciation. Heredity 83:363–372
20. Dittrich-Reed DR, Fitzpatrick B (2013) Transgressive hybrids as hopeful monsters. Evol Biol 40:310–315
21. Neme R, Amador C, Yildirim B et al (2017) Random sequences are an abundant source of bioactive RNAs or peptides. Nat Ecol Evol 1(6):0217. https://doi.org/10.1038/s41559-017-0127
22. Wilson BA, Foy SG, Neme R et al (2017) Young genes are highly disordered as predicted by the preadaptation hypothesis of de novo gene birth. Nat Ecol Evol 1(6):0146–0146. https://doi.org/10.1038/s41559-017-0146
23. Gu S, Szafranski P, Akdemir ZC (2016) Mechanisms for complex chromosomal insertions. PLoS Genet 12(11):e1006446. https://doi.org/10.1371/journal.pgen.1006446
24. Symer DE, Connelly C, Szak ST et al (2002) Human L1 retrotransposition is associated with genetic instability in vivo. Cell 110:327–328
25. Beck CR, Garcia-Perez JL, Badge RM et al (2011) LINE-I elements in structural variation and disease. Annu Rev Genomics Hum Genet 12:187–215
26. Valton AL, Dekker J (2016) TAD disruption as oncogenic driver. Curr Opin Genet Dev 36:34–40
27. Lupianez DG, Spielmann M, Mundlos S (2016) Breaking TADs: how alterations of chromatin domains result in disease. Trends Genet 33(4):225–237
28. Middelkamp S, van Heesch S, Braat AK et al (2017) Molecular dissection of germline chromothripsis in a developmental context using patient-derived iPS cells. Genome Med 9(1):9. https://doi.org/10.1186/s13073-017-0399-z
29. Collins RL, Brand H, Redin CE et al (2017) Defining the diverse spectrum of inversions, complex structural variation, and chromothripsis in the morbid human genome. Genome Biol 18(1):36. https://doi.org/10.1186/s13059-017-1158-6
30. Tan EH, Henry IM, Ravi M et al (2015) Catastrophic chromosomal restructuring during genome elimination in plants. elife 4:e06516. https://doi.org/10.7554/eLife.06516
31. Froment JV, Kaidi A, Jackson SP (2012) Chromothripsis and cancer: causes and consequences of chromosome shattering. Nat Rev Cancer 12(10):663–670
32. Notta F, Chang-Seng-Yue M, Lemire M et al (2016) A renewed model of pancreatic cancer evolution based on genomic rearrangement patterns. Nature 538(7625):378–382
33. de Pagter MS, van Roosmalen MJ, Baas AF et al (2015) Chromothripsis in healthy individuals affects multiple protein-coding genes and can result in severe congenital abnormalities in offspring. Am J Hum Genet 96(4):651–656
34. Bertelsen B, Nazaryan-Petersen L, Sun W et al (2016) A germline chromothripsis event stably segregating in 11 individuals through three generations. Genet Med 18(5):494–500. https://doi.org/10.1038/gim.2015.112
35. Fukami M, Shima H, Suzuki E et al (2017) Catastrophic cellular events leading to complex chromosomal rearrangements in the germline. Clin Genet 91:653–660
36. McDermott DH, Gao JL, Liu Q et al (2015) Chromothriptic cure of WHIM syndrome. Cell 160:686–699
37. Sheldon PR (1990) Shaking up evolutionary patterns. Nature 345:772
38. Dutrillaux B (1979) Chromosomal evolution in primates: tentative phylogeny from microcebus murinus (prosimian) to man. Hum Genet 48:251–314
39. Britton-Davidian J, Catalan J, Ramalhinho M et al (2000) Rapid chromosomal evolution in island mice. Nature 403:158

40. Noor MAF, Grams KL, Bertucci LA et al (2001) Chromosomal inversions and the reproductive isolation of species. Proc Natl Acad Sci U S A 98:12084–12088
41. Yunis JJ, Sawyer JR, Dunham K (1980) The striking resemblance of high-resolution g-banded chromosomes of man and chimpanzee. Science 208:145–1148
42. Newman TL, Tuzun E, Morrison VA et al (2005) A genome-wide survey of structural variation between human and chimpanzee. Genome Res 15:1344–1356
43. Pevzner P, Tesler G (2003) Genome rearrangements in mammalian evolution: lessons from human and mouse genomes. Genome Res 13:37–45
44. Dennis MY, Harshman L, Nelson BJ et al (2017) The evolution and population diversity of human-specific segmental duplications. Nat Ecol Evol 1:0069. https://doi.org/10.1038/s41559-016-0069
45. Carbone L, Harris RA, Gnerre S et al (2014) Gibbon genome and the fast karyotype evolution of small apes. Nature 513:195–201
46. Lässig M, Mustonen V, Walczak AM (2017) Predicting evolution. Nat Ecol Evol 1:0077. https://doi.org/10.1038/s41559-017-0077

Part II

The Identification of Chromothripsis

Check for updates

Chapter 5

Analysis of Chromothripsis by Combined FISH and Microarray Analysis

Ruth N. MacKinnon

Abstract

Fluorescence in situ hybridization (FISH) to metaphase chromosomes, in conjunction with SNP array, array CGH, or whole genome sequencing, can help determine the organization of abnormal genomes after chromothripsis and other types of complex genome rearrangement. DNA microarrays can identify the changes in copy number, but they do not give information on the organization of the abnormal chromosomes, balanced rearrangements, or abnormalities of the centromeres and other regions comprised of highly repetitive DNA. Many of these details can be determined by the strategic use of metaphase FISH. FISH is a single-cell technique, so it can identify low-frequency chromosome abnormalities, and it can determine which chromosome abnormalities occur in the same or different clonal populations. These are important considerations in cancer. Metaphase chromosomes are intact, so information about abnormalities of the chromosome homologues is preserved. Here we describe strategies for working out the organization of highly rearranged genomes by combining SNP array data with various metaphase FISH methods. This approach can also be used to address some of the uncertainties arising from whole genome or mate-pair sequencing data.

Key words Chromothripsis, Metaphase FISH, Chromosome, Cytogenetics, M-FISH, M-BAND, Microarray, SNP array, Cytogenomics

1 Introduction

Whole genome sequencing, mate-pair sequencing, and DNA microarray analysis are ideal for studying complex genome reorganization, including chromothripsis, at high resolution. Metaphase fluorescence in situ hybridization (FISH) is a valuable complementary tool [1], as has been demonstrated in several recent publications describing highly complex genome remodeling, including chromothripsis [2–5]. It can provide essential information and confirm deductions about the structure of the abnormal chromosomes.

In microarray data arranged against chromosome idiograms, chromothripsis can be identified as a series of chromosome segments in a distinctive pattern of two or three alternating copy

Franck Pellestor (ed.), *Chromothripsis: Methods and Protocols*, Methods in Molecular Biology, vol. 1769,
https://doi.org/10.1007/978-1-4939-7780-2_5,

number states [4]. However, microarray data do not reveal how these chromosome fragments are organized in the abnormal chromosomes. Much of this information can be provided by the strategic use of metaphase FISH.

FISH is especially suitable for localizing regions of highly repetitive DNA, such as centromeres, that are intractable to analysis by both microarray and sequencing. Centromeres play an important role in chromosome stability and cancer evolution. Knowing the identity and location of centromeres and other repetitive elements can help infer the steps in the formation of abnormal chromosomes, as we have demonstrated (Fig. 1) [6, 7], yet this information is often overlooked. The entire short arms of the five acrocentric human chromosomes (chromosomes 13, 14, 15, 21, and 22) are not represented on DNA microarrays. They are comprised of the poorly characterized nucleolar organizer region (NOR), which includes significant repetitive DNA regions [8].

Metaphase FISH is a single-cell technique, which can discern the differences between different clones, and, as the chromosomes are intact, information on chromosomal organization is preserved. In microarray data and most sequencing applications, it is often not clear whether nonoverlapping complex aberrations of a chromosome involve one or both homologues. This can be resolved with the help of metaphase FISH.

A major advantage of using a microarray platform that includes SNP (single-nucleotide polymorphism) data is that it identifies changes in heterozygosity (e.g., *see* Fig. 2).

A change in copy number is usually matched by a change in the B allele frequency of heterozygous loci, and these data can be used to calculate the percentage of each clonal population. In examples of chromothripsis where there are mixed populations of normal cells and affected cells (as is typically seen in cancer), all the regions of copy number loss or gain have identical B allele frequencies, as is expected for copy number aberrations that occur simultaneously (*see*, e.g., Fig. 3). This supports the hypothesis that chromothripsis is a single event [7].

Chromothripsis can arise from breakpoints clustered together on the parent chromosome [4, 9]. Here we describe our method for using a combination of SNP array and different metaphase FISH techniques to work out the structure of chromosomes created by chromothripsis and other complex genome rearrangements. We include other complex rearrangements, because these are likely to be part of a genome with chromothripsis and because complex rearrangements provide useful illustrations of approaches for solving difficult patterns.

SNP array, multicolor FISH (M-FISH), and multicolor banding (M-BAND) are first used to define and order segments in the abnormal chromosomes. The finer detail and ambiguities may be resolved by careful selection of single-locus and region-specific FISH probes.

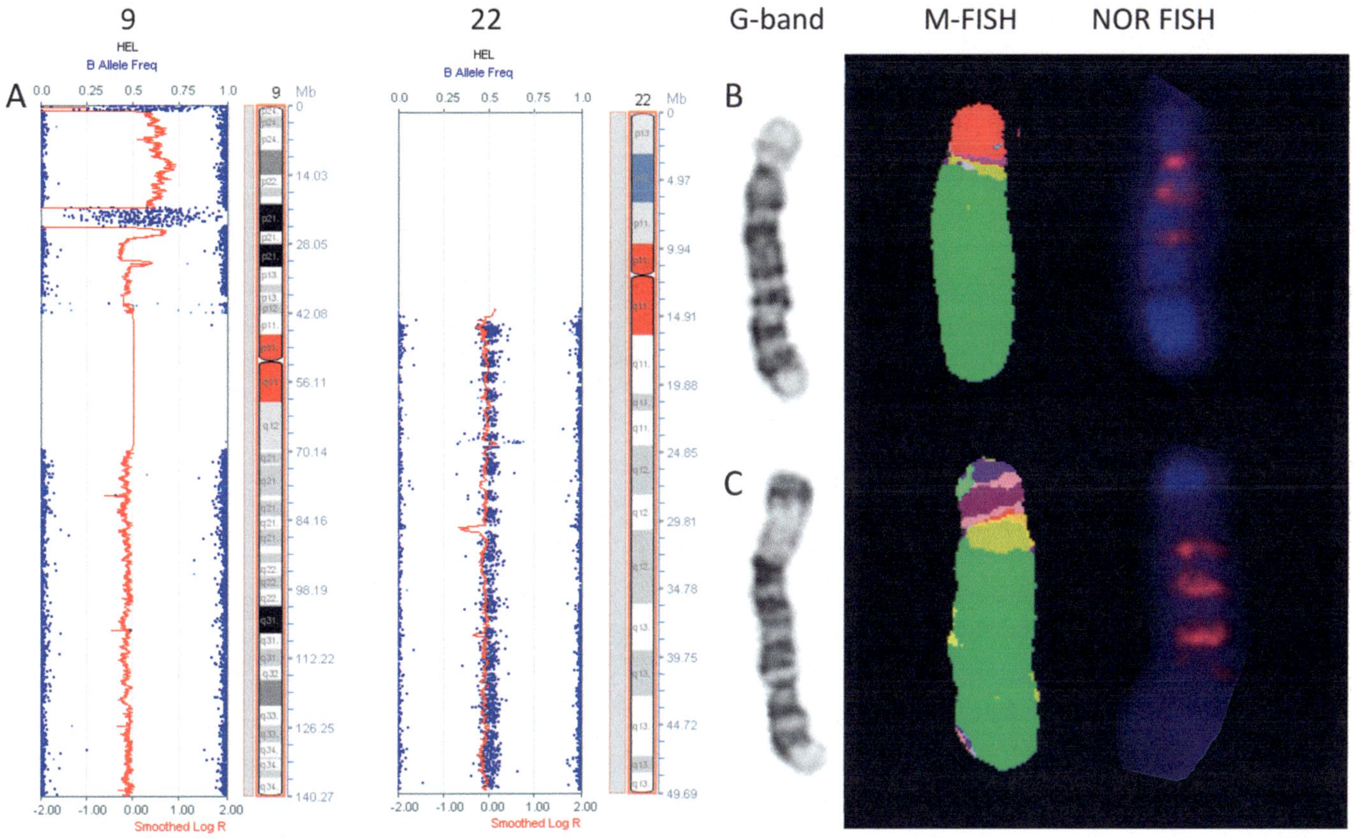

Fig. 1 Oncogene amplification by breakage-fusion-bridge cycles involving the 22p NOR. This example shows how a probe for the NOR was used to help determine the sequence of events in the evolution of the pseudotriploid myeloid cell line HEL. (**a**) SNP array shows that there is a uniparental disomy of chromosome 9 (two copies of the same homologue) with *CDKN2A* homozygous deletion and *JAK2* amplification, and two nearly normal copies of chromosome 22. (**b**, **c**) are the two abnormal chromosomes derived from chromosome 9. These both consist of a normal 9q (as shown by the SNP array) with deletion and amplification of 9p and other chromosome materials joined to the deleted end of 9p. One of the copies of chromosome 22 is at the top of **b** (labeled red by M-FISH). The NOR probe detected bands of NOR from the short arm of chromosome 22 on both **b** and **c** (red signal), but M-FISH did not detect any chromosome 22 material in this region of either **b** or **c;** it is just labeled with the chromosome 9 false color. (Note that this is the region devoid of probes above the centromere on the chromosome 22 SNP array.) Therefore, the likely course of events was the formation of a dicentric chromosome by joining of 9p and 22p; breakage-fusion-bridge cycles between the centromeres on 22p and 9p causing amplification and deletion of 9p and amplification of 22p (9q and 22q are relatively normal); and duplication of this chromosome with replacement of the chromosome 22q material with material from chromosomes 11, 18, 19, and 20 to form **c** (Modified from MacKinnon et al. [6])

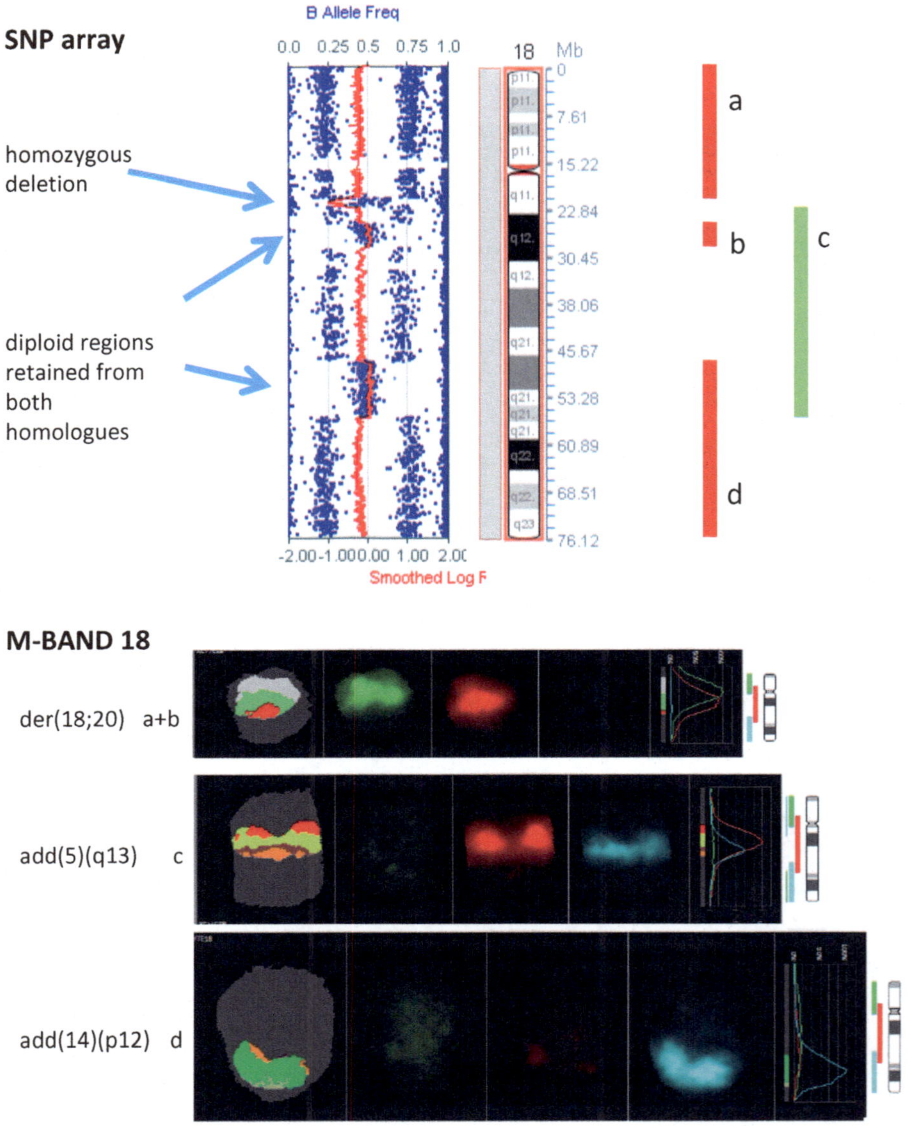

Fig. 2 Integration of SNP-A and FISH results to explain complex rearrangements. A complex SNP array pattern for chromosome 18 that was clarified with the help of M-BAND. SNP array: there are alternating retained and deleted regions raising the possibility of a chromothripsis event. Note however that the B allele frequency patterns for these deletions, while similar, are not identical (therefore occurring in different proportions of cells), and there is also a region of homozygous deletion. The complex AML karyotype had no normal 18, so we can conclude that both chromosome 18 homologues are abnormal. The red (left) and green vertical bars are aligned with the retained segments consistent with the B allele frequency and M-BAND pattern. The diploid regions are retained from both homologues. The cells containing section c of chromosome 18 have slightly tighter B allele frequency values for the heterozygous loci; therefore, there is deletion of the green homologue in fewer cells than have deletion of the red homologue. This suggests that the events causing rearrangement of the red homologue occurred before events causing rearrangement of the green homologue. M-BAND for chromosome 18 clarifies the deletion patterns in each homologue and the location of the SNP array segments. The M-BAND images show the false-color banding pattern for regions of the chromosome containing chromosome 18 material on the left and the individual fluorochrome hybridization pattern.

We used SNP array as a high-resolution genome analysis tool, but our approach is also useful for filling in the gaps after sequencing. With whole genome sequencing, the order and sequence of chromosome segments can be worked out in finer detail, and mate-pair sequencing can be used to work out chromosome recombination events and breakpoints. However, sequencing is also problematic at the centromeres and other regions of highly repetitive DNA. Metaphase FISH is useful for filling in these gaps and for confirming the details obtained by these methods.

2 Materials

2.1 G-Banded Karyotype

1. It is assumed that the reader has access to a G-banded karyotype or a cytogenetics laboratory (*see* **Note 1**).

2.2 DNA Microarray Data

1. A microarray platform—one that includes SNP analysis is preferable (SNP array or CGH + SNP array) (CGH, comparative genomic hybridization).
2. A karyotype-based SNP array analysis tool. DNA copy number and allele frequency are visualized against chromosome idiograms.
3. We used CytoSNP 12 microarrays and KaryoStudio software (Illumina, San Diego, CA). The processing of SNP arrays can be outsourced to a commercial provider. The data are returned and processed by the abovementioned software in-house.

2.3 Metaphase Preparation

1. Dividing cells in culture.
2. 10 μg/mL colcemid.
3. 0.075 M KCl.
4. Carnoy's fixative: 3 parts methanol to 1 part glacial acetic acid.
5. Water bath, centrifuge.

Fig. 2 (continued) The idiogram at the right of each image shows the fluorochrome labeling scheme for the XCyte 18 probe. M-BAND identified three abnormal chromosomes comprising parts of chromosome 18. Segments a and d contain a telomere and can be placed at the end of an abnormal chromosome. The green homologue (section c) has lost both telomeres and is flanked on both sides by parts of other chromosomes. Section a, from the red homologue, spans the 18 centromere. The green homologue appears to have lost its centromere and has joined to other chromosome segments that can provide a centromere. The short segment labeled b is positioned with segment a because the der(18;20), and not the add(14)(p12), has the red signal expected from this segment. Note that the changes to the B allele frequency values are much more distinct and easier to quantify than the changes to the smoothed log R ratio (the vertical red line) relating to copy number. Note the strong M-BAND signals produced by these 25–30 Mb regions

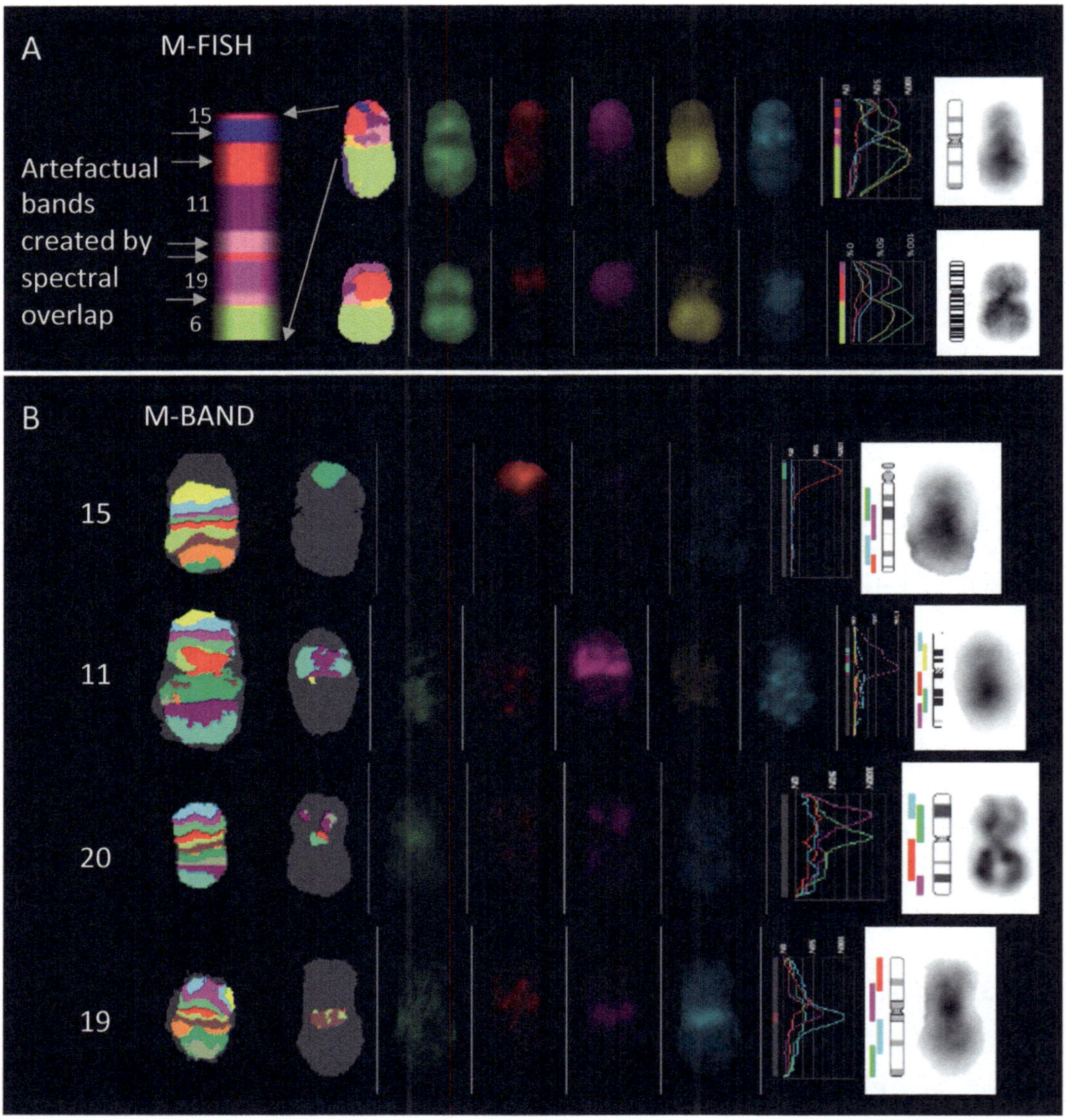

Fig. 3 Comparing the resolution of M-FISH and M-BAND. (**a**) Two examples of M-FISH labeling of an abnormal chromosome, a der(19)t(6;11;15;19;20), in the myeloid cell line HEL [6], so named because it has a 19 centromere and elements of each of the listed chromosomes (8 Mb from chromosome 15, 25 Mb including a partial duplication from chromosome 11, sections of about 2 and 1 Mb from chromosome 20, and the centromere and a further 6 Mb from chromosome 19). Two examples are given to illustrate the variability of the labeling when the chromosome segments are in this size range. The long arm is comprised of chromosome 6 material only, and the short arm is derived from the other chromosomes. On the left is a schematic representation of the false colors of the section of the upper image between the arrows. The true false colors are labeled according to the chromosome they represent, and there are several artifactual false colors due to overlapping fluorochromes, which are arrowed. The false-color pattern is not helpful for working out the composition of the abnormal chromosome. The single-color galleries are more helpful but still open to interpretation. The take-home message is that the false-color image of the short arm is of little value when there are several joined segments of this size range. However, the chromosome composition can be solved by interpretation of the SNP array segments together with M-BAND. (**b**) M-BAND images of the chromosomes in this abnormal chromosome, which are clearly positive for each of the chromosomes comprising

2.4 FISH

1. Fixed metaphase suspensions.
2. Clean glass microscope slides.
3. Phase-contrast microscope for examination of spreading quality.
4. Rubber cement: e.g., Fixogum (Marabu).
5. Coverslips of a size to match the area of metaphases to be probed.
6. Sealed humidified chamber (*see* **Note 2**).
7. 37 °C incubator.
8. One to three water baths.
9. Coverslips for mounting (24 × 50–60 mm).
10. Fluorescence microscope with filter sets appropriate to the fluorochromes used.
11. Low-fluorescence immersion oil.
12. FISH capture and analysis station with M-FISH capacity (e.g., Isis, MetaSystems).

2.4.1 M-FISH and M-BAND (see Note 3)

1. 0.07 M NaOH.
2. 20 × SSC saline-sodium citrate buffer: 3.0 M NaCl and 0.3 M $C_6H_5Na_3O_7{\bullet}2H_2O$ (tri-Sodium citrate dihydrate) for preparation of 0.1 × SSC, pH 7.0–7.5, and 2 × SSC, pH 7.0–7.5. Adjust pH with concentrated NaOH or HCl solutions.
3. Water bath set at 70 °C.
4. Coplin jars.
5. Probe cocktail: Metasystems 24XCyte (mFISH) or XCyte (mBAND) (*see* **Note 4**).
6. Water baths set at 75 °C and 37 °C and ice or thermocycler.
7. Tween 20 Polyoxyethylenesorbitan monolaurate syrup (e.g., Sigma P-1379) (Tween™ is a trademark of ICI America, Inc.) (*see* **Note 5**).

Fig. 3 (continued) the short arm. On the left of each set is a normal chromosome showing the M-BAND pattern for that chromosome (15p is unlabeled as it consists of repeats) and on the right (left to right) the der(19) showing the false-color M-BANDs, single-color galleries, fluorescence intensity profile, labeling scheme, and DAPI-banded chromosome. Top to bottom: M-BAND 15 (8 Mb segment), M-BAND 11 (25 Mb including a partial duplication), M-BAND 20 (2 Mb from the long arm in gray and purple false colors and a paracentric segment of about 1 Mb in green and red false colors), M-BAND 19 (the centromere and about 6 Mb). Note that there are two segments which each contribute a telomere, as expected (from chromosomes 15 and 6). The centromere of this chromosome was difficult to identify. SNP array identified a paracentric segment from chromosome 19, allowing us to postulate that it was present in this chromosome and with it the adjacent 19 centromere. M-BAND confirmed that this paracentric segment was on this abnormal chromosome (B, M-BAND 19 image), and FISH with a 1/5/19 centromere probe was positive

8. 1 × SSC, pH 7.0–7.5 (*see* **Note 6**).
9. 4 × SSCT pH 7.0–7.5: add 0.05% Tween 20 to 4 × SCS, pH 7.0–7.5.
10. PBS (phosphate-buffered saline solution): 120 mM NaCl, 7 mM Na_2HPO_4, 3 mM NaH_2PO_4, and 2.7 mM KCl; adjust to pH 7.4 with HCl or NaOH.
11. DAPI/antifade for XCyte probes (available from Metasystems).
12. Software specific to the probe (Isis, Metasystems).

2.4.2 Other Metaphase FISH

1. Hot plate with variable temperature up to at least 80 °C.
2. Hybridization buffer: use the one recommended by the manufacturer.

   Alternatively, prepare the following multipurpose hybridization buffer [10]:

   (a) 20 × SSCP: 20 mL 7.01 g NaCl, 4.41 g sodium citrate, 40 mL 0.5 M phosphate buffer (0.5 M $NaHPO_4$ mixed with 0.5 M NaH_2PO_4 to pH 5.6).

   (b) Hybridization buffer: 20 mL 20 × SSCP at pH 6, 50 mL 20% dextran sulfate in deionized formamide (*see* **Note 4**).

   (c) To make 10 μL probe, mix 7 μL hybridization buffer, required amount of probe, and make up to 10 μL with autoclaved deionized water.
3. FISH probes for the region of interest (*see* **Notes 4, 7–12**).
4. Mounting medium: DAPI/antifade (e.g., from Metasystems or Vectashield).
5. Water bath set at 73 °C.
6. 0.4 × SSC/0.3% NP40 (*see* **Note 5**).
7. 2 × SSC/0.1% NP40.

3 Methods

3.1 Karyotype

1. From the G-banded karyotype, determine the normal and rearranged chromosomes and karyotype complexity. If G-banding is not available, the SNP array and M-FISH steps described below will give a comprehensive picture of rearrangements, with the exception of balanced internal rearrangements. These can be detected by adding M-BAND with probes to all of the 23 (female) or 24 (male) chromosomes (*see* below).

3.2 Identify SNP Array Segments

1. For any departure from the normal diploid pattern, note copy number and breakpoints.
2. Identify the retained segments that therefore form part of an abnormal chromosome (*see* **Note 13**).

3. It is helpful to make a diagram of the retained sections alongside the chromosome idiograms (*see* examples in Figs. 2, 4, and 7), bearing in mind that both homologues may be affected (*see* **Notes 14** and **15**).

3.3 Metaphase Preparation

Adapted from [11]

1. Add colcemid to a final concentration of 0.2 μg/mL.
2. Incubate at 37 °C for 20 min (*see* **Note 16**).
3. Pellet cells by centrifugation, 5 min at 310 × *g*.
4. Discard supernatant.
5. Add 8 mL KCl.
6. Incubate at 37 °C for 20 min.
7. Centrifuge for 5 min at 310 × *g*.
8. Discard supernatant, and resuspend pellet thoroughly.
9. Add 5 mL Carnoy's fixative dropwise while mixing gently.
10. Centrifuge for 5 min at 310 × *g*.
11. Add 5 mL Carnoy's fixative.
12. Repeat **steps 10** and **11**.
13. Resuspend pellet in Carnoy's fixative to the desired cell concentration.
14. Store at −20 °C.

3.4 M-FISH and M-BAND

The following method is based on the user instructions for the XCyte probes (Metasystems) [12]:

1. Slide preparation. Pipette an appropriate volume of metaphase suspension onto a clean microscope slide. Check metaphases under a phase-contrast microscope for morphology and spreading (*see* **Note 17**).
2. Mark the area to be hybridized, for example, with a diamond pencil, or spread onto pre-marked slides.
3. Prewarm 2 × SSC, pH 7.0–7.5, in a Coplin jar to 70 °C.
4. Prepare Coplin jars containing 0.07 M NaOH and 0.1 × SSC, pH 7.0–7.5 at room temperature.
5. Cool 0.1 × SSC, pH 7.0–7.5 and 2 × SSC, pH 7.0–7.5 in Coplin jars to 4 °C.
6. Incubate slides in 2 × SSC, pH 7.0–7.5 for 30 min.
7. Remove Coplin jar from the water bath, and cool at room temperature down to about 37 °C (this takes about 20 min).
8. Start probe denaturation (**step 15**) in parallel with the following steps [9–14].

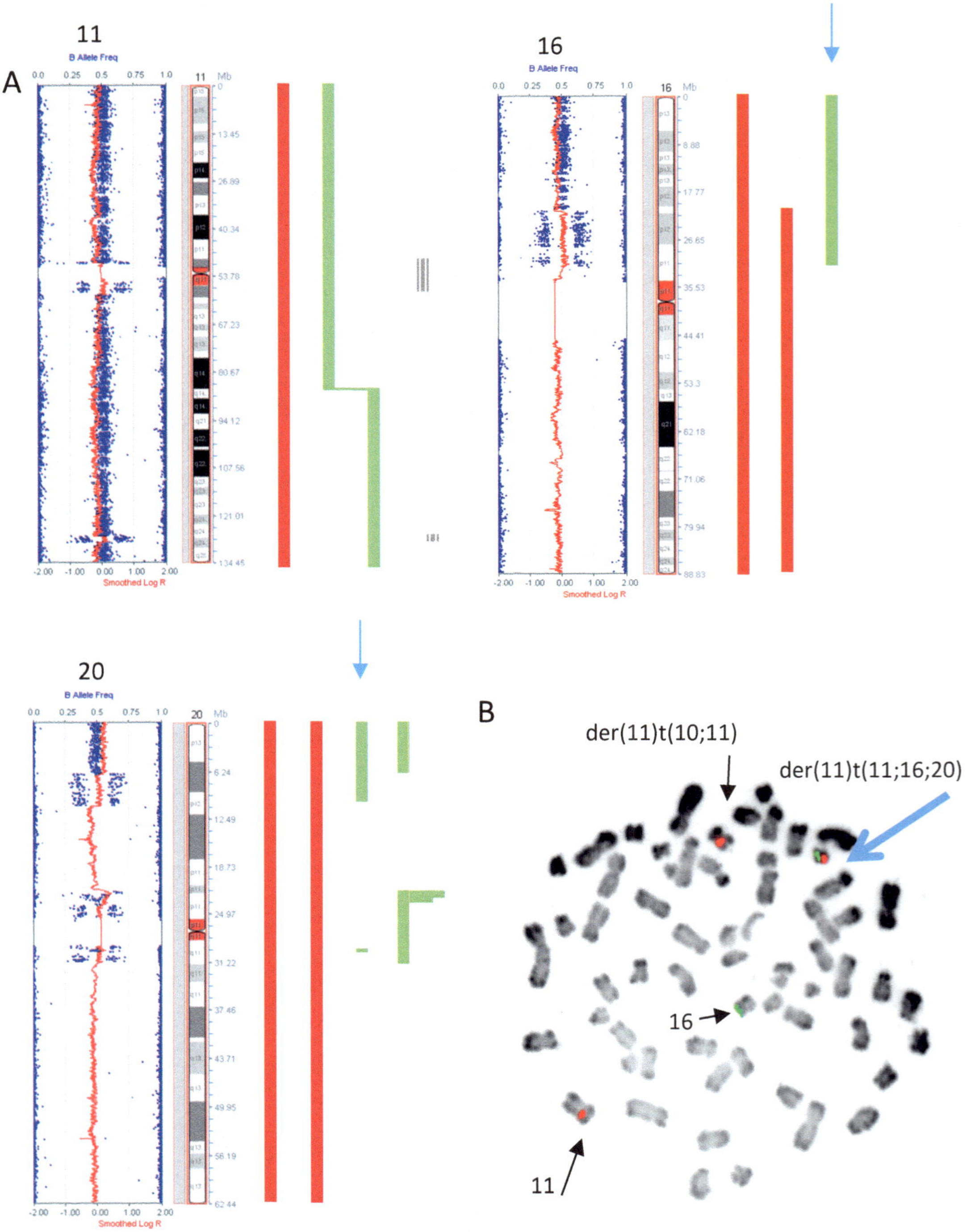

Fig. 4 Centromere capture. Images from the cell line U937, a triploid myeloid cell line with a complex karyotype. (**a**) There is gain of a 4 Mb section of chromosome 11 that spans the centromere. The two vertical green bars connected by a horizontal bar beside the chromosome 11 idiogram represent a balanced translocation involving chromosome 11. The two alternative homologues are represented by red and green vertical bars; it is not known which homologue is duplicated (gray striped vertical bar). To identify the abnormal chromosome containing the extra copy of the 11 centromere, we searched for a chromosome with no obvious centromere.

9. Immerse slides in 0.1 × SSC, pH 7.0–7.5, room temperature, 1 min.
10. Denature in 0.07 M NaOH, room temperature for exactly 1 min.
11. Immerse slides in 0.1 × SSC, pH 7.0–7.5, 4 °C, 1 min.
12. Immerse slides in 2 × SSC, pH 7.0–7.5, 4 °C, 1 min.
13. Dehydrate through ethanol series: 1 min each in 70, 95, and 100% ethanol.
14. Air-dry.
15. Spin the probe stock tube briefly in a microfuge to ensure the probe is at the bottom of the tube.
16. Pipette the required amount of probe into a microfuge tube (*see* **Note 18**).
17. Denature at 75 °C for 5 min.
18. Place briefly on ice.
19. Incubate at 37 °C for 30 min to reduce background by pre-annealing labeled and unlabeled repeat sequences in the probe cocktail (*see* **Note 19**).
20. Spin the tube briefly in a microfuge to ensure the denatured probe is at the bottom of the tube.
21. Pipette the probe onto the marked area of the slide containing the metaphases to be probed.
22. Carefully place a coverslip over the probe, avoiding air bubbles, which can be gently pressed out if necessary.
23. Seal around the coverslip using rubber cement.
24. Incubate in the sealed humidified chamber at 37 °C overnight (M-BAND probes) or for 2–4 days (M-FISH probes).
25. Prewarm 1 × SSC, pH 7.0–7.5 to 75 °C, and prepare a Coplin jar containing 4 × SSCT, pH 7.0–7.5.
26. Remove rubber cement and coverslips carefully.
27. Wash slides in 1 × SSC, pH 7.0–7.5, 75 °C, 5 min.
28. Incubate slides in 4 × SSCT, pH 7.0–7.5, 5 min.
29. Wash slide briefly in distilled water to avoid crystal formation.

Fig. 4 (continued) M-FISH showed that one abnormal chromosome contained the sections of chromosomes 16 and 20 indicated by arrows. Both telomeres are accounted for, but in these segments, there is no centromere-spanning segment. (**b**) Metaphase FISH confirmed that this chromosome contains the 11 centromere: a *MYH11* probe (green signal) present in the green SNP array segment of 16p was used to identify this abnormal chromosome, which was also positive for the co-hybridized 11 centromere probe (red signal). The thin black arrows point to the other chromosomes identified with these probes

30. Drain fluid off and let the slides air-dry.
31. Apply 20 μL of DAPI/antifade and coverslip.
32. Slides are best stored in the dark at −20 °C.

3.5 M-FISH and M-BAND Capture and Analysis

Detailed instructions on image capture and analysis are provided with the software specific to the probe (Isis, Metasystems).

1. Metaphases can be found by viewing the DAPI counterstain with a low-power objective.
2. Locate a spread, apply immersion oil, and view under high power (*see* **Note 20**).
3. Assess for hybridization by viewing different fluorochrome signals.
4. Ensure the spectrum symbol is selected before capture and any of the following image processing steps.
5. Select the experiment type (M-FISH or a specific M-BAND), and capture the DAPI image and each fluorochrome in succession using automatic integration control (*see* **Note 21**).
6. Define the region of interest (select labeled chromosomes) (*see* **Note 22**).
7. Apply background correction.
8. Apply automatic upper and lower thresholding.
9. Correct pixel shift if necessary. Do not use the automatic register color function for M-BAND.
10. Maximize the metaphase as desired.
11. Adjust the object threshold to delineate the chromosomes, and separate the chromosomes (*see* **Notes 23** and **24**).
12. Enter karyotype mode.
13. For M-FISH: apply automatic karyotyping, and then make any adjustments to allow for incorrectly classified chromosomes (*see* **Note 25**).
14. The first false-color image is produced using the built-in classifier. The chromosomes in the karyogram may be used to train a new classifier. For M-FISH place translocated chromosomes in the buffer and untranslocated chromosomes in their correct position, and select "Train M-FISH Classifier." This will result in a more homogeneous false-color image and can result in the correction of incorrect false colors.
15. For M-BAND classifier training: place the abnormal labeled chromosomes at the correct position and the normal chromosomes at the chromosome 1 position where they will be used to train the multicolor banded chromosomes.
16. Select "Train M-BAND Classifier" and select the number of bands. A classifier from another metaphase can be used (*see* **Note 26**).

17. For M-FISH and M-BAND: DAPI banding can be enhanced using the "Adjust Filter Power" function.
18. The minimum classification intensity can be adjusted so that the unlabeled areas are gray (it determines the amount of fluorescence that is classified as true signal and converted into a false color).
19. Single-color galleries can be copied to compare with other metaphases or experiments.

3.6 Identify M-FISH and M-BAND Segments

1. Carry out M-FISH (*see* above).
2. From the M-FISH results, note the sections of different chromosome origin in the abnormal chromosomes (*see* **Note 27**).
3. Carry out M-BAND with a probe specific for each rearranged chromosome for which the pattern is not already clear (e.g., *see* Figs. 2 and 3). Rearrangements within a chromosome (deletion, inversion, or duplication) may be characterized with M-BAND. Determine the location, orientation, and approximate breakpoints (*see* **Note 28**).
4. Locate these M-FISH and M-BAND segments on the abnormal chromosomes (*see* **Notes 29** and **30**).

3.7 Match SNP Array and M-FISH/M-BAND Segments

1. Reconstruct the abnormal chromosomes by matching SNP array and M-FISH/M-BAND segments.
2. As a general rule, each linear chromosome should have a telomere-containing segment at each end, and consequently it should be possible to place each telomeric segment at one end of a chromosome (*see* **Note 31**). Identify a centromere for each abnormal chromosome. Each chromosome should have at least one functional centromere. Centromere placement will be simplest if there is one centromere-spanning segment for each abnormal chromosome (Fig. 4). Exceptions include centromere capture [6, 7], dicentric or multicentric chromosomes, dicentric chromosomes which have subsequently lost a centromere [13–16] (Fig. 5), and rearrangements occurring at the centromere including short paracentric segments—i.e., the copy number is different either side of the centromere so that it is not clear from the SNP array data whether the centromere is retained or which chromosome it is on (Figs. 3 and 6).

3.8 FISH

The following is our default method, based on the Vysis protocol (Abbott Molecular):

1. Drop suspension containing metaphase chromosomes onto a microscope slide. *See* Subheading 3.4, **step 1** (*see* **Note 32**).
2. Make a probe mix containing probe(s) at an appropriate final concentration (*see* **Note 33**).

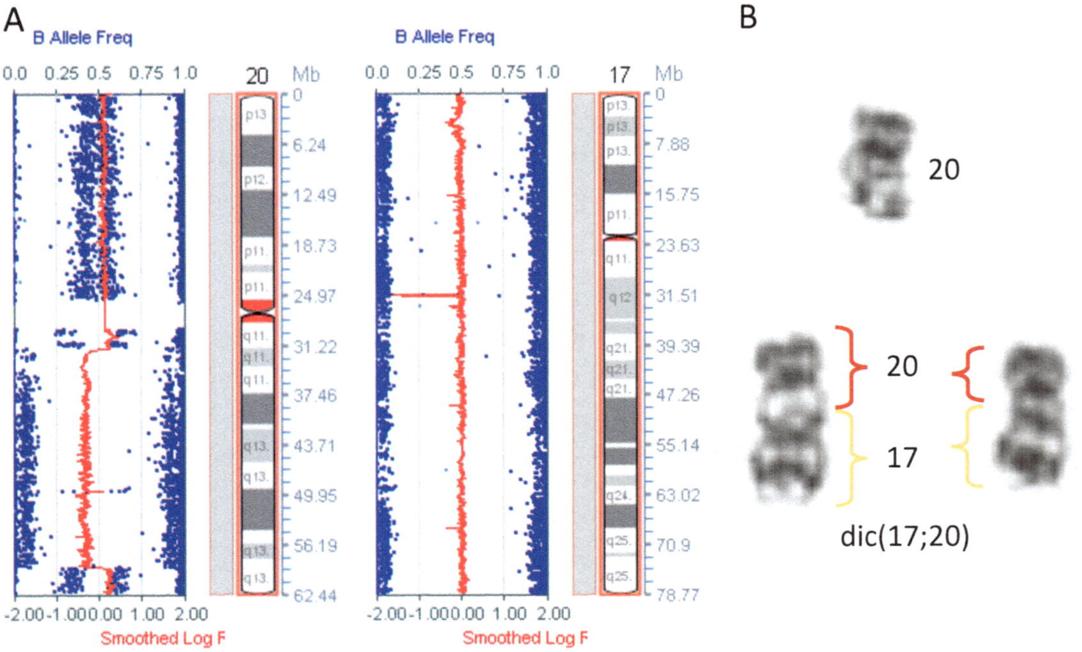

Fig. 5 Fusion events restricted to repetitive sequence at telomeres are not detected by SNP array. (**a**) The SNP array patterns for chromosomes 17 and 20 in a case of AML with a complex karyotype. The SNP array representation is limited to non-repetitive sequence and suggests that the ends of both chromosomes are retained and not involved with any other chromosome. (**b**) However, cytogenetics showed that there was a dicentric chromosome formed by fusion between 17p and 20q. The image shows the normal chromosome 20 (there was no normal chromosome 17) and indicates the parts of chromosomes 17 and 20 in the two copies of the dic(17;20) as in this instance of probable telomere fusion, end-to-end joining can occur without loss of the terminal SNP array probes of the chromosome. In these dicentric chromosomes, the 17 or 20 centromere was often deleted. This is not shown on the SNP array but was determined by metaphase FISH [16]

3. Add a volume of probe mix to the microscope slide, cover with a coverslip, and seal around the edge with rubber cement (*see* **Note 32**).
4. Co-denature the chromosome, and probe DNA by placing the slide on a hot plate at the recommended temperature for the required time (*see* **Note 34**).
5. Incubate overnight at 37 °C in a sealed humidified container.
6. Prepare Coplin jars containing 0.4 × SSC/0.3% NP40 at 73 °C and 2 × SSC/0.1% NP40 at room temperature.
7. Remove coverslips.
8. Wash slides in 0.4 × SSC/0.3% NP40 for 2 min at 73 °C.
9. Wash slides in 2 × SSC/0.1% NP40 for 1 min at room temperature.
10. Air-dry in the dark.
11. Mount in DAPI/antifade.

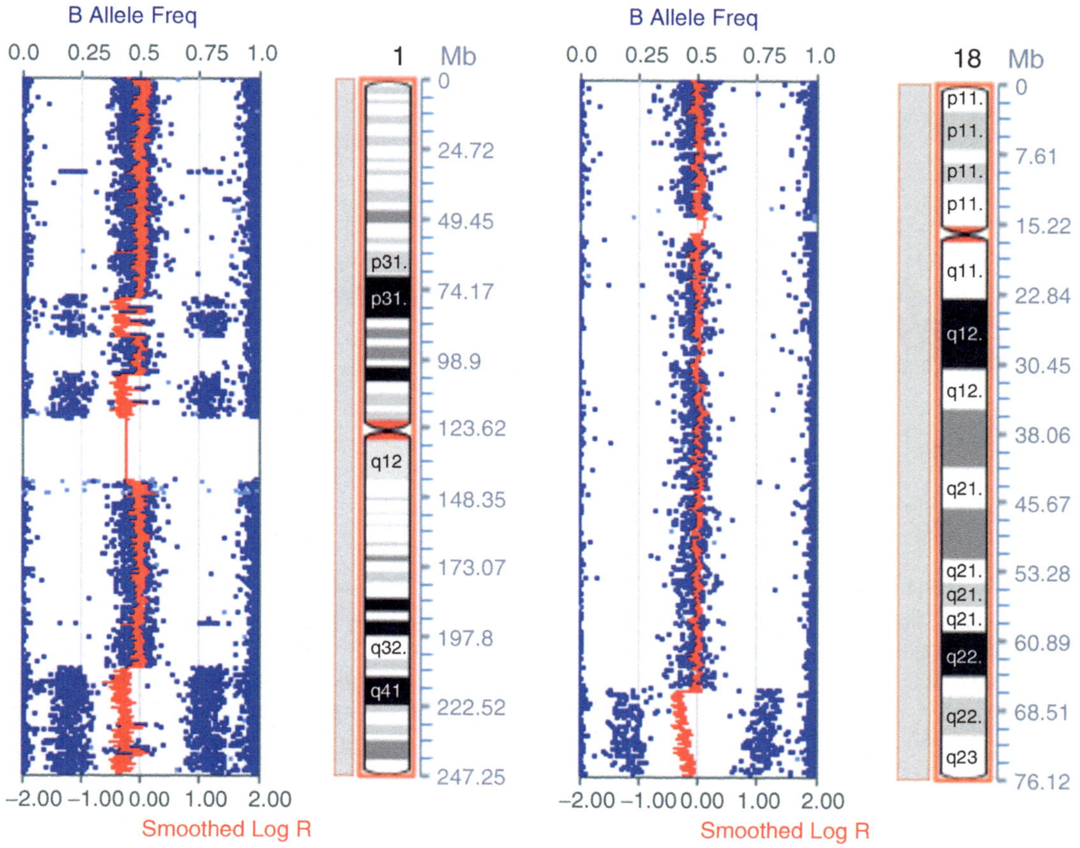

Fig. 6 The SNP array pattern for chromosomes 1 and 18 from a case of AML with chromothripsis. The telomere-containing terminal ends of 1q and 18q have been lost. There is a deletion of 1p adjacent to the 1 centromere, so the SNP array data do not show whether the 1 centromere is retained. Metaphase FISH with a centromere probe showed that it had been lost. The retained sections of chromosome 1 therefore lack both a centromere and a telomere. M-FISH showed that they had obtained these by joining to chromosome 18 at the site of the terminal 18q deletion. In this mixture of cells with normal and abnormal genomes, all the deletions have the same B allele frequency pattern, which shows that each deletion is present in the same proportion of cells, a feature that supports the rearrangements characteristic of chromothripsis having occurred at the same time rather than in a stepwise process [7]

12. Proceed to visualization with a fluorescence microscope using appropriate filters for DAPI and the fluorochromes used (*see* **Note 35**).

3.9 Disambiguation

Ambiguities can be resolved by the following measures (*see* **Note 36**):

1. To localize a SNP array, M-FISH, or M-BAND segment where there is still ambiguity, choose a probe from that segment, and use it for FISH to metaphase chromosomes.
2. To investigate whether a SNP array segment consists of multiple rearranged segments, select probes at different points along it. Subdivide the SNP array segments in this way as appropriate to achieve the desired level of detail and resolution.

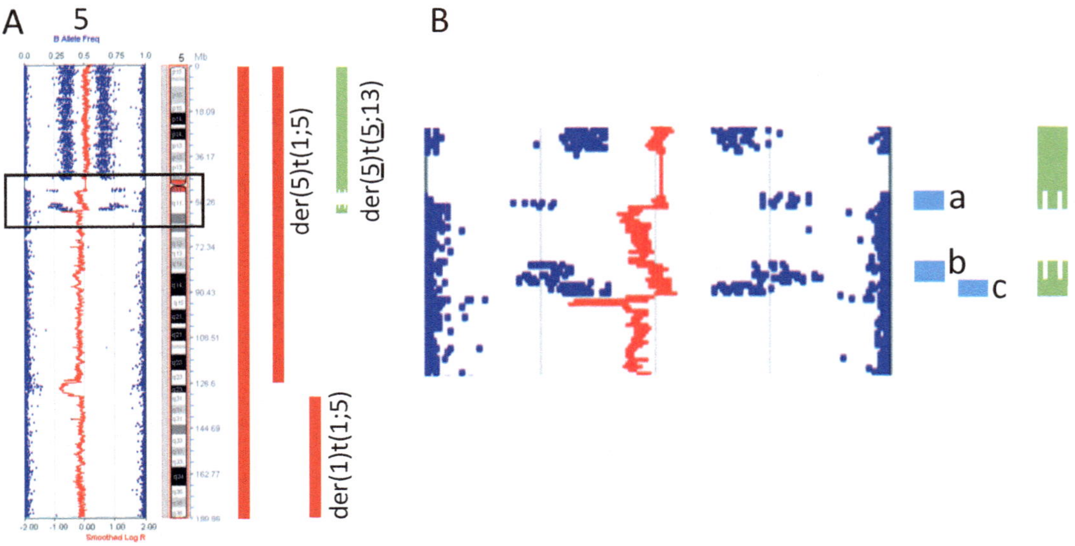

Fig. 7 BACs can be used to map small retained segments. This SNP array image shows a complex pattern for chromosome 5 below the centromere in the myeloid cell line U937. (**a**) Whole chromosome image showing the content of the four abnormal chromosomes from two homologues (red and green) as vertical bars to the right. (**b**) Close-up of the boxed area in (**a**). BACs for each of three segments (*a–c* represented by blue bars) were used to localize these segments and explain the SNP array pattern. The results are represented on the right with a green vertical bar. These segments were on the same abnormal chromosome, but the sections represented by vertical stripes were present in only 70% of cells

3. Co-hybridization: FISH can be performed with a mixture of probes labeled with different fluorochromes in the following situations (*see* **Notes 37** and **38**).
4. If the identity of an abnormal chromosome will not be clear by morphology after FISH, identify it by hybridizing a marker probe from the candidate chromosome with your test probe. Use the SNP array, G-band, M-FISH, and/or M-BAND results to identify suitable marker probes for the abnormal chromosomes (*see* **Notes 39** and **40**; Fig. 4).
5. If it is suspected that two regions have become separated or joined, check this by co-hybridizing probes for these loci.
6. If probes with different optimal conditions are to be combined, first test a range of conditions (pretreatment, minimum co-denaturation temperature, hybridization buffer, and dilution ratio) on a normal cytogenetic preparation to find those that give the optimal signal and chromosome morphology (*see* **Notes 41** and **42**).
7. If the identity of the centromere on the abnormal chromosome is ambiguous, this can be tested with specific centromere probes. Sets of centromere probes labeled with different fluorochromes can be tested together, to speed up this search, if

there is no obvious centromere candidate. An alternative is to use a multicentromere FISH probe [17]. A method for making such a probe can be found in Liehr et al. [18] (*see* **Notes 43** and **44**).

8. The short arms of the acrocentric chromosomes (chromosomes 13, 14, 15, 21, or 22) contain highly repetitive DNA, including the nucleolar organizer region (NOR). Metaphase FISH with a probe for the NOR can help resolve uncertainties involving these regions [6, 7]. A short arm of one of these chromosomes, which is "invisible" by SNP array, may provide one of the telomeres of an anachromosome (chromosome produced by chromothripsis and also chromosomes made by complex rearrangements in the context of this chapter) [7]. We use a commercial probe for the NOR (*see* Fig. 1).
9. Distinguish between abnormalities of the two different chromosome homologues. The B allele frequency data can help with this if the abnormalities of the two homologues overlap. Otherwise use G-band, M-FISH, or M-BAND data to determine whether abnormalities are in the same or different homologues (*see* **Note 45** and **46**; Fig. 2).
10. Calculate the proportion of each genotype in a population with two different genotypes as follows [7] (*see* **Note 47**):

    In a mixture of two cell populations with genotypes 1 and 2, each with different B allele frequencies, the proportion of cells with genotype 1 (F) can be calculated using the formula

    $$F = (\mathrm{b}_2 - \mathrm{B}a_2)/[\mathrm{B}(a_1 - a_2) - b_1 + b_2]$$

    where b_1 is the number of B alleles in genotype 1 and b_2 is the number of B alleles in genotype 2, a_1 is the total number of alleles in genotype 1 and a_2 is the total number of alleles in genotype 2, and B is the B allele frequency value for heterozygous loci (this works using either of the two inner B allele frequency columns) (*see* **Notes 48** and **49**).

4 Notes

1. G-banding will help recognize balanced internal rearrangements of a chromosome (e.g., inversions), which won't be picked up by M-FISH or SNP array. Alternatively, M-BAND can detect these types of rearrangement—carry out all 23 (female) or 24 (male) M-BAND experiments. *See* Subheading 2.4.1.
2. A cloth saturated with 2 × SSC or water and laid flat in a sealable plastic container is suitable. The slides are placed faceup on the wet cloth.

3. M-FISH and M-BAND probes are complex. For simplicity and reliability, we recommend purchasing commercial probes. The method given is for XCyte mFISH and mBAND, kits for which are available from MetaSystems [12].
4. FISH probes are light-sensitive. Hybridization buffers and premixed probes contain formamide, which is a suspected teratogen.
5. Igepal CA-630 replaces Nonidet P-40 (available from Sigma-Aldrich) and can be substituted for Tween 20. Tween 20 (or Igepal CA-630) is added after autoclaving.
6. Check the pH of the solutions at room temperature; correct pH is important.
7. Single-locus FISH probes for many diagnostically relevant probes are available from a variety of suppliers and can be convenient to use if they are available. Region-specific (chromosome paints, chromosome arm paints, and chromosome band) and repetitive-locus (e.g., centromere) probes are also available. Custom-made locus-specific or region-specific probes can be ordered (*see* below) already labeled with the desired fluorochrome. If probes will potentially be mixed and co-hybridized, have them labeled with fluorochromes that can be distinguished with the available filters. Suitable suppliers include Vysis (Abbott Molecular), Cytocell (Oxford Gene Technology), MetaSystems Probes, Empire Genomics, and Kreatech (Leica Biosystems). The Aquarius LPE NOR probe (Cytocell) is a suitable probe for the nucleolar organizer region (NOR) on the acrocentric chromosome short arms, and for the telomere, the LNA (locked nucleic acid) telomere FISH probe available from Exiqon is suitable. Whole chromosome, chromosome arm, and chromosome band-specific paints can be co-hybridized with single locus and repeat sequence probes.
8. Labeled BAC clones can be used as locus-specific FISH probes. They can be chosen from a suitable genome location in the Ensembl genome browser archive, Ensembl 75 GRCh37 (http://grch37.ensembl.org/index.html) or earlier. To find the chromosome region in question, search for a gene or locus, or click on the "view karyotype" icon and center on the desired location. View RP11-BAC clone sets by selecting "Configure this page" on the left, and select the 32k clone set and Tilepath from "Clones & misc. regions." BACs marked with black triangles in this genome browser have been mapped by FISH. Labeled RP11-BACs can be purchased from Empire Genomics. Empire Genomics also provides an online tool to search their BAC database by genome position (https://www.empiregenomics.com/CloneCentral/region_search). Alternatively, if the necessary expertise is available, BACs can be purchased as clones or DNA (BACPAC Resources or Thermo Fisher) and

labeled in-house. Human FISH probes are normally prepared with a carrier DNA, e.g., salmon sperm DNA, and human Cot-1 DNA if they contain unwanted repeats.

(a) Another option for single-locus probes is to purchase custom-labeled oligonucleotide probes at the desired genome position (e.g., SureFISH, Agilent). These have the advantage of being repeat-free, eliminating background noise from cross-hybridization.

9. It is best to use concentrated probes if possible and make them up to the required concentration so that they can be combined as desired without diluting the probe more than what is optimal.
10. The position of BAC clones should be confirmed by FISH, as sometimes their location does not match that given on the human genome build (Elizabeth Baker, personal communication).
11. With each new genome build, the position of a locus changes. Be sure to note which genome build you are using and adjust the position if you are comparing to a different genome build, for example, in the SNP array analysis software. In Ensembl the position of a locus in a previous build can be found by going to the archived site (the "view in archive site" link is at the bottom right of the page). To convert a position to a more recent genome build, add a string to the Ensembl address as follows. For example, the *ABL1* gene in Ensembl build NCBI36 is at http://may2009.archive.ensembl.org/Homo_sapiens/Location/View?db=core;g=ENSG00000097007;r=9:132578987-132752883. In the current version of Ensembl, you would remove the first part of the address up to and including "archive" and add ";a = ncbi36" at the end, i.e., http://ensembl.org/Homo_sapiens/Location/View?db=core;g=ENSG00000097007;r=9:132578987-132752883;a=ncbi36, to go to the location in the current genome build (http://www.ensembl.org/Help/Faq?id=286). Check the surrounding loci and BACs in the Ensembl archive site to check that your conversion is correct (e.g., going from *ABL1* to the NCBI36 build in the archive gives an incorrect conversion at the time of writing). One way to check the conversion is to click on the gene tab if there is one (in this case the *ABL1* tab, which gives the correct location). Alternatively, check the surrounding landscape, and if it does not match your original, search for a locus in the build you are converting to.
12. The variety of probe types and the option of ordering customized probes mean that there is usually a solution for any situation.
13. A SNP array segment is a series of retained SNP loci from a contiguous stretch of the original chromosome with the same copy number. Balanced rearrangements within a SNP array

segment will not produce a change in copy number, so a SNP array segment may consist of more than one rearranged chromosome segment. Mate-pair sequencing or whole genome sequencing has the advantage of being able to detect most of these rearrangements.

14. This is addressed in Subheading 3.9, **step 9**.
15. If using SNP arrays, it is best to look at the B allele frequency data to identify a change in copy number, as these are more sensitive to copy number change.
16. Colcemid concentration and incubation time may need to be adjusted depending on the cell type.
17. Good morphology is indicated by well-defined chromosomes clear of cytoplasm, neither phase bright nor pale. Well-spread chromosomes with minimal overlaps are best for the interpretation of M-FISH, whereas more overlaps can be tolerated with M-BAND.
18. It is important to use at least the volume of probe specified in the user instructions for the coverslip used, as M-FISH and M-BAND probes are complex.
19. A thermocycler can be used instead of **steps 17–19**. Set at 75 °C for 5 min, 10 °C for 30 sec, and 37 °C for 30 min, and hold at 10 °C or on ice until used.
20. Avoid overexposure of the slides to light by blocking the filter when not viewing the slide.
21. After selecting "Capture," use the up and down arrows to select the field used to determine exposure time. Any high-intensity background should be excluded from this field, or the signal of interest will be underexposed. Ensure a positive signal for each fluorochrome is included in this field (particularly for M-BAND), or exposure times will be too high.
22. First check all fluorochromes by viewing the images in each fluorochrome channel; otherwise some labeled chromosomes may be deleted at this step. This is because some fluorochromes are not visible in the triple-color images.
23. Check that small chromosomes are not being lost by over-thresholding.
24. Separating the chromosomes while viewing the DAPI image makes it easier to identify whole chromosomes.
25. View the single-color galleries to check the accuracy of the false colors and to check for small segments that do not produce a false color (Fig. 3).
26. M-BAND (and M-FISH) classifiers can be saved. The same classifier should be used on M-BAND images from different metaphases if the banding patterns are to be compared.

27. M-FISH can identify balanced chromosome rearrangements that microarrays do not detect (that have no change in copy number).
28. M-FISH can identify segments down to the order of about 5–10 Mb, depending on the fluorochromes used for the chromosome probes. Junctions of chromosomes, which have probes with overlapping combinations of fluorochromes, can be harder to identify. Shorter rearranged sections may produce a misleading false color due to overlap from fluorochromes in adjoining regions. Look at the individual fluorochrome signal patterns to check the chromosome identity if the regions are 20 Mb or less (Fig. 3).
29. A single M-FISH false color for a region that has undergone chromothripsis suggests that segments from the same chromosome have rejoined, i.e., that these are all intrachromosomal rearrangements. However, M-FISH is a low-resolution technique. M-BAND is more sensitive than M-FISH for chromosome identification because each M-BAND probe is specific for only one chromosome. Chromosome material down to about 1–2 Mb may be detected by M-BAND (Fig. 3). There is a better chance of detecting material from the same chromosome on different anachromosomes [7] or from different chromosomes on the same anachromosome, with M-BAND compared to M-FISH. Internal rearrangements are a different matter, and those in the order of 10 Mb or more may be detected by M-BAND. Single-locus FISH using probes from each section provides the best resolution for localizing short chromothripsis segments (*see* below).
30. Bear in mind the possibility that an M-BAND probe cross-hybridizes with another chromosome. We found this with the XCyte 5 probe. Check the M-BAND probe on a normal control specimen if this is suspected.
31. A shattered and repaired chromosome that has lost its telomeres may use a telomere-containing segment from a different chromosome [7] (Fig. 6). Exceptions may include ring chromosomes, which have no telomere, and chromosomes joined by telomere fusion (Fig. 5) [19]. Telomere erosion associated with telomere fusion may not be obvious on SNP array data. The probes representing the ends of the chromosomes on the SNP arrays are not telomeres but rather the subtelomeric regions. Ideally there will be twice as many telomeric segments as there are rearranged chromosomes, so that segment matching is straightforward.
32. Plan ahead of time the volume of metaphase suspension, coverslip size, and volume of probe. 3 μL of chromosome suspension spotted on a microscope slide covers an area suitable for a

9 mm round coverslip or 7 μL for a 12 mm round coverslip. The smaller coverslip leaves open the option of carrying out multiple FISH assays on a single microscope slide to help preserve precious probe or specimen. As a precaution against confusing specimens, do not spread different specimens on the same slide. FISH can be performed in volumes of 1 μL or less under appropriately sized coverslips. We routinely use 0.8 μL probe under a 9 mm round coverslip. Other probe volumes can be calculated based on the area of the coverslip.

33. When working with fluorescent probes, keep exposure to light to a minimum.
34. Carry out test hybridizations to optimize these if they are not known.
35. Store probed slides at −20 °C in the dark.
36. In genomes with chromothripsis and other highly rearranged genomes, there will probably be ambiguities after the segment identification and matching outlined above, particularly if there is chromothripsis with clustering of the breakpoints close together. In chromothripsis, SNP array segments may be below the resolution of M-BAND and may contain multiple actual rearranged segments (where there is no copy number change on either side of the breakpoint).
37. Performing multiple FISH tests by co-hybridization at once also improves efficiency.
38. M-FISH and M-BAND probes are not suitable for co-hybridization.
39. DAPI banding can also help identify chromosomes. In the Isis software (Metasystems), G-bands are often discernible as regions of different staining intensities. The DAPI is at a reduced concentration in the Metasystems DAPI/antifade solution compared to some other preparations, which helps achieve DAPI banding (Ilse Chudoba, personal communication). The contrast can be improved by using the "Adjust filter power" function.
40. A whole chromosome paint can also be used for this purpose.
41. Probes from different manufacturers can have different denaturation temperatures or different wash conditions, and some require pretreatment of the chromosomes. If the temperature is too low, the probe will not denature. The higher the denaturation temperature, the greater the risk of compromising chromosome morphology.
42. When combining probes for co-hybridization, the ideal situation is that the probe stocks are in concentrated form, so they can be combined without diluting them further than intended, and that their intended hybridization conditions are the same.

Some commercial probes are already at the final concentration in hybridization buffer. Robust probes (that produce a strong signal) that are diluted when mixed together should still produce an analyzable signal.

43. Some centromere probes have cross-specificity, e.g., probes for the 1, 5, and 19 centromeres.
44. Highly repetitive DNA is not assayed by SNP array. In contrast, as FISH signals are most easily obtained from repetitive DNA, FISH can be used to identify whether these regions are present and where, in the following situations:
    (a) If a SNP array pattern shows different copy numbers on either side of a centromere, this centromere may or may not be retained on the higher copy number segment (Figs. 3 and 6). Whole arm translocations with breakpoints at the centromeres of both involved chromosomes can produce a chromosome with both centromeres [6].
    (b) Centromeres can be deleted without a change to the SNP array pattern [6, 20]. An abnormal chromosome that contains two regions that flank a centromere in the parent chromosomes suggests that it is dicentric, but on the other hand, one of the centromeres in a dicentric chromosome can be deleted without loss of the flanking probes (Fig. 5) [15].
    (c) If an anachromosome does not contain any segment that spans a centromere, it is possible that a centromere has been captured from another centromere [6]. A centromere-spanning segment from another chromosome, which has not been localized or accounted for, is an obvious candidate (Figs. 4 and 6).
    (d) If no centromere can be identified, it is possible that a neo-centromere has been created [21]. If so, antibodies to the centromere protein CENP-C could be used to test for a functional centromere [22], if desired. Double minute chromosomes, containing highly amplified segments, have no centromere. Double minutes can be produced as a result of chromothripsis [4].
45. This information is preserved in intact metaphase chromosomes. FISH probes could be chosen for each interval if the breakpoints are close together (Fig. 7).
46. When material on both homologues is present in equal proportions, the B allele frequency (BAF) for heterozygous loci should sit at 0.5. If a trisomic chromosome is deleted, the BAF for heterozygous loci will show whether the deletion is in the duplicated homologue (resulting in values closer to or equal to 0.5, if the deletion is in less than or 100% of cells, respectively) or in the

single-copy homologue (producing a further shift toward homozygosity of the heterozygous loci) (Figs. 4 and 7) [6].

47. In Illumina SNP array data, B allele frequency values for the homozygous loci are represented by the columns of blue dots at 0 (AA loci) and 1 (BB loci). A normal diploid karyotype also has a column of dots at 0.5 representing the heterozygous (AB) loci. In abnormal karyotypes, this central column splits into two columns equidistant from 0.5 when there are unequal numbers of both homologues, including when there are mixtures of cell populations with different genotypes. In a mixed cell population where there is imbalance of the two chromosome homologues, the B allele frequency of heterozygous loci is the number of B alleles in each clone weighted by the proportion of cells in that clone, divided by the total number of alleles in each clone weighted by the proportion of cells in that clone. In this scenario there are two B allele frequency columns for heterozygotes, one with less than 50% B alleles and one with more than 50% B alleles.

48. For example, if a proportion (*F*) of cells has a deletion of the B allele at a heterozygous locus, the genotype is A in cells with the deletion and AB in undeleted cells. The value of b_1 is 0, b_2 is 1, a_1 is 1, and a_2 is 2.

    The first example given in footnote a in Table 1 of MacKinnon et al. [7], is incorrect and is for the A vs. AB combination of alleles, not A vs. AAB as stated.

Acknowledgments

Many thanks to Meaghan Wall for critical reading of the manuscript.

References

1. Palumbo E, Russo A (2016) Chromosome imbalances in cancer: molecular cytogenetics meets genomics. Cytogenet Genome Res 150:176–184
2. Garsed DW, Marshall OJ, Corbin VDA et al (2014) The architecture and evolution of cancer neochromosomes. Cancer Cell 26:653–667
3. Li Y, Schwab C, Ryan SL et al (2014) Constitutional and somatic rearrangement of chromosome 21 in acute lymphoblastic leukaemia. Nature 508:98–102
4. Stephens PJ, Greenman CD, Fu B et al (2011) Massive genomic rearrangement acquired in a single catastrophic event during cancer development. Cell 144:27–40
5. Landry JJ, Pyl PT, Rausch T et al (2013) The genomic and transcriptomic landscape of a HeLa cell line. G3 (Bethesda) 3:1213–1224
6. MacKinnon RN, Wall M, Zordan A et al (2013) Genome organization and the role of centromeres in evolution of the erythroleukaemia cell line HEL. Evol Med Publ Health 2013:225–240
7. MacKinnon RN, Campbell LJ (2013) Chromothripsis under the microscope: a cytogenetic perspective of two cases of AML with catastrophic chromosome rearrangement. Cancer Genet 206:238–251
8. Floutsakou I, Agrawal S, Nguyen TT et al (2013) The shared genomic architecture of

human nucleolar organizer regions. Genome Res 23:2003–2012
9. Korbel JO, Campbell PJ (2013) Criteria for inference of chromothripsis in cancer genomes. Cell 152:1226–1236
10. Donlon TA, Litt M, Newcom SR et al (1983) Localization of the restriction fragment length polymorphism D14S1 (pAW-101) to chromosome 14q32.1 -> 32.2 by in situ hybridization. Am J Hum Genet 35:1097–1106
11. Moorhead PS, Nowell PC, Mellman WJ et al (1960) Chromosome preparations of leukocytes cultured from human peripheral blood. Exp Cell Res 20:613–616
12. MacKinnon RN, Chudoba I (2011) The use of M-FISH and M-BAND to define chromosome abnormalities. In: Campbell LJ (ed) Cancer cytogenetics methods and protocols. Humana Press, New York, pp 203–218
13. MacKinnon RN, Campbell LJ (2007) Dicentric chromosomes and 20q11.2 amplification in MDS/AML with apparent monosomy 20. Cytogenet Genome Res 119:211–220
14. MacKinnon RN, Campbell LJ (2010) Monosomy 20 in the karyotypes of myeloid malignancies is usually the result of misclassification of unbalanced chromosome 20 abnormalities. In: Urbano KV (ed) Advances in genetic research. Nova Science Publishers, pp 57–72
15. MacKinnon RN, Campbell LJ (2011) The role of dicentric chromosome formation and secondary centromere deletion in the evolution of myeloid malignancy. Genet Res Int 2011:643628
16. MacKinnon RN, Patsouris C, Chudoba I et al (2007) A FISH comparison of variant derivatives of the recurrent dic(17;20) of myelodysplastic syndromes and acute myeloid leukemia: obligatory retention of genes on 17p and 20q may explain the formation of dicentric chromosomes. Genes Chromosom Cancer 46:27–36
17. Camps J, Mrasek K, Prat E et al (2004) Molecular cytogenetic characterisation of the colorectal cancer cell line SW480. Oncol Rep 11:1215–1218
18. Liehr T, Weise A, Kosyakova N (2016) cenM-FISH approaches. In: Liehr T (ed) Fluorescence in situ hybridization (FISH) application guide. Springer, Berlin, Heidelberg, pp 249–255
19. MacKinnon RN, Duivenvoorden HM, Campbell LJ et al (2016) The dicentric chromosome dic(20;22) is a recurrent abnormality in myelodysplastic syndromes and is a product of telomere fusion. Cytogenet Genome Res 150:262–272
20. MacKinnon RN, Selan C, Wall M et al (2010) The paradox of 20q11.21 amplification in a subset of cases of myeloid malignancy with chromosome 20 deletion. Genes Chromosom Cancer 48:998–1013
21. Choo KH (1997) Centromere DNA dynamics: latent centromeres and neocentromere formation. Am J Hum Genet 61:1225–1233
22. Sullivan BA, Schwartz S (1995) Identification of centromeric antigens in dicentric Robertsonian translocations: CENP-C and CENP-E are necessary components of functional centromeres. Hum Mol Genet 4:2189–2197

Check for updates

Chapter 6

Chromothripsis Detectable in Small Supernumerary Marker Chromosomes (sSMC) Using Fluorescence In Situ Hybridization (FISH)

Thomas Liehr

Abstract

The formation of small supernumerary marker chromosomes (sSMC) still remains enigmatic. However, it is suggested that majority of all kinds of de novo sSMC (inverted duplication-, ring-, and centric-minute-shaped ones) are products of incomplete trisomic rescue during early embryogenesis. Recent work, based on molecular cytogenetics, suggests that these trisomic rescue events are going together with chromothripsis, directed against the supernumerary chromosome to be degraded. Here we present a protocol how to characterize so-called discontinuous sSMC by means of fluorescence in situ hybridization (FISH).

Key words Small supernumerary marker chromosomes (sSMC), Incomplete trisomic rescue, Early embryogenesis, Chromothripsis, Molecular cytogenetics fluorescence in situ hybridization (FISH)

1 Introduction

Nowadays it is suggested that the early embryo has a huge capacity to correct many kinds of numerical and even structural chromosomal aberrations [1]; evidence for this came, among others, from studying cases with uniparental disomy (UPD) [2] and small supernumerary marker chromosomes (sSMC) [3]. Interestingly, UPD and sSMC can also appear together in de novo sSMC cases [4].

Even though sSMC are rare, there are about three million carriers of an sSMC worldwide, and in ~70% of the cases, the 47th (derivative) chromosome appeared de novo [3]. sSMC can have three basic shapes; they can be inverted duplication-, ring-, or centric-minute-shaped. A de novo sSMC is suggested to be a product of incomplete trisomic rescue during early embryogenesis.

To provide better possibilities for studying sSMC, numerous fluorescence in situ hybridization (FISH)-based probe sets were developed. The most recent one is the so-called pericentromeric-critical region FISH (PeCR-FISH) probe set,

Franck Pellestor (ed.), *Chromothripsis: Methods and Protocols*, Methods in Molecular Biology, vol. 1769,
https://doi.org/10.1007/978-1-4939-7780-2_6, © Springer Science+Business Media, LLC, part of Springer Nature 2018

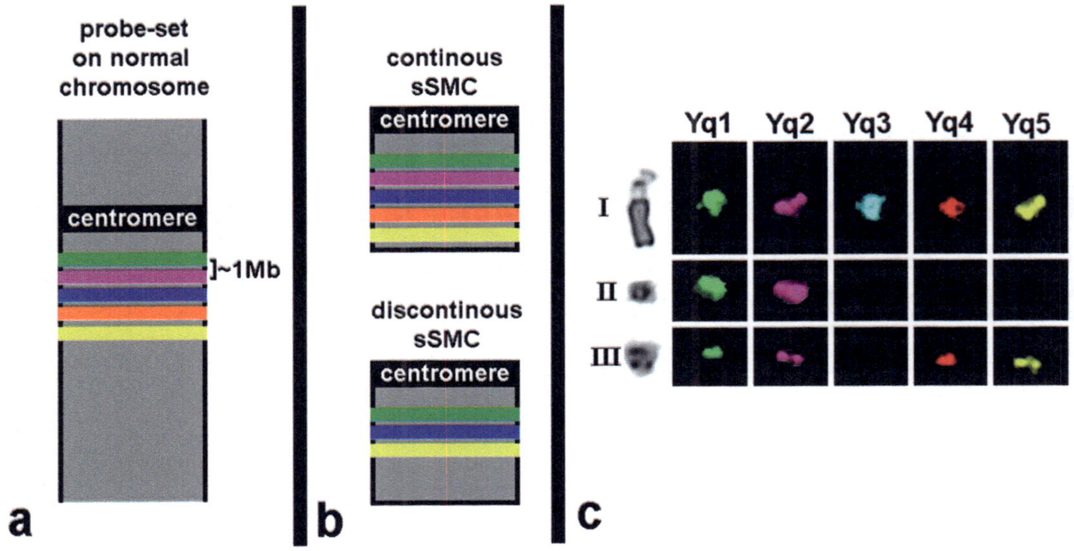

Fig. 1 (**a**) Scheme how chromosome arm-specific probe sets for an sSMC with known origin may look like. (**b**) Schematic depictions of a continuous and discontinuous sSMC. (**c**) Results of a Yq-pericentromeric region-specific probe set consisting of five BAC probes in about 1 Mb distance obtained in a (*I*) normal Y-chromosome, (*II*) continuous sSMC derived from a Y-chromosome der(Y)(:p11.1→q11.21:), and (*III*) discontinuous sSMC derived from a Y-chromosome der(Y)(pter→q11.21::q11.221→q11.221:)

consisting of three to five locus-specific probes spanning about 3–5 Mbp (Fig. 1a) [5]. Using PeCR-FISH identified 26% of the randomly chosen cases to be discontinuous sSMC (Fig. 1b) [5]. This finding led to the idea that sSMC carriers are not only models for individuals attributed to an embryo established originally with a trisomy but also may be formed by chromothripsis, at least in a subset of the cases.

To enable more laboratories to study an sSMC of known origin for a possible "chromothriptic origin," we present here a protocol how to characterize so-called discontinuous sSMC by means of FISH.

2 Materials

Standard cell biological and molecular cytogenetic equipment, including standard solutions (e.g., ethanol, methanol, formamide, formaldehyde, etc.), is required.

As this protocol comprises environmental toxins (like formamide and formaldehyde), proper collection of and treatment of these substances as hazardous waste are required.

2.1 Processing of BACs

1. QIAprep® Miniprep Kit (Qiagen).
2. Antibiotics: use and set up the antibiotics as required by the special BAC clone resistance; corresponding information can be obtained from the individual BAC clone suppliers.

2.2 Locus-Specific Probe Preparation

1. Glycerol.
2. Nick translation kit.
3. Peptone.
4. tRNA.
5. Yeast extract.
6. 0.5 M EDTA (pH 8).
7. LB (Luria–Bertani) medium: to 800 ml aqua dest, add 10 g of peptone, 5 g yeast extract, and 10 g NaCl. Adjust the pH to 7.5 with NaOH (*see* **Note 1**). Adjust the volume to 1 l with aqua dest. Sterilize by autoclaving. Unopened bottles can be stored at room temperature (RT).
8. 3 M sodium acetate: 24.6 g sodium acetate is added to 100 ml aqua dest; the pH is adjusted to 5.2 with acetic acid (HCl) (*see* **Note 2**).

2.3 Fluorescence In Situ Hybridization (FISH)

1. Antifade Vectashield (store at +4 °C).
2. Cot1 DNA human (store at −20 °C).
3. Cytogenetically worked up cell pellet of one or more patient(s) with an sSMC (provided by a cytogenetic lab and not part of this protocol).
4. DAPI (4,6-diamidino-2-phenylindol•2HCl) stock solution (store at −20 °C).
5. Ethanol 100%, 90%, and 70%, 100 ml, each (store at RT).
6. Formamide: aliquot and store at −20 °C; please remember to discard the formamide solution as hazardous waste.
7. Saline-sodium citrate (SSC 20×) (Gibco BRL). Set up 1× and 2× SSC before use (store at RT).
8. Hybridization buffer: dissolve 2 g dextran sulfate in 10 ml 50% deionized formamide/2 × SSC/50 mM phosphate buffer for 3 h at 70 °C. Adjust pH to 7 with phosphate buffer; hydrochloric acid destabilizes buffer solution. Aliquot and store at −20 °C.
9. Phosphate-buffered saline (PBS 1×) (Biochrom) (store at RT).
10. Rubber cement: Fixogum™ (Marabu) (store at RT).
11. DAPI solution: dissolve 1.5 μL of 1 M DAPI stock solution in 1 ml of Vectashield antifade (store at +4 °C; can be used at least for 3 months).

12. Denaturation buffer: 70% (v/v) formamide (*see* **Note 3**), 20% (v/v) filtered double distilled water, 10% (v/v) 20 × SSC; make fresh as required.
13. Formamide solution: 2 × SSC/50% formamide, pH 7.0. Please remember to discard the formamide solution as hazardous waste.
14. Phosphate buffer: prepare 0.5 M Na_2HPO4 and 0.5 M NaH_2PO4, mix these two solutions (1:1) to get pH 7.0, and then aliquot and store at −20 °C.
15. Three to five color FISH probe sets to be created from commercially available bacterial artificial chromosome (BAC) probes according to chromosomal origin and size of the sSMC in question (Fig. 1a–c).

3 Methods

3.1 Selection and Processing of BACs

To study an sSMC of known chromosomal origin and known to contain euchromatic content from short and/or long arm, specific BAC probes can be selected and ordered from corresponding browsers (e.g., UCSC: http://genome.ucsc.edu/cgi-bin/hgGateway).

When selecting BAC clones, cross hybridization caused by sequence homologies should be avoided; the latter can be checked online using eFISH website (http://projects.tcag.ca/cgi-bin/efish/index.cgi). For the here suggested protocol, five BACs, each in a distance of ~1 Mb, should be selected.

3.1.1 Cultivation of E. coli with BACs

BAC clones are provided as *E. coli* clones on LB agar stabs and can be stored for several weeks (*see* **Note 4**). For BAC/plasmid isolation, several methods and kits are available, e.g., QIAprep® Miniprep Kit. The following protocol is described for a so-called Miniprep resulting in ~20 μg plasmid DNA from 1 to 5 ml overnight cultures of *E. coli.*

1. Add the appropriate antibiotics at correct concentration to 5 ml LB medium in a suited sterile jar.
2. Pick *E. coli* by a sterile pipette tip (*see* **Note 5**) from the LB agar stab, and transfer into the LB medium.
3. Cover the sterile jar in a way that air can be exchanged; cultivate for ~16 h on a shaker with 200 g at 37 °C (*see* **Note 6**).

3.1.2 Plasmid DNA Extraction from E. coli

To preserve the corresponding BAC clones, mix 0.5 ml of the *E. coli* culture with 0.5 ml glycerol, and store at −80 °C. For plasmid DNA extraction, follow the details given in the corresponding instruction manual.

3.1.3 Labeling of BAC-DNA

To label BAC-DNA (directly with fluorochromes or indirectly by biotin or digoxigenin), there are several options. Easiest way is labeling by Nick translation, which can be done in each fluorochrome or hapten wanted by corresponding kits (e.g., Roche) (*see* **Note 7**).

3.2 Fluorescence In Situ Hybridization (FISH)

1. 10 μg of each labeled BAC is precipitated together with 2 μg COT1 DNA.
2. After drying of the pellet (e.g., in a SpeedVac machine), the pellet is diluted in 10 μl hybridization buffer.
3. Denature the probe in a 0.5 or 1.5 ml reaction cup at 75 °C for 5 min (*see* **Note 8**) and do a prehybridization at 37 °C for 25 min (*see* **Note 9**).
4. Store the probe set on ice, until being applied to a denatured slide (**step 10**) (*see* **Note 8**).
5. Dehydrate slides with metaphases of an sSMC carrier in an ethanol series (70%, 90%, 100%, 3 min each), and air-dry.
6. Apply 100 μl of denaturation buffer on the slide surface, and cover with 24 × 50 mm coverslip.
7. Move the slides to a heating plate for 2–4 min at 75 °C.
8. To conserve target DNA as single strands, remove the coverslip, and place slide(s) in a Coplin jar filled with 70% ethanol (−20 °C; 3 min) immediately.
9. Repeat **step 5**.
10. Use the denatured probe set from **step 4**, and add onto denatured slides; put suited coverslips on the drop, and seal with rubber cement.
11. Incubate slides for 8–16 h at 37 °C in a humid chamber.
12. Remove the slides from the chamber, and scratch rubber cement with forceps from the slide; then carefully remove coverslips. Optional: let coverslips swim off in washing buffer (RT, 100 ml Coplin jar).
13. Subject slides for 5 min to post-washing in 1 × SSC solution (62–64 °C) with gentle agitation; then wash slides briefly in 1 × PBS (RT).
14. Dehydrate slides in an ethanol series (70%, 90%, 100%, 3 min each), and air-dry.
15. Counterstain the slides with 20 μl of DAPI solution (antifade already included), cover with a coverslip 24 × 60, and evaluate under a fluorescence microscope. If five fluorochromes are used, corresponding evaluation software is necessary. In Fig. 1b, schematic examples indicate how a continuous and a discontinuous sSMC look like.

4 Notes

1. Take care when handling with NaOH—it might be mordant on skin similar to acids!
2. Take care when handling with HCl—it might be mordant on skin!
3. As formamide is teratogen, pregnant women are not allowed to work in a molecular cytogenetic laboratory!
4. Stabs need to be stored in a way that they cannot dry out, e.g., at 4 °C.
5. In our hands sterile pipette tips are more appropriate here than otherwise often-suggested sterile toothpicks.
6. Different countries may have specific rules when working with genetically modified organisms; in Germany, e.g., one must work in an S1 laboratory.
7. Direct labeling leads to shorter working times of the whole FISH procedure, as no detection and less washing steps are necessary. However, indirect labeling may be used in case a probe needs signal amplification.
8. Denaturation of a probe can be done in water bath or, if available, more easily in a thermocycler.
9. Prehybridization time may be adjusted (i.e., be longer) in case of BACs producing more background.

References

1. Pellestor F (2014) Chromothripsis: how does such a catastrophic event impact human reproduction? Hum Reprod 29:388–393
2. Liehr T (2014) Uniparental disomy (UPD) in clinical genetics. A guide for clinicians and patients. Springer, Heidelberg
3. Liehr T (2012) Small supernumerary marker chromosomes (sSMC). A guide for human geneticists and clinicians; with contributions by UNIQUE (the rare chromosome disorder support group). Springer, Heidelberg
4. Liehr T, Ewers E, Hamid AB et al (2011) Small supernumerary marker chromosomes and uniparental disomy have a story to tell. J Histochem Cytochem 59:842–848
5. Hamid AB, Pekova S, Fan X et al (2018) Small supernumerary marker chromosome may provide information on dosage-insensitive pericentric regions in human. Curr Genomics; (in press)

Chapter 7

Identification of Chromothripsis in Biopsy Using SNP-Based Microarray

Veronica Ortega, Christina Mendiola, and Gopalrao V. N. Velagaleti

Abstract

One of the well-known hallmarks of cancer is genomic instability. Although gradualism is a well-established process of cancer evolution, recent studies have shown that chromothripsis or chromoanasynthesis can result in complex genomic rearrangements by a single catastrophic event rather than several incremental steps. These two novel phenomena suggest an evolutionary modality for cancer cells to circumvent individual mutational events with one simultaneous shattering of chromosomes or chromosome regions resulting in the random reassembling of shattered genetic material to form complex derivative chromosomes. Although sequencing methods are ideal for the detection of chromothripsis, single-nucleotide polymorphism (SNP)-based microarray methods are also useful in detecting chromothripsis in biopsy samples. Issues related to sample collection, storage, and transport, especially with tumor biopsies, may limit the options for sequencing studies, and in such cases, SNP-based microarray may be a viable alternative for detecting chromothripsis.

Key words Chromothripsis, SNP microarray, Copy number variants (CNVs), DNA, Cancer, Biopsy

1 Introduction

One of the well-described distinctive features of cancer is genomic instability. As a result, methods to uncover pathways accelerating such instability have become important in cancer studies. Until recently, genomic instability was thought to arise in a process known as "gradualism" wherein sequential accumulation of independent genomic lesions occurs over a span of several years in a multistep process [1–3]. Such genomic lesions include point mutations, CNVs in the form of gains and losses, balanced structural rearrangements, and inversions [4–7]. Recent studies identified novel phenomena called "chromothripsis, chromoanasynthesis, and chromoplexy" which led to the hypothesis that cancer cells might acquire all genomic lesions at once to circumvent the defensive responses for the host genome [8–11]. Although the underlying mechanisms of these novel phenomena are not known and

Franck Pellestor (ed.), *Chromothripsis: Methods and Protocols*, Methods in Molecular Biology, vol. 1769,
https://doi.org/10.1007/978-1-4939-7780-2_7,

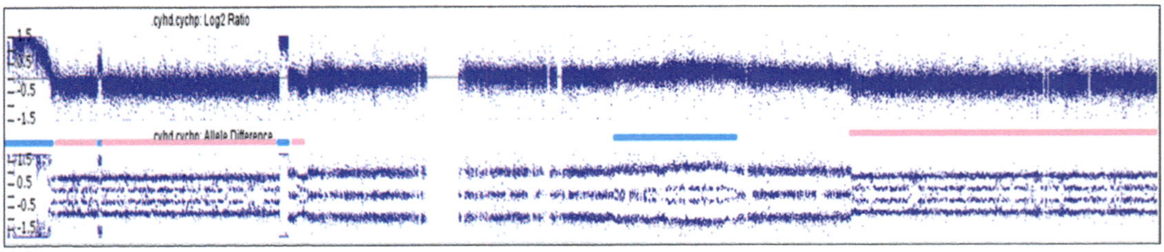

Fig. 1 Plot showing a typical chromothripsis like pattern on SNP array

were thought to be different, the resulting complex genomic rearrangements are believed to be derived from a single catastrophic event rather than by several incremental steps [12–16]. The term "chromoanagenesis or chromosomal rebirth" was coined to group these one-step catastrophic events together [17]. Chromothripsis is described in most cancer types, and the prevalence of chromothripsis across tumor entities varies widely [18]. The discrepancy in the prevalence rates is attributed to both the types of methodology used and the definitions of chromothripsis [18, 19]. Although the initial articles describing chromothripsis used SNP-based microarrays for CNV profiling, whole genome sequencing (WGS)-based methods became the choice for detection of chromothripsis as the WGS-based methodologies were able to not only accurately characterize the breakpoints but increase the sensitivity in detecting smaller CNVs that are beyond the resolution of SNP-based microarrays [19–21]. In spite of the limitations, SNP-based microarrays continue to be a useful tool for detecting chromothripsis (Fig. 1).

2 Materials

2.1 DNA Extraction from Biopsy Material

DNeasy Tissue Kit (Qiagen, Germantown, MD) (*see* **Note 1**)
Store the contents according to manufacturer's suggestions.

1. Sterile surgical scissors.
2. Sterile surgical forceps.
3. Centrifuge tubes, 15 mL.
4. Cell culture petri dishes, 60 × 15 mM.

2.2 Microarray Equipment and Reagents (*See* Note 2)

1. GeneChip™ Scanner 3000Dx (Thermo Fisher Scientific).
2. GeneChip™ Hybridization Oven 645 (Thermo Fisher Scientific).
3. GeneChip™ Fluidics Station 450 (Thermo Fisher Scientific).
4. Chromosome Analysis Suite (ChAS) software, version 2.1 or higher (Thermo Fisher Scientific).

5. CytoScan HD Array and Reagent Kit Bundle (Thermo Fisher Scientific):
   (a) CytoScan™ Module 1: Pre-Lab Restriction and Ligation (store at −25 to −15 °C).
      NSP I, 10× NSP I buffer, 100× BSA, adaptor NSP I (50 μM), 10× T4 DNA ligation buffer, T4 DNA ligase, Reference DNA 103 (50 ng/μL), PCR primer.
   (b) CytoScan™ Module 2: Pre-Lab TE Buffer and Nuclease-Free Water (store at 2–8 °C).
   (c) CytoScan™ Module 3: Post-Lab Fragmentation, Labeling, and Hybridization (store at −25 to −15 °C).
      Fragmentation reagent, 10× fragmentation buffer, TdT enzyme, 5× TdT buffer, 30mM DNA labeling reagent, oligo control reagent, hyb. buffer part 1, hyb. buffer part 2, hyb. buffer part 3, hyb. buffer part 4.
   (d) Cytoscan™ Module 4: Post-Lab Stain, Holding Buffer, Beads, and Water (store at 2–8 °C).
      Stain buffer 1, stain buffer 2, array holding buffer, purification beads, nuclease-free water.
   (e) CytoScan™ Module 5: Post-Lab Elution Buffer and Purification Wash Buffer (store at 15–30 °C).
      Elution buffer, purification wash buffer, wash A, and wash B.
6. Bleach, sodium hypochlorite: working concentration of 0.615% (v/v).
7. PCR gel DNA ladder: 25 bp.
8. Fragmentation gel ladder: 25–2000 bp.
9. Ethanol, absolute.
10. Gel-loading buffer.
11. Laminar flow or PCR cabinet (*see* **Note 3**).
12. Benchtop cooler, with the capacity to hold 8–32 tubes (1.5 mL) and ability to maintain temperature below −15 °C for 2 h.
13. Centrifuge plate, multipurpose, 330 μL of capacity.
14. Cooling chamber, double-block, with the capacity to hold 96-well plates with a maximum capacity of 330 μL.
15. Freezer, −20 to −15 °C, manual defrost.
16. Microfuge (for tubes and strip tubes).
17. Pipettors.
   (a) 12-channel, 2–20 μL.
   (b) 12-channel, 20–200 μL.
   (c) Single channel, 100–1000 μL.

   (d) Single channel, 20–200 μL.
   (e) Single channel, 2–20 μL.
18. Rectangular ice tray, large (9-L capacity).
19. Storage racks, tube, 96-well.
20. Thermal cycler: capable of holding 200-μL volume and 96-well plate; heat block capable of holding temperature of 4–99.9 °C; temperature accuracy of ±0.25 °C (at 35–99.9 °C); average heating and cooling rate of 2.6 °C per s; thermal uniformity of ±0.5 °C.
21. Vortexer, 60 Hz, 75 W, 600–3200 g.
22. Adhesive film applicator (hard plastic) (*see* **Note 4**).
23. Adhesive films, clear, PCR-certified, 96-well plates (Life Technologies).
24. Adhesive film applicator (hard plastic).
25. Anti-vibration pad, used with vortexer to prevent movement during operation.
26. Benchtop cooler, with the capability to hold 8–32 tubes (1.5 mL) and ability to maintain temperature below −15 °C for 2 h.
27. Cooling chamber, double-block, with the capacity to hold 96-well plates with a maximum capacity of 330 μL.
28. Electrophoresis supplies.
29. Freezer, −20 to −15 °C, manual defrost.
30. Gel imager.
31. Magnetic rack with magnet on the side and capable of holding 8–12 tubes of 1.5–2-mL capacity.
32. Microcentrifuge, non-refrigerated with capacity to hold 24 tubes and maximum rotation speed of 16,200 × *g*.
33. Microcentrifuge tubes, nuclease-free, sterile.
   (a) 1.5 mL.
   (b) 2.0 mL.
34. Microtube foam insert.
35. Pipettors.
   (a) 12-channel, 2–20 μL.
   (b) 12-channel, 20–200 μL.
   (c) Single channel, 100–1000 μL.
   (d) Single channel, 20–200 μL.
   (e) Single channel, 2–20 μL.
36. Plate centrifuge, refrigerated multipurpose, plate carriers for 4 × 96-well assay plates.

37. Platform head, 6 in. for microtube foam insert.
38. Rectangular ice tray, large (9 L of capacity).
39. Refrigerator, 2–8 °C.
40. Spectrophotometer, single or multichannel, UV/VIS.
41. Storage racks, tube, 96-well with lid, 1.5–2-mL tubes.
42. Thermal cycler: capable of holding 200-μL volume and 96-well plate; heat block capable of holding temperature of 4–99.9 °C; temperature accuracy of ±0.25 °C (at 35–99.9 °C); average heating and cooling rate of 2.6 °C per s; thermal uniformity of ±0.5 °C.
43. Vortexer, 60 Hz, 75 W, 600–3200 g.
44. Adhesive films, clear, PCR-certified, 96-well plates (Life Technologies).
45. Adhesive label dots, 1/2 to 3/8 in.
46. TBE precast gels, 4% (Sigma-Aldrich).
47. TBE precast gels, 2% (Sigma-Aldrich).
48. Microcentrifuge tubes, nuclease-free, sterile.
    (a) 1.5 mL.
    (b) 2.0 mL.
49. Microcentrifuge tubes, nuclease-free, sterile, nonstick.
    (a) 1.5-mL amber polypropylene.
    (b) 1.5-mL blue polypropylene.
    (c) 1.5-mL natural polypropylene.
50. Pipette tips with aerosol barriers, 20, 200, 1000 μL.
51. PCR plates, unskirted with a maximum volume of 330 μL.
52. Reagent reservoir, 25 mL.
53. Strip tubes, nuclease-free, sterile, 8-well, 0.2-mL polypropylene.
54. Titanium DNA amplification kit (Clontech).

3 Methods (Fig. 2)

3.1 DNA Isolation from Biopsy Tissue

1. Cut biopsy tissue into small pieces.
2. Place the pieces into a 1.5-mL microcentrifuge tube.
3. Add 100-μL buffer ATL.
4. Add 20-μL Proteinase K.
5. Mix by vortexing for 30 s.
6. Incubate at 56 °C until all pieces are completely lysed.
7. Add 200-μL buffer AL.

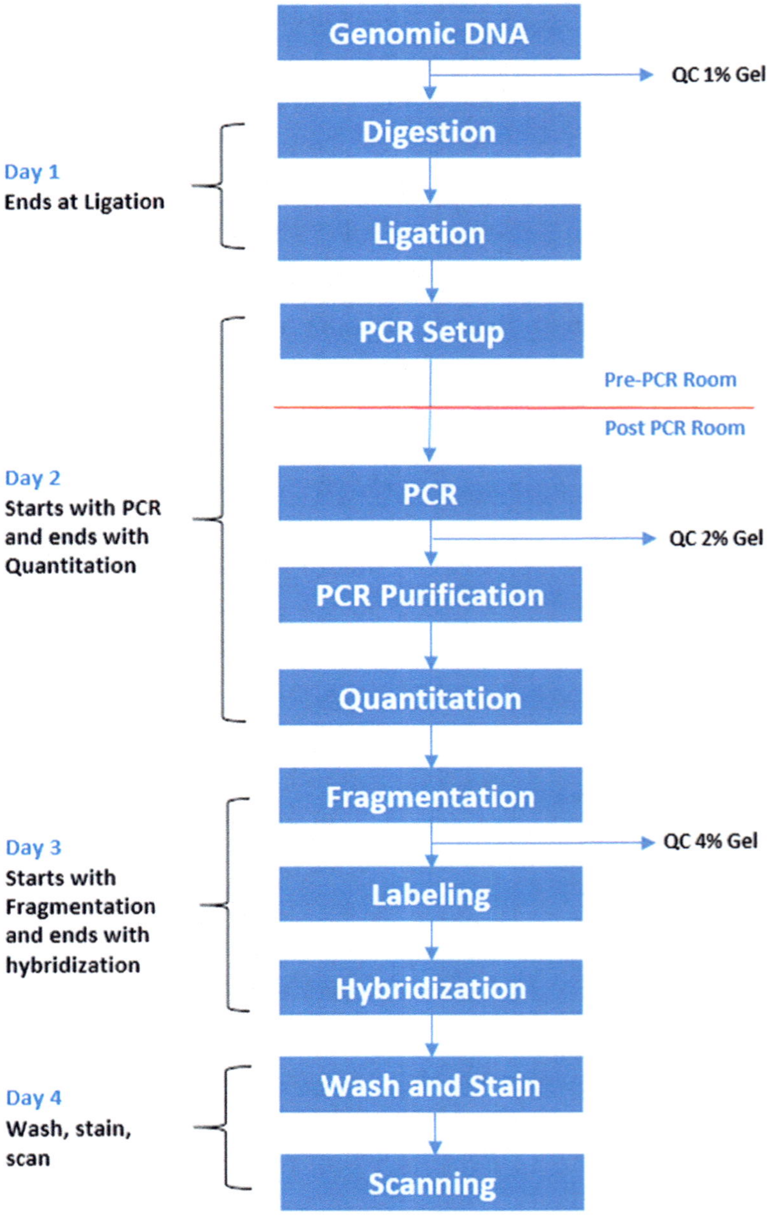

Fig. 2 Schematic showing the procedure

8. Mix thoroughly by vortexing for 30 s.
9. Incubate sample at 56 °C for 10 min.
10. Add 200-μL ethanol (96–100%).
11. Mix thoroughly by vortexing for 30 s.
12. Transfer the solution to a DNeasy mini spin column placed in a 2-mL collection tube.
13. Centrifuge at ≥6000 × *g* for 1 min.
14. Discard the flow-through and the collection tube.

15. Place the spin column in a new 2-mL collection tube.
16. Add 500 μL of buffer AW1.
17. Centrifuge at ≥6000 × *g* for 1 min.
18. Discard the flow-through and the collection tube.
19. Place the spin column in a new 2-mL collection tube.
20. Add 500 μL of buffer AW2.
21. Centrifuge at ≥20,000 × *g* for 3 min.
22. Discard the flow-through and the collection tube.
23. Place the spin column in a new 1.5- or 2-mL pre-labeled microcentrifuge tube.
24. Add 200-μL buffer AE to the center of the spin column to elute the DNA.
25. Incubate for 1 min at room temperature.
26. Centrifuge at ≥6000 × *g* for 1 min (*see* **Note 5**).
27. Measure the DNA concentration and store in the freezer.

3.2 Digestion and Ligation (Day 1: Pre-PCR Lab)

3.2.1 Set up Work Area

1. Fill ice bucket with ice. Place cooling chamber on ice.
2. Decontaminate hood with UV light for at least 15 min.
3. Turn on thermal cycler to preheat lid.

3.2.2 Prepare gDNA (Genomic DNA) Samples

1. Remove stock gDNA samples from freezer. Place gDNA at 15–30 °C until thawed (≤30 min); place in cooling block on ice.
2. Remove TE buffer from refrigerator. Each sample must be diluted to 250 ng/5 μL with TE buffer in 1.5-mL or 2-mL microcentrifuge tubes.
3. Pulse vortex each gDNA sample three times and centrifuge 5 s. Place back in ice bucket.
4. Use OD concentration value to determine dilution volumes for a final concentration of 50 ng/μL for each sample. Adjust the gDNA volume to bring the final volume to 5 μL.

   For example:

Required final quantity:	250 ng
OD measurement:	121.4 ng/μL
Volume of gDNA needed:	2.1 μL(250 ng/121.4 ng = 2.1 μL)
Final volume:	2.10-μL sample
	±2.90 TE buffer
	5.00 μL (working concentration of 250 ng/5 μL)

   When OD measurement is 50 ng/μL, no dilution is required—remove 5 μL of the sample, and place directly to a ppro-priate assigned well designation.

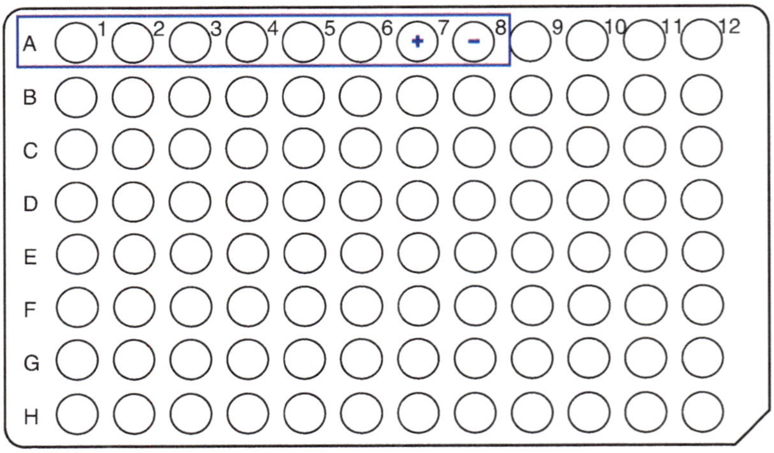

Fig. 3 Labeling of a 96-well plate with the number of samples including the positive (+) and negative (−) controls

3.2.3 Aliquot Prepared gDNA into Assay Plate

1. Mark a 96-well plate with a black marker around the number of samples to be tested along with the positive (+) and negative (−) controls (Fig. 3).
2. Place plate on the lower half of cooling block.
3. Vortex gDNA for 3 s; centrifuge.
4. Transfer **5 μL** of the first sample from gDNA microcentrifuge tube to well A1 of DIG-LIG (digestion-ligation) plate; transfer all samples the same way.
5. Tightly seal DIG-LIG plate with a new seal, and centrifuge at 650 × *g* for 1 min. Place sealed plate on the lower half of cooling block.

3.2.4 Restriction Enzyme Digestion

1. Place nuclease-free water (NFW) on ice.
2. Label a 1.5-mL tube "dig" (digestion) and place in cooling block.
3. Thaw to room temperature (≤30 min) both the 10× NSP1 buffer and 100× BSA (*see* **Note 6**).
4. Once thawed, vortex three times for 1 s each.
5. Quick spin for 3 s.
6. Place in cooling block on ice.
7. Prepare digestion master mix (MM) by adding the following reagents per table below except for the enzyme (prepare and store all reagents and tubes in a cooling block placed on ice) (*see* **Notes 7–10**).

Reagent	1 sample (μL)	8 samples (μL)	16 samples (μL)	24 samples (μL)
NFW	11.55	110.9	221.8	332.6
10× NSP1 buffer	2.00	19.2	38.4	57.6
100× BSA	0.20	1.9	3.8	5.8
NSP1	1.00	9.6	19.2	28.8
Total	14.75	141.6	282.2	424.8

8. Vortex the "dig" tube three times for 1 s each.
9. Quick spin for 3 s.
10. Place the "dig" tube in a chilled cooling block (2–8 °C).
11. Remove NSP 1 enzyme from the freezer.
12. Quick spin NSP 1 enzyme for 1 s.
13. Vortex for 1 s.
14. Quick spin for 3 s.
15. Return to −15 to −25 °C cold storage chamber.
16. Immediately add NSP 1 enzyme to digestion MM.
17. Return the enzyme to −15 to −25 °C cold storage chamber.
18. Vortex digestion MM three times for 1 s each.
19. Quick spin 3 s.
20. Place in a chilled cooling block (2–8 °C).
21. Add **14.75-μL** digestion MM to each sample in row A (Fig. 4).
22. Seal plate tightly with a new seal.
23. Vortex plate 1 s all corners and center.
24. Centrifuge plate at 650 × *g* for 1 min.
25. Load plate on thermal cycler, ensure thermal cycler lid is preheated, and run CytoScan® Digest program.

CytoScan Digest program	
Temperature (°C)	Time
37	2 h
65	20 min
4	5 min

26. Discard remaining digestion MM and return reagents to the freezer.
27. Digest takes 2 h and 25 min to process; when done, leave plate in thermal cycler. Process plate within 2.5 h (*see* **Note 11**).

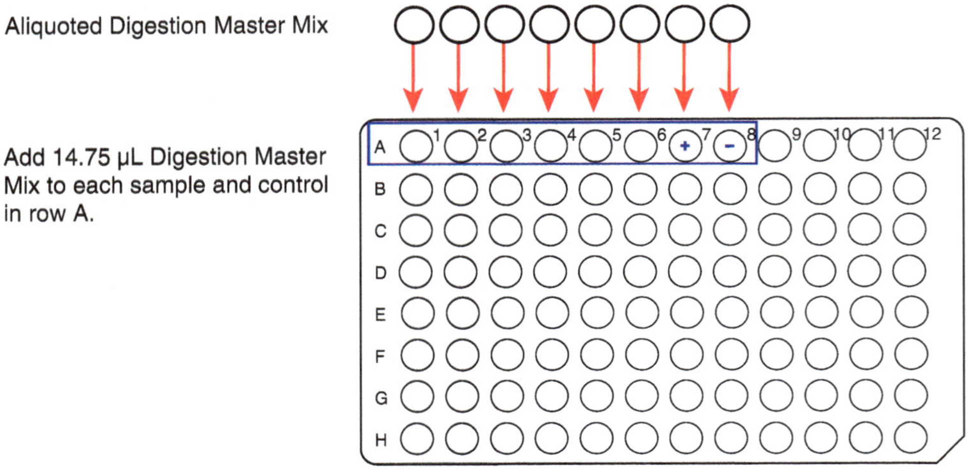

Fig. 4 Adding digestion master mix to the each sample in row A of a 96-well plate

3.2.5 Ligation

1. Thaw reagents and digested samples.
2. Label a 1.5-mL tube "lig" (ligation).
3. Thaw to room temperature (≤30 min) both the 50-μM adaptor for NSP 1 and 10× T4 DNA ligation buffer.
4. Once thawed, vortex three times for 1 s each.
5. Quick spin for 3 s.
6. Place in cooling block on ice and use within 1 h.
7. Remove the "dig" plate from thermal cycler, and centrifuge at 650 × *g* for 1 min.
8. Place the "dig" plate in the lower half of cooling block on ice.
9. Prepare ligation master mix by adding the following reagents per table below except for enzyme (prepare and store all reagents and tubes in a cooling block placed on ice) (*see* **Note 12**).

Reagent	1 sample (μL)	8 samples (μL)	16 samples (μL)	24 samples (μL)
10× T4 DNA ligation buffer	2.50	25.0	50.0	75.0
50-μM NSP 1 adaptor	0.75	7.5	15.0	22.5
T4 DNA ligase	2.00	20.0	40.0	60.0
Total	5.25	52.5	105.0	157.5

10. Vortex the "lig" tube for three times for 1 s each.
11. Quick spin for 3 s.
12. Place the "lig" tube in a chilled cooling block (2–8 °C).

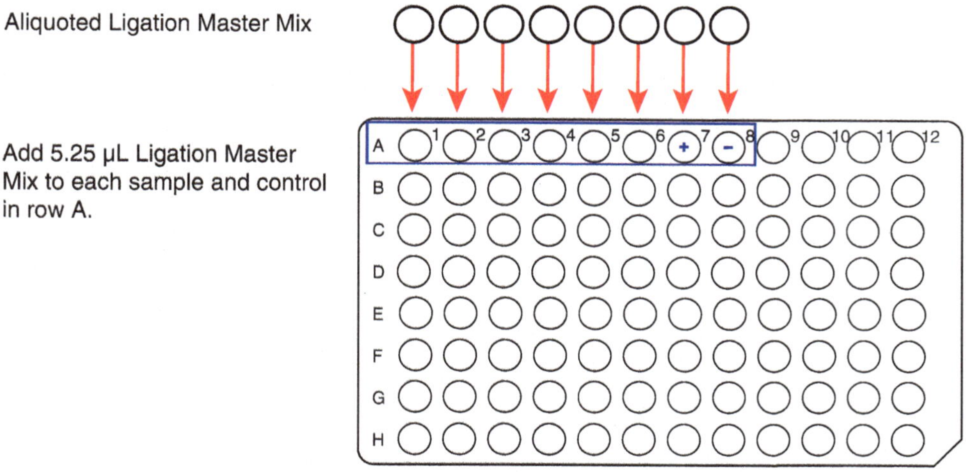

Fig. 5 Adding ligation master mix to the each sample in row A of a 96-well plate

13. Remove the cold storage chamber containing the T4 DNA ligase enzyme from freezer.
14. Quick spin SNP 1 enzyme for 1 s.
15. Vortex for 1 s.
16. Quick spin for 3 s.
17. Return to −15 to −25 °C chamber.
18. Immediately add enzyme to ligation MM, and return the cold storage chamber to freezer.
19. Vortex ligation MM three times for 1 s each.
20. Quick spin 3 s.
21. Place ligation MM in cooling block on ice.
22. Unseal digested sample plate and discard seal.
23. Add **5.25-µL** ligation MM to each digested sample (Fig. 5).
24. Seal plate tightly with a new seal.
25. Vortex plate 1 s all corners and center.
26. Centrifuge plate at 650 × *g* for 1 min.
27. Load plate onto thermal cycler, ensure thermal cycler lid is preheated, and run CytoScan® ligate program.

CytoScan ligate program

Temperature (°C)	Time
16	3 h
70	20 min
4	5 min
4	Hold

28. Discard the remaining ligation MM and return reagents to freezer.
29. Ligation takes 3 h and 25 min to complete. When done, leave plate in thermal cycler. The "lig" plate can be left in the thermal cycler at 4 °C for up to 60 h (*see* **Note 13**).

3.3 PCR (Day 2: Pre-PCR Lab and Post-PCR Lab)

3.3.1 Set Up Work Area

1. Fill ice bucket with ice. Place cooling chamber on ice.
2. Decontaminate hood with UV light for at least 15 min.
3. Have someone turn on thermal cycler to preheat lid in the post-PCR lab.

3.3.2 Dilute Ligated Samples

1. Twenty minutes before use, place NFW on ice.
2. Remove plate from thermal cycler and centrifuge at 650 × *g* for 1 min (*see* **Note 14**).
3. Place plate in the upper half of cooling block on ice.
4. Unseal the ligated sample plate and discard seal.
5. Use P200 pipette to add **75-μL** NFW to each reaction.
6. Tightly seal plate with a new seal.
7. Vortex 1 s in all corners and center.
8. Centrifuge at 650 × *g* for 1 min.

Ligated DNA	**25 μL**
Chilled NFW	75 μL
Total	100 μL

3.3.3 Run PCR

1. Place the diluted ligated sample plate on the upper half of cooling block.
2. Label the new plate "PCR" and place in the lower half of cooling block.
3. Unseal the ligated and diluted sample plate and discard seal.
4. Use the multichannel P20 pipette to transfer 10 μL of each sample to the corresponding well of PCR plate (transfer four times). Start with row D1–D8, then C1–C8, until A1–A8 is reached (Fig. 6).
5. Store the leftover samples at −15 to −25 °C.
6. Seal plate tightly with a new seal and centrifuge at 650 × *g* for 1 min.
7. Thaw to room temperature (≤30 min) both the 10× Titanium Taq PCR buffer, dNTP mixture, and PCR primer.
8. Once thawed, vortex three times for 3 s each.
9. Quick spin for 3 s.
10. Place in cooling block (2–8 °C) on ice and use within 1 h.

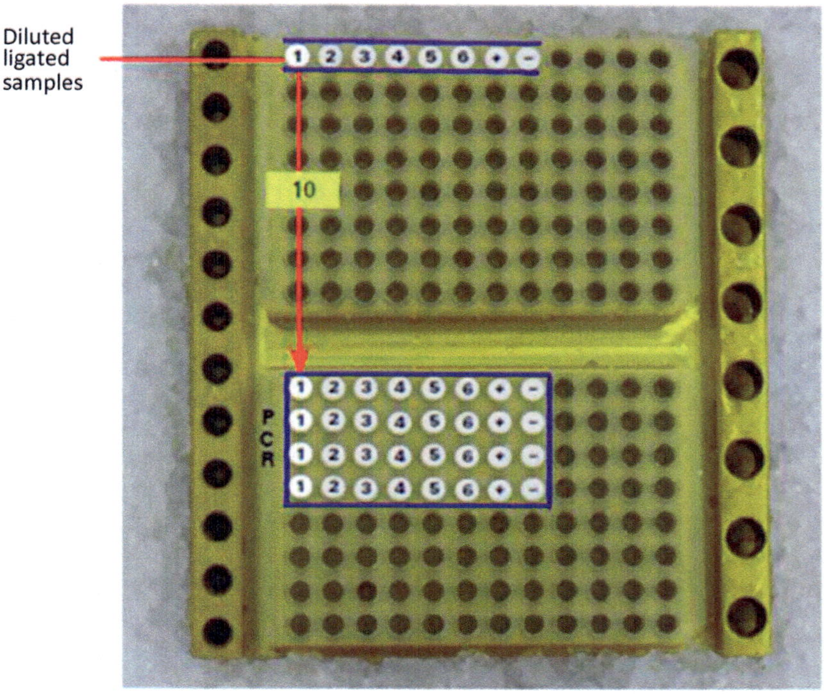

Fig. 6 Transferring the sample to a new 96-well plate (four aliquots for each sample)

11. Label a 15-mL centrifuge tube "PCR" and place on ice.
12. Place on ice both NFW and GCM.
13. If a precipitate is present in GCM, vortex until precipitate is dissolved.
14. Immediately place plate in the lower half of chamber.
15. To the 15-mL "PCR" centrifuge tube, add the reagents in the following order as shown in the table below (*see* **Notes 15** and **16**).

Reagent	1 sample (μL)	8 samples (μL)	16 samples (μL)	24 samples (μL)
NFW	39.5	1453.6	2907.2	4360.8
10× Titanium Taq PCR buffer	10.0	368.0	736.0	1104.0
GC-melt reagent	20.0	736.0	1472.0	2208.0
dNTP mixture (0.25 mM each)	14.0	515.2	1030.4	1545.6
PCR primer (100 μM)	4.5	165.6	331.2	496.8
50× Titanium Taq DNA polymerase (*see* **Note 17**)	2.0	73.6	147.2	220.8
Total	90.0	3312.0	6624.0	9936.0

16. Vortex the 15-mL "PCR" tube for 3 s.
17. Remove cold storage chamber containing the Taq polymerase from freezer.
18. Mix Taq polymerase.
19. Quick spin for 1 s.
20. Vortex for 1 s.
21. Quick spin for 1 s.
22. Immediately add to the PCR MM, and then return the enzyme to the cold storage chamber.
23. Vortex PCR MM three times for 1 s each.
24. Quick spin for 3 s.
25. Unseal PCR sample plate and discard seal.
26. Add **90-μL** PCR MM to each sample on PCR plate (*see* **Note 18**).
27. Tightly seal the plate with a new seal.
28. Vortex for 1 s in all corners and center.
29. Repeat vortex one more time.
30. Centrifuge at 650 × *g* for 1 min.

Ligated and diluted DNA	**10 μL**
PCR master mix	90 μL
Total	100 μL

31. Move plate on ice to post-PCR lab.
32. Load the plate on thermal cycler.
33. Run CytoScan® PCR program

CytoScan PCR program		
Temperature (°C)	Time	Cycles
94	3 min	1×
94	30 s	30×
60	45 s	
68	15 s	
68	7 min	1×
4	5 min	1×
4	Hold	(Can be overnight)

34. PCR program takes 1 h. When done, leave plate in thermal cycler. Plate must be processed within 2.5 h.
35. Decontaminate hood with UV light for at least 15 min.
36. When PCR program is completed, ensure plate is tightly sealed, and centrifuge at 650 × *g* for 1 min.

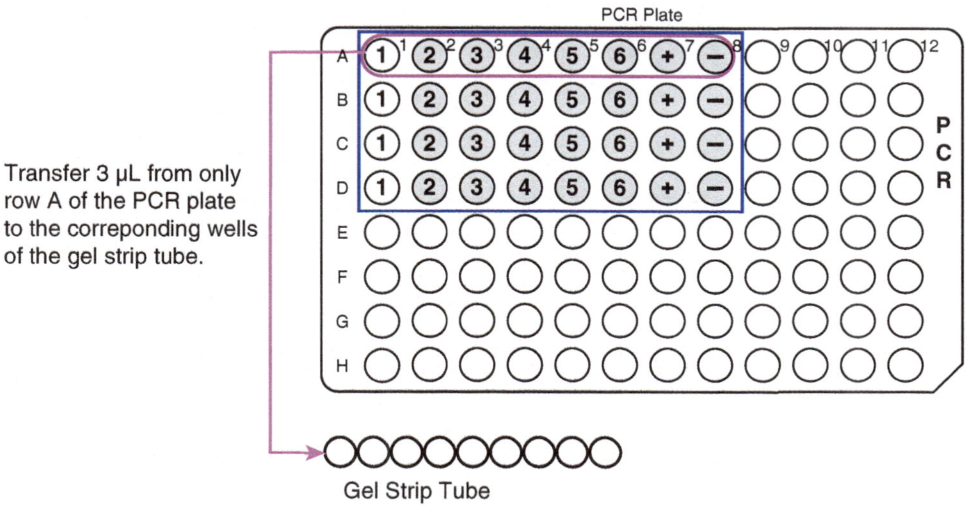

Fig. 7 Transferring PCR product to a strip tube to check PCR reaction

3.3.4 Checking PCR Reaction

1. Unseal PCR plate and discard seal.
2. Transfer 3-μL PCR product from each well of row A to corresponding wells of a strip tube (Fig. 7).
3. Set strip tube aside, start purification, and run gel during the 10 min RT incubation.
4. Run samples on 2% agarose gel to resolve bands.
5. Requirement: PCR target distribution is 150–2000 bp (note higher intensities at 300–2000 bp) (Fig. 8).

3.3.5 PCR Product Purification

1. Prepare purification wash buffer.
   (a) Add 45-mL absolute ethanol to PW BUF bottle.
   (b) Cap bottle tightly, and mix by inverting the unopened PW BUF bottle ten times.
   (c) Enter the date of ethanol addition on the bottle label.
2. Prepare PCR products.
   Pool PCR products (*see* **Note 19**).
   (a) Write sample number on a 1.5-mL centrifuge tube with a permanent marker for each sample.
   (b) Use P200 pipette to transfer all four well aliquots of each sample to each of the marked 1.5-mL centrifuge tube (Fig. 9) (*see* **Note 20**).
   (c) Ensure total volume in each well is transferred and pooled.
   (d) PCR wells [4] 100 μL from each well = 400 μL.
   (e) Total volume in each 1.5-mL tube = 400 μL/tube −3 μL aliquoted for PCR gel.

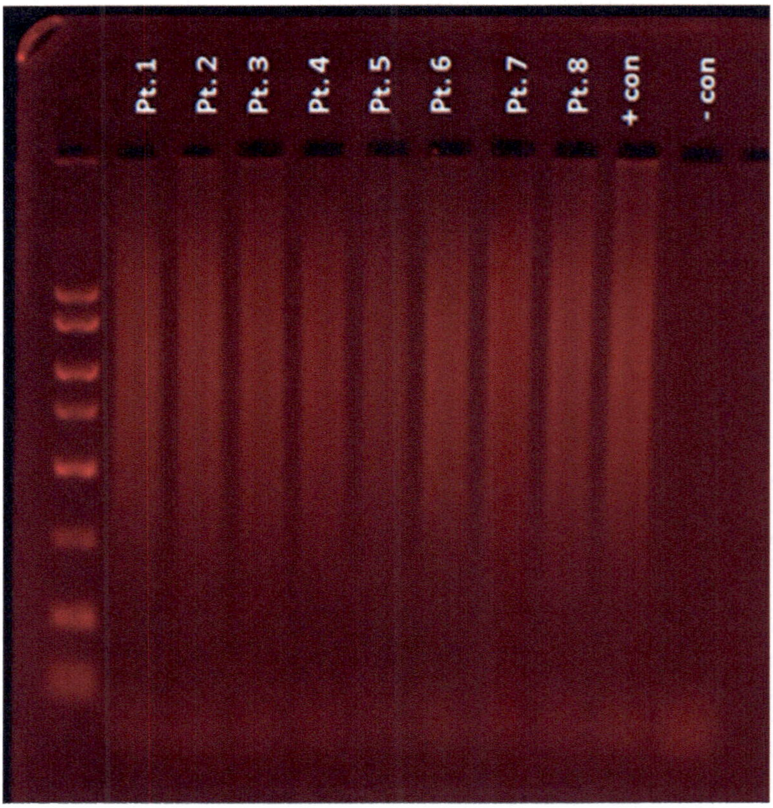

Fig. 8 PCR gel showing the target distribution (150–2000 bp)

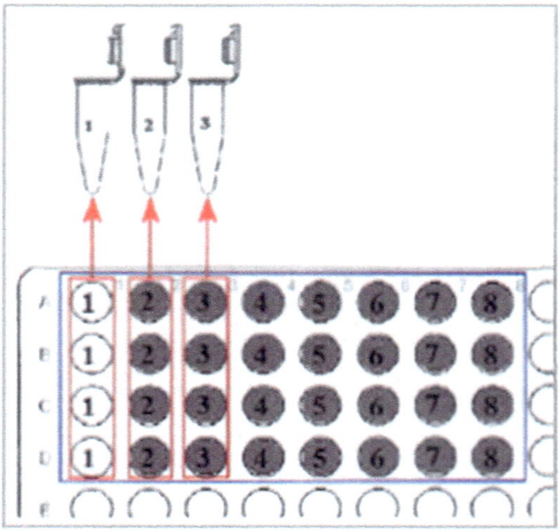

Fig. 9 Transferring the four aliquots to a labeled 1.5-mL centrifuge tube

3. Purify PCR products.
   Change tips between pipetting steps.
   (a) Thoroughly mix PURIF BDS stock—invert the bottle ten times. Examine the bottom of the bottle to ensure solution appears homogeneous.
   (b) Slowly open tube caps to prevent spill.
   (c) Using a P1000 pipette, add 720 μL of PURIF BDS to each pooled sample (*see* **Note 21**).
   (d) Securely cap each tube and invert ten times.
   (e) Incubate at RT for 10 min.
   (f) Load tubes onto centrifuge with cap hinge facing out (Fig. 10), and centrifuge at 16,000 × *g* for 3 min.
   (g) Place tube on magnetic stand so cap hinge is over the magnet. Ensure pellet is completely pulled toward the magnet.
   (h) Use P1000 pipette to remove supernatant without disturbing bead pellet. Discard supernatant.
4. Add purification wash buffer.
   (a) Use P1000 pipette to add 1-mL PW BUF to each tube.
   (b) Cap tubes and load into foam tube adapter. Fully insert tubes into foam to secure. Space sample tubes evenly. Use balance if necessary.
   (c) Vortex 2 min (*see* **Note 22**).
   (d) Centrifuge tubes, cap hinges facing out, at 16,000 × *g* for 3 min.
   (e) Place tubes on magnetic stand, and visually verify that beads are completely pulled toward the magnet with no beads present in PW BUF supernatant at the bottom of tubes.

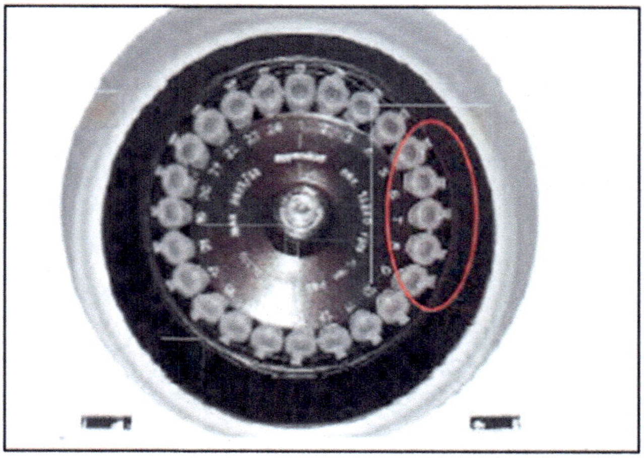

Fig. 10 Loading the tubes onto a centrifuge (keep the cap hinge facing out)

(f) Visually verify that bead pellet is adjacent to the magnet when removing supernatant.

(g) Use P1000 pipette to remove supernatant without disturbing bead pellet.

(h) Discard supernatant.

(i) Centrifuge tubes, cap hinges facing out, at 16,000 × *g* for 30 s; place tubes back on magnetic stand, so cap hinge is over the magnet. Ensure pellet is completely pulled toward the magnet.

(j) Use P20 pipette to remove remaining drops of PW BUF from each tube bottom, one sample at a time (*see* **Note 23**).

(k) Take tubes off magnetic stand, and allow the remaining PW BUF to evaporate by leaving tubes uncapped at RT for 10 min.

5. Add elution buffer.

(a) Use P100 pipette to add 52-μL elution buffer to each tube directly onto beads.

(b) Cap tubes and load into foam tube adapter. Balance tubes.

(c) Vortex 10 min to resuspend beads.

(d) Examine each tube to be sure beads are resuspended in homogeneous slurry (*see* **Note 24**).

(e) With cap hinges facing out, centrifuge tubes at 16,000 × *g* for 3 min.

(f) Place tubes on magnetic stand for 10 min, so beads are pulled to the side of the tube (*see* **Note 25**).

(g) Transfer 47 μL of the eluted sample to the correct well on a new 96-well plate (Fig. 11) (*see* **Notes 26** and **27**).

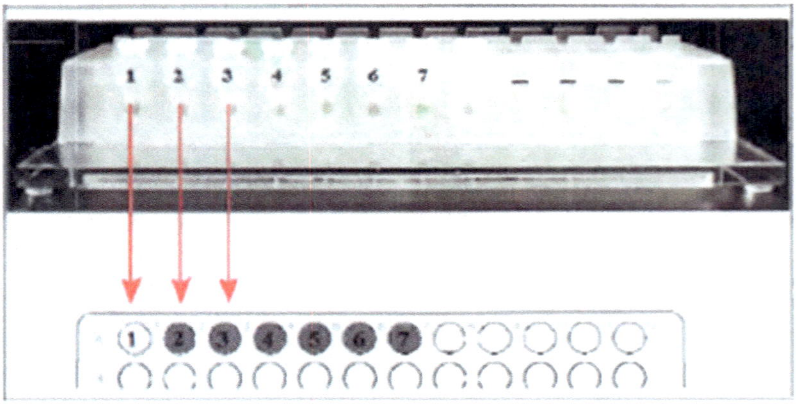

Fig. 11 Transferring eluted sample to the correct well on a new 96-well plate

(h) Remove 2 μL from each well and pipet into a labeled strip tube. Set aside for quantitation step.

(i) Tightly seal plate, and vortex 1 s in all corners and center.

(j) Centrifuge at 650 × *g* for 1 min.

(k) Ensure plate is tightly sealed.

(l) Store plate at −15 to −25 °C.

(m) Proceed to quantitation.

3.3.6 Quantitation

This assay was optimized using a UV spectrophotometer for quantitation. OD measurement accuracy is critical (*see* **Note 28**).

1. Prepare reagents, equipment, and consumables.
   (a) Place NFW on benchtop.
2. Prepare diluted aliquots of purified samples.
   (a) To the transferred 2 μL of each purified sample in the strip tube, add 18 μL of NFW.
   (b) Pipette up and down in each tube to ensure NFW and sample are thoroughly mixed.
3. Quantitate diluted purified PCR product.
   (a) Determine concentration of each PCR product.
   (b) Reading should be above 250 ng/μL (*see* **Note 29**).

3.4 Fragmentation (Day 3: Post-PCR Lab)

Perform all additions, dilutions, and mixing on ice. Ensure all reagents reach equilibrium before use.

3.4.1 Precautions

1. Enzyme is temperature sensitive.
2. Handle tube by cap or base only. Do not touch tube sides.
3. Keep at −15 to −25 °C except when aspirating enzyme volume.
4. Centrifuge so contents of tube are uniform.
5. Perform all steps rapidly and without interruption.
6. Add enzyme to fragmentation MM last.
7. Enzyme is viscous; pipette slowly.
8. Avoid excess solution outside pipette tip.

3.4.2 Set up Work Area (Post-PCR Lab)

1. Fill ice bucket with ice. Place cooling chamber on ice.
2. Decontaminate hood with UV light for at least 15 min.
3. Turn on thermal cycler to preheat lid.
4. Set plate centrifuge to 2–8 °C at least 15–20 min before fragmentation step, and close lid. Keep lid closed at all times before and after centrifugation to ensure proper temperature at every step.

3.4.3 Thaw and Prepare Reagents

1. Thaw plate of purified, quantitated samples at RT (≤30 min).
2. Ensure tight seal and centrifuge plate at 650 × *g* for 1 min.
3. Place plate on the lower half of cooling block on ice for at least 10 min (*see* **Note 30**).
4. Thaw fragmentation buffer at RT (≤30 min); place on cooling block on ice.
5. Label a 1.5-mL centrifuge tube "frag" (fragmentation) and keep chilled in cooling block (*see* **Note 31**).
6. Prepare fragmentation buffer by vortexing three times for 1 s each.
7. Quick spin 3 s.
8. Place in chilled cooling block (2–8 °C).

3.4.4 Prepare Fragmentation Master Mix

Prepare and store all reagents, tubes, and cooling block on ice (*see* **Note 32**).

Always prepare fragmentation MM per table below, regardless of the number of samples.

Fragmentation master mix	
NFW	123.8 μL
10× fragmentation buffer	158.4 μL
Fragmentation reagent	5.8 μL
Total	288.0 μL

1. Add NFW and 10× fragmentation buffer per table above.
2. Vortex the tube three times for 1 s.
3. Quick spin 3 s.
4. Remove cold storage chamber containing FRAG RGT from freezer.
5. Immediately quick spin 3 s to bring down reagent from the top of tube.
6. Vortex 1 s.
7. Quick spin 3 s.
8. Immediately return to −15 to −25 °C cold storage chamber.
9. Add fragmentation reagent per table above.
10. Immediately return to cold storage chamber.
11. Vortex fragmentation MM three times for 1 s each.
12. Quick spin 3 s and immediately place in cooling block.
13. Use a multichannel P20 pipette to transfer **10-μL** fragmentation MM to each sample (Fig. 12) (*see* **Note 33**).
14. Seal plate tightly with a new seal.

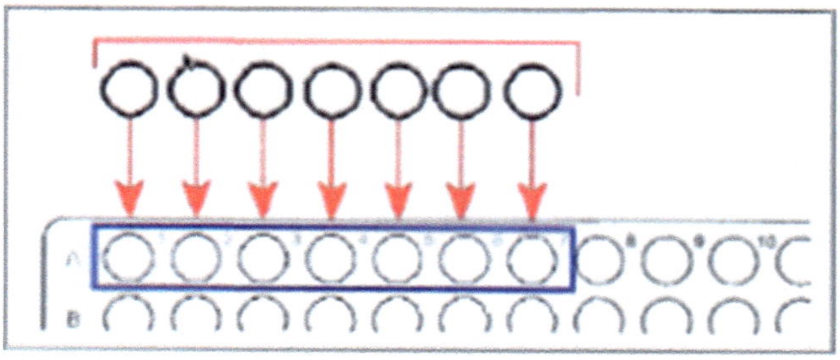

Fig. 12 Transferring fragmentation master mix to each sample

15. Vortex 1 s in all corners and center.
16. Bring sample plate to centrifuge on cooling block in ice bucket (2–8 °C).
17. Centrifuge plate in prechilled centrifuge at 650 × *g* for 1 min. Quickly remove plate from centrifuge, and place in cooling block in ice box.
18. Carry sample plate on cooling block in ice bucket, and immediately load fragmentation plate on thermal cycler with the preheated lid.
19. Run the CytoScan® fragment program.

CytoScan fragment program	
Temperature (°C)	Time
37	35 min
95	15 min
4	5 min
4	Hold

20. The fragment program takes 55 min. When done, leave plate in thermal cycler (*see* **Note 34**).
21. After removing from thermal cycler, ensure plate is sealed tightly, and centrifuge at 650 × *g* for 1 min, and transfer to chilled cooling block on ice.
22. Remove and discard any remaining fragmentation MM (*see* **Note 35**).

Purified PCR product	**45 μL**
Fragmentation master mix	10 μL
Total	55 μL

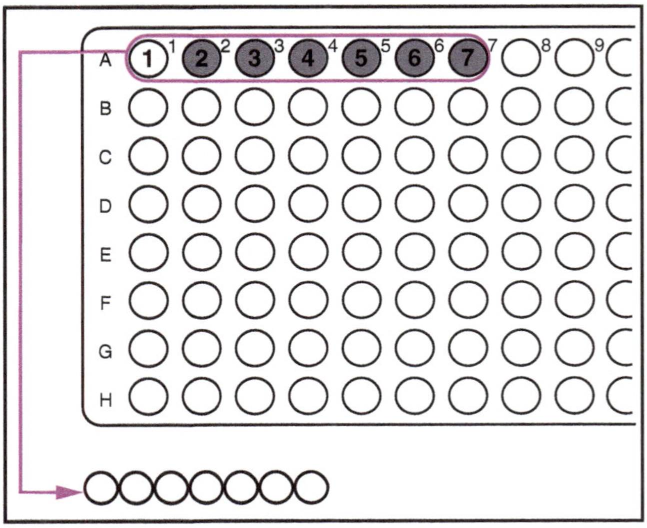

Fig. 13 Transferring fragmented products to strip tube

3.4.5 Checking Fragmentation Reaction

1. Remove seal from plate. Add 4 μL of fragmented samples from plate to corresponding wells of a strip tube. Set aside strip tube to run 4% gel after labeling.
2. Transfer the fragmented product to strip tube (Fig. 13).
3. Seal the fragmented DNA plate with a new seal, and keep on the lower half of cooling block on ice.
4. Majority of fragment distribution must be 25–125 bp (Fig. 14).

3.5 Labeling

Turn on thermal cycler to preheat lid.

3.5.1 Set Up Work Area

1. Label a 1.5-mL centrifuge tube "lbl" (labeling), and place in cooling block on ice.
2. Thaw at RT (≤30 min) both the TDT buffer and DNA label RGT, and then immediately place on ice (*see* **Note 36**).
3. Prepare TDT buffer and DNA label RGT by vortexing both three times for 1 s each.
4. Quick spin 3 s; place in cooling block.

3.5.2 Prepare Labeling Master Mix

Prepare and store all reagents, tubes, and cooling block on ice (*see* **Note 37**).

1. Add to the 1.5-mL tube on ice both 5× TDT buffer and 30 mM DNA labeling buffer per table below.

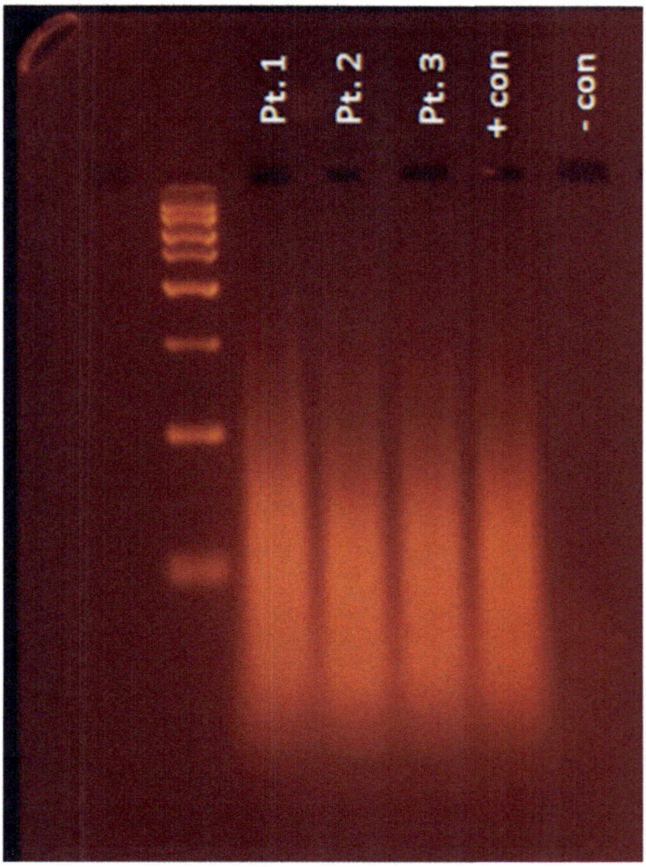

Fig. 14 Gel image showing fragment distribution (majority should be between 25 and 125 bp)

Reagent	1 sample (μL)	8 samples (μL)	16 samples (μL)	24 samples (μL)
TDT buffer	14.0	134.4	268.8	403.2
DNA label RGT	2.0	19.2	38.4	57.6
TDT enzyme	3.5	33.6	67.2	100.8
Total	19.5	187.2	374.4	561.1

2. Vortex three times for 1 s each.
3. Quick spin 3 s.
4. Remove the cold storage chamber containing the TDT enzyme from freezer.
5. Quick spin TDT enzyme 1 s.
6. Mix TDT enzyme by vortexing three times for 1 s each.
7. Quick spin 3 s, and place in cooling block.

8. Return TDT enzyme to cold storage chamber.
9. Add TDT ENZ to labeling MM.
10. Vortex labeling MM three times for 1 s each.
11. Quick spin 3 s. Return cold storage chamber to freezer.

3.5.3 Labeling

Keep samples in cooling block and all tubes on ice when making additions.

1. Use P20 pipette to add **19.5-μL** labeling MM to each sample.

Fragmented DNA (less 4.0 μL for gel analysis)	51.0 μL
Labeling master mix	19.5 μL
Total	70.5 μL

2. Use a new seal and tightly seal plate.
3. Vortex 1 s each in all corners and center.
4. Centrifuge at 650 × *g* for 1 min.
5. Place labeling plate in thermal cycler and run CytoScan® label program.

CytoScan® label program	
Temperature (°C)	Time
37	4 h
95	15 min
4	5 min
4	Hold

6. Labeling program takes 4 h and 20 min to complete.
7. During this time, run 4% gel to check fragmentation reaction.
8. When program finishes, leave plate in thermal cycler, or transfer to chilled cooling block on ice.
9. Ensure plate is sealed tightly and centrifuge at 650 × *g* for 1 min (*see* **Note 38**).

3.6 Target Hybridization

1. Log in to Affymetrix Molecular Diagnostic Software (AMDS).
2. At least 1 h before hybridization, turn oven on to 50 °C and set rpm to 60 (*see* **Note 39**).

3.6.1 Create Test Request and Register Array

1. Left workflow panel: select Active Worklist.
2. In toolbar, click Create.
3. Go to Assay Name drop-down list and select CytoScan HD (*see* **Notes 40** and **41**).
4. Manually enter lab number or specimen ID.
5. In toolbar, click Submit.

3.6.2 Register Probe Array

1. Unwrap arrays and place on clean, lint-free surface at room temperature no more than 30 min before hybridization.
2. Using a silver permanent marker, label the backs of each array with lab number. Label the front as well but any color marker may be used.
3. Left workflow panel: click Registration.
4. Move curser to array ID box next to the corresponding specimen ID (*see* **Note 40**).
5. Use wireless barcode scanner to scan barcode on array chip.
6. In Registration Worklist toolbar, click Complete Step.

3.6.3 Prepare Arrays

1. Place arrays in a clean area at RT.
2. On array back, clean any excess fluid around septa.
3. Insert 200-μL pipette tip into the upper right septum of each array.
4. Apply two 1/2-inch adhesive label dots to top edges of array (Fig. 15).

3.6.4 Prepare for Hybridization

1. Place reagent reservoir on the upper half of cooling block on ice.
2. Label a 15-mL centrifuge tube "Hyb" and place on ice.
3. Prepare and store all reagents, tubes, and cooling block on ice (*see* **Note 42**).
4. Place plate in the lower half of cooling block on ice.
5. Thaw at room temperature (≤30 min), and immediately place the hybridization buffer 1, hybridization buffer 2, hybridization buffer 3, hybridization buffer 4, and OCR reagent on ice. Use within 1 h.
6. Vortex each reagent three times for 1 s each.
7. Quick spin 3 s, and then place in cooling block.

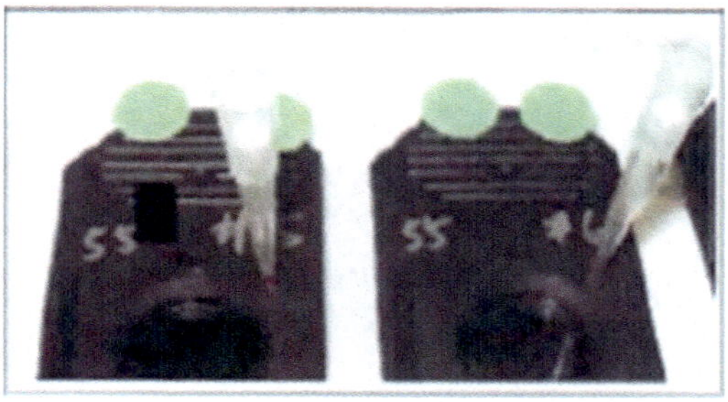

Fig. 15 Applying two 1/2-inch adhesive label dots to top edges of array

3.6.5 Prepare Hybridization Master Mix (see **Note 43***).*

1. To the 15-mL tube labeled "Hyb," add each reagent per table below (*see* **Notes 44** and **45**).

Reagent	1 sample (μL)	8 samples (μL)	16 samples (μL)	24 samples (μL)
Hyb. buffer 1	165.0	1584.0	3168.0	4752.0
Hyb. buffer 2	15.0	144.0	288.0	432.0
Hyb. buffer 3	7.0	67.2	134.4	201.6
Hyb. buffer 4	1.0	9.6	19.2	28.8
OCR	2.0	19.2	38.4	57.6
Total	190.0	1824.0	3648.0	5472.0

3.6.6 Add Hybridization MM and Denature Samples

1. Remove and discard plate seal.
2. Pour hybridization MM into reagent reservoir placed on the upper half of cooling chamber on ice.
3. Add **190 μL** of hybridization MM to samples.
4. Use new seal and tightly seal plate.
5. Vortex 1 s each in all corners and center.
6. Repeat vortexing to ensure complete mixing.
7. Centrifuge at 650 × *g* for 1 min (*see* **Note 46**).
8. Place plate in thermal cycler and run CytoScan® Hyb program.

CytoScan Hyb program	
Temperature (°C)	Time
95	10 min
49	3 min
49	∞

9. When thermal cycler reaches 49 °C forever (∞), open lid, and load samples onto arrays immediately (*see* **Note 47**). Remove plate seal while plate is still sitting in the thermal cycler. CAUTION: Thermal cycler is hot!
10. Use a P200 pipette to remove 200 μL of the first sample from plate, and immediately inject whole amount into bottom left septa on array.
11. Remove pipette tip covering top right septa, and cover both septa with 1/2-inch adhesive label dots and press firmly. Allow edge of dot to overlap the circle window for ease of removal (Fig. 16) (*see* **Notes 48** and **49**).
12. Continue this process with four samples at a time. Load arrays evenly spaced into oven trays. Complete all sample and hybridization oven loading within 30 min.

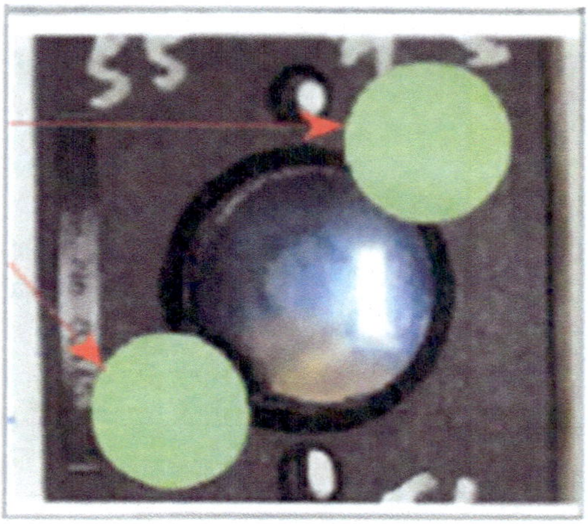

Fig. 16 Covering septa with 1/2-inch adhesive label dots

13. Immediately place trays back into hybridization oven (*see* **Note 50**).
14. In workflow panel, click Hybridization Oven (*see* **Notes 51** and **52**).
15. Place cursor into selected oven number field. Associate test requests with oven by manually entering “1.” Click “Select All” to highlight all samples. Click Start.
16. Allow arrays to rotate at 60 rpm and 50 °C for 16–18 h.
17. Discard leftover hybridization solution and plate.

3.7 Wash, Stain, and Scan Arrays (Day 4: Post-PCR Lab)

3.7.1 Prime Fluidics Station

Prime Fluidics Station before processing array chips

1. Add HD wash solution A and wash solution B to fluidics station (*see* **Note 53**).
2. Place three empty 1.5-mL tubes in sample holders of module to prime.
3. In workflow panel, click Fluidics. Click Station Setup. In Assay drop-down list, select CytoScan HD.
4. Highlight stations to prime.
5. Click Prime Station.
6. Provide e-signature.
7. Follow instructions in Fluidics Station LCD window.
8. In Fluidics Worklist toolbar, click Close Setup Screen.

3.7.2 Wash and Stain Arrays

1. Setup.
   (a) Vortex stain and AH buffer bottles before use.
   (b) Add 500 μL of SB 1 to an "amber"-colored 1.5-mL microfuge tube, and place it in the first slot on the fluidics (*see* **Note 54**).
   (c) Add 500 μL of SB 2 in a "clear-/natural"-colored 1.5-mL microfuge tube, and place it in the 2nd slot on the fluidics.
   (d) Add 800 μL of AH buffer in a "yellow"-colored 1.5-mL microfuge tube, and place it in the third slot on the fluidics.
   (e) In Hybridization Oven Worklist toolbar, "Select All" test requests.
   (f) Click End.
   (g) Click Complete Step.
   (h) When done, remove arrays from Hybridization Oven (*see* **Note 55**).
   (i) Remove 1/2-inch adhesive label dots from arrays.
2. Wash and Stain.
   (a) In left workflow panel, click Fluidics.
   (b) On the fluidics, keep the cartridge lever in down (eject) position.
   (c) Place probe array into the correct module of fluidics station.
   (d) Return cartridge lever to up (engaged) position.
   (e) Scan each probe array with wireless barcode reader.
   (f) Immediately scan fluidics station module for each probe array (*see* **Note 56**).
   (g) When status is "Ready," select test requests in AMDS.
   (h) In Fluidics Worklist panel toolbar, click Start.
   (i) Follow instructions and view status in Fluidics Station LCD window.
   (j) Turn off oven.
   (k) When done, press cartridge lever down (eject).
   (l) Remove array, and check array window for bubbles or air pockets. If air bubbles are present, return array to fluidics station. Pull lever up to remove bubbles. If air bubbles are still present, repeat the process a few times. If still unsuccessful, air bubbles must be removed manually.
   (m) Insert 200-μL pipette tip into the upper right septum of array.
   (n) Use a pipette to remove half the solution from bottom left septa.

(o) Completely fill array with AH BUF.

(p) Remove all arrays without bubbles from fluidics station, and pull up cartridge lever to engage wash block.

(q) When prompted, replace stain tubes with three empty tubes.

(r) In fluidics step of the workflow panel, click Complete Step.

3.7.3 Scan Arrays

Turn on scanner at least 10 min before use.

1. Prepare Arrays.
   (a) Use nonabrasive tissue to clean array glass surface before scanning. Do not use alcohol.
   (b) On array cartridge back, clean excess fluid around septa.
   (c) Carefully cover both septa with 3/8-inch adhesive label dots.
   (d) Press to flatten. If spots do not apply smoothly, remove and apply new. This time it is important NOT to allow dots to overlap circle window.
2. Scan Arrays.
   (a) In the left workflow panel, select Scanner.
   (b) Load probe arrays clockwise into AutoLoaderDx v.2—start with position 1 (red slot).
   (c) In Scanner Worklist toolbar, click Start.
   (d) Wait for status to show Complete.
   (e) In Scanner Worklist toolbar, click Complete Step.

3.7.4 Shut Down Fluidics Station

Shut down when done. Do not keep fluidics station on if you will not use it again within 12 h.

1. Prepare three 1.5-mL (empty) tubes for each module.
2. Place wash lines into a bottle filled with deionized water.
3. In Fluidics step of workflow panel, select Fluidics Station Setup.
4. Select fluidics station to shut down and click Shutdown Station.
5. Follow instructions on fluidics station LCD window.
6. When done, turn switch off.
7. After all chips are scanned and the fluidics is shut down, remove chips from the scanner, and shut down the computer.

4 Notes

1. This protocol is written using the Qiagen DNA extraction kit as recommended by the Affymetrix (Thermo Fisher Scientific) microarray protocol. Although we have not tried other DNA extraction methods or kits, any kit and method should work

equally well as long as the DNA is of high quality and not degraded.

2. This protocol is exclusive and proprietary for Affymetrix (Thermo Fisher Scientific), and all the reagents and equipment are specific for Affymetrix arrays and protocols. The method described here works only with Affymetrix, and we have not applied or tried using this methodology with any other array products or companies.
3. In the equipment list, item numbers 11–21 are to be placed in the pre-PCR clean room. Do not mix or transfer the equipment from pre- to post-PCR rooms.
4. In the equipment list, items 22–43 are to be placed in post-PCR room. Do not mix or transfer the equipment from pre- to post-PCR rooms.
5. In our experience, repeating this elution step increases the DNA yield by about 25%.
6. Set up the reaction within 1 h for best results.
7. For best result, dispense within 10 min.
8. For more than one sample, the calculations include 20% overage.
9. Do not remove enzymes from freezer until ready to add to MM.
10. Always transport enzymes in cold storage chamber.
11. Delay beyond the 2.5 h limit will result in quality control failure in our experience. If for any reason, the samples cannot be processed within 2.5 h of digestion, do not proceed with digestion.
12. For best results, dispense within 10 min.
13. This is an optional stopping point. The ligation plate can be stored in the freezer at −15 to −25 °C for up to 1 week without affecting the quality of results.
14. If samples are frozen, thaw at RT (≤30 min). Centrifuge 650 × *g* for 1 min. Immediately place plate on the upper half of cooling block. Process plate within 1 h.
15. For more than one sample, the calculations include 15% overage.
16. Since there are four replicates for each sample during PCR step, the volumes are multiplied by four.
17. Do not add until ready to aliquot master mix to ligated sample.
18. Avoid contamination, and change pipette tips after each transfer.
19. Requires transfer of assay intermediate from plate to independent tubes done at room temperature.

20. Avoid cross-contamination—use new pipette for each transfer, and take care with pipette tips when pooling samples for purification.
21. Eight samples or less: use single-channel P1000 pipette to add 720-μL PURIF BDS directly from bottle.
22. Bead pellet may not be completely resuspended—this is acceptable.
23. Do not disturb or break off any bead pellet.
24. If beads are not fully resuspended, flick tube to dislodge pellet, and vortex 2 min. Reexamine tubes, and dislodge and vortex pellet until beads are resuspended.
25. Visually verify that bead pellet is completely pulled toward magnet with no beads present in ELU BUF at the bottom of tubes.
26. Ensure correct volume is pipetted—fragmentation step is sensitive to input mass.
27. Eluate may appear brownish. Sometimes brown residue is seen at the pipette tip but usually remains behind on tip when sample is pipetted out.
28. Be sure OD measurement is within instrument linear range.
29. In our experience when the concentration is <250 ng/mL, the results fail the quality control measures.
30. Process within 1 h.
31. Use within 1 h.
32. Dispense within 10 min.
33. Do not pipette up and down. Do not introduce air bubbles at the bottom of tubes.
34. Process within 2.5 h.
35. Never reuse fragmentation master mix.
36. Use within 1 h.
37. Dispense within 10 min.
38. You can hold plate on the thermal cycler at 4 °C for up to 60 h.
39. Remove array from storage 1 h before hybridization to equilibrate to room temperature. Do not remove arrays from pouch.
40. Specimen IDs do not have to be unique and can have 40 characters.
41. Ensure there are no unintentional specimen ID duplications.
42. If the same specimen ID is used multiple times, system prompts you to select the correct one to associate with the corresponding array cartridge. Follow instructions on the AMDS screen.
43. Dispense within 10 min.

44. Use within 1 h.
45. Any precipitation in hybridization master mix after mixing dissolves during 50 °C hybridization.
46. Optional stopping point—seal plate tightly, and store with samples in hybridization master mix at −15 to −25 °C for up to 1 week. To start the hybridization, follow these steps:
    (a) Thaw at room temperature.
    (b) Ensure plate is sealed tightly.
    (c) Centrifuge 650 × *g* for 1 min.
    (d) Remove and replace seal with a new seal.
    (e) Seal tightly.
    (f) Vortex to ensure complete mixing.
    (g) Centrifuge 650 × *g* for 1 min.
    (h) Place plate on cooling block on ice.
47. Hybridization mix is viscous. Pipette slowly for accuracy.
48. Do not allow loaded arrays to sit at RT for more than 1 min.
49. Ensure oven is balanced as you load trays, and trays are rotating at 60 rpm at all times.
50. You cannot change any parameter in Time/Rotation/Duration field displayed on Hybridization Oven Worklist.
51. Ensure wash buffers are correctly positioned with appropriate wash lines in respective buffers.
52. SB 1 is light sensitive, and hence use the "amber"-colored or any other light-sensitive-colored microfuge tubes for best results.
53. Remove only four arrays at a time from oven, and transfer all arrays from hybridization oven to fluidics station(s) within 20 min for best results.
54. Array barcode on probe array identifies proper test request registered to that array ID.
55. Scanner uses a laser and is equipped with a safety interlock system. Do not defeat interlock system, or you may be exposed to hazardous laser light.
56. Warning: the door is locked while the instrument is scanning. Do not attempt to open the door.

Acknowledgments

The authors express their sincere gratitude to Affymetrix (Thermo Fisher Scientific), especially to Dr. Clint VanValkenburgh, Alan Silverman, Dr. Xiaowen Rudner-Hobden, and Corina Nikolof for helpful suggestions, for assistance in validating and establishing the assay, and for continued technical support.

References

1. Greaves M, Maley CC (2012) Clonal evolution in cancer. Nature 481:306–313
2. Nowell PC (1976) The clonal evolution of tumor cell populations. Science 194:23–28
3. Yates LR, Campbell PJ (2012) Evolution of the cancer genome. Nat Rev Genet 13:795–806
4. Mitelman F, Johansson B, Mertens F (2007) The impact of translocations and gene fusions on cancer causation. Nat Rev Cancer 7: 233–245
5. Rabbits TH (2009) Commonality but diversity in cancer gene fusions. Cell 137:391–395
6. Stephens PJ, McBride DJ, Lin ML et al (2009) Complex landscapes of somatic rearrangement in human breast cancer genomes. Nature 462:1005–1010
7. Stratton MR, Campbell PJ, Futreal PA (2009) The cancer genome. Nature 458:719–724
8. Baca SC, Prandi D, Lawrence MS et al (2013) Punctuated evolution of prostate cancer genomes. Cell 153:666–677
9. Liu P, Erez A, Nagamani SC et al (2011) Chromosome catastophes involve replication mechanisms generating complex genomic rearrangements. Cell 146:889–903
10. Rausch T, Jones DT, Zapatka M et al (2012) Genomic sequencing of pediatric medulloblastoma links catastrophic DNA rearrangements with TP53 mutations. Cell 148:59–71
11. Stephens PJ, Greenman CD, Fu B et al (2011) Massive genomic rearrangements acquired in a single catastrophic event during cancer development. Cell 144:27–40
12. Foremont JV, Kaidi A, Jackson SP (2012) Chromothripsis and cancer: causes and consequences of chromosome shattering. Nat Rev Cancer 12:663–670
13. Jone MJ, Jallepalli PV (2012) Chromothripsis: chromosomes in crisis. Dev Cell 23:908–917
14. Maher CA, Wilson RK (2012) Chromothripsis and human disease: piecing together the shattering process. Cell 148:29–32
15. Meyerson M, Pellman D (2011) Cancer genomes evolve by pulverizing single chromosomes. Cell 144:9–10
16. Tubio JM, Estivill X (2011) Cancer: when catastrophe strikes a cell. Nature 470:476–477
17. Holland AJ, Cleveland DW (2012) Chromoanagenesis and cancer: mechanisms and consequences of localized, complex chromosomal rearrangements. Nat Med 18: 1630–1638
18. Rode A, Maass KK, Willmund KV et al (2016) Chromothripsis in cancer cells: an update. Int J Cancer 138:2322–2333
19. Ortega V, Chaubey A, Mendiola C et al (2016) Complex chromosomal rearrangements in B-cell lymphoma: evidence of chromoanagenesis? A case report. Neoplasia 18:223–228
20. Kloosterman WP, Koster J, Molenaar JJ (2014) Prevalence and clinical implications of chromothripsis in cancer genomes. Curr Opin Oncol 26:64–72
21. Kinsella M, Patel A, Bafna V (2014) The elusive evidence for chromothripsis. Nucleic Acids Res 42:8231–8242

Chapter 8

Detection of Chromothripsis in Plants

Isabelle M. Henry, Luca Comai, and Ek Han Tan

Abstract

Chromothripsis, or chromosome shattering, occurs after chromosomes missegregate, are pulverized and subsequently repaired erroneously, leading to highly complex structural rearrangements. In plants, chromothripsis has been observed as a result of mitotic malfunction connected with the incomplete loss of haploid inducer chromosomes during uniparental genome elimination. Uniparental genome elimination, a process that results in haploid induction, is a phenomenon that typically results in the loss of an entire parental chromosome set in early embryos, resulting in haploid plants. In *Arabidopsis thaliana*, genome elimination can be achieved via the manipulation of the centromere-specific histone H3 variant, CENH3. Genomic characterization of F1 progeny resulting from CENH3-mediated genome elimination crosses in *Arabidopsis* revealed haploids (~39%), diploids (~25%), and aneuploids (~37%). Within the aneuploid class, ~11% show evidence for chromothripsis. Here, we present a protocol to identify *Arabidopsis* aneuploids that have inherited chromothriptic chromosomes during genome elimination crosses and describe in detail how to perform in silico reconstructions for individuals with chromothripsis using the somatic mutation finder (SMuFin) tool.

Key words CENP-A, Chromatin, Haploid induction, Epigenetics, Genome instability, Embryonic development, Chromosome missegregation, Aneuploidy

1 Introduction

The chromothripsis phenomenon challenges the tenet that highly complex rearrangements of chromosomal structures are the result of gradual accumulation of changes over time [1]. This phenomenon was first described in cancer from studies aimed at reconstructing tumor genomes [2]. A classic feature of chromothripsis is the observation of oscillating copy number changes that are clustered along a single chromosome and the ability to reconstruct or walk along the breakpoints of this shattered chromosome [3]. Deeper understanding of chromothripsis is especially important for work aimed at preventing mitotic chromosomal instability that can potentially become tumorigenic [4]. However, chromothripsis is not necessarily a terminal event. Human subjects whose karyotype includes chromothriptic chromosomes have also been reported [5, 6], and chromothripsis has even been suggested to have cured a rare genetic disease [7].

Franck Pellestor (ed.), *Chromothripsis: Methods and Protocols*, Methods in Molecular Biology, vol. 1769, https://doi.org/10.1007/978-1-4939-7780-2_8, © Springer Science+Business Media, LLC, part of Springer Nature 2018

In plants, chromothripsis was first detected during a survey of *Arabidopsis thaliana* aneuploids [8], which are individuals that deviate from a balanced, euploid chromosome number, i.e., they carry more copies or segments of some chromosomes than others. These aneuploids that carry chromothriptic chromosomes were the by-product of CENH3-mediated genome elimination crosses using the *GFP-tailswap* haploid inducer [9, 10], which produces haploid, diploid, and aneuploid progeny. About ~11% of aneuploid plants harbor chromosomes that underwent chromothripsis. Since then, chromothriptic chromosomes have been documented while examining progeny from a different type of haploid inducer, one based on wide complementation of a *cenh3* mutation [11]. CENH3 genome elimination has the potential to become a common method to generate haploids and accelerate crop breeding efforts [12, 13]. Therefore, it is important to study the implications of chromothripsis from CENH3-mediated genome elimination as this process becomes more widely adopted.

The evidence from plants adds weight to the theory that chromothripsis is the result of a single catastrophic event associated with chromosome missegregation. This happens in dividing cells such as those that can drive carcinogenesis, or in very early embryogenesis [14], resulting in aberrant chromosomes shared by all or most cells, such as those observed in *Arabidopsis* as a consequence of CENH3-mediated genome elimination. In *Arabidopsis*, chromothripsis was associated with missegregated chromosomes during embryo formation. However, when missegregation was triggered during meiosis by triploidy, chromothripsis was not observed [8], raising the question of tissue specificity or of a specific role for altered centromeric chromatin. An important cellular aberration that is observed when chromothripsis occurs is the presence of micronuclei, nuclear bodies that form around missegregated chromosomes during mitosis [15, 16]. Micronuclei formation had been reported from haploid induction crosses in various plant species [17, 18] as well as from the *Arabidopsis* centromere-mediated elimination crosses [8]. The chromosome(s) that are enclosed in a micronucleus are much more likely to undergo double-stranded DNA breaks, improper replication, and DNA repair occurs primarily through the nonhomologous end-joining (NHEJ) pathway.

The discovery of chromothripsis in plants has opened new avenues to the investigation of the mechanisms behind and the consequences of this phenomenon, both of which can potentially be applied to cancer research. For example, in *Arabidopsis thaliana*, chromosomes that have undergone chromothripsis can be passed on to the next generation [8]. In addition, a recent genomic analysis of a population of gamma-irradiated mutants in poplar [19] resulted in the identification of several individuals carrying dosage variation consistent with chromothripsis (Henry I., personal communication—*see* **Note 1**), suggesting that this phenomenon might be more common than originally expected. Therefore, in order to promote the investigation of chromothripsis in plants, tools need to be available to describe it, which is the purpose of this protocol. In all documented cases of chromothripsis, the complex rearrangements that arise result in dosage imbalance and are accompanied by radical oscillating copy number changes within a single or a few chromosomes [3]. This makes chromothripsis easily distinguishable at the genomic level. Here, we present a detailed bioinformatics approach to detect chromothriptic individuals in *Arabidopsis thaliana* generated using the *GFP-tailswap* centromere-mediated elimination cross, along with strategies to reconstruct chromothriptic chromosomes in silico, and to represent the data as a Circos plot. We will use a previously undescribed individual with chromothripsis (FRAG00133) in *Arabidopsis* to illustrate this process.

2 Materials

2.1 Genome Elimination Cross and DNA Extraction

This protocol can be easily adapted to other species provided that a chromosome or close to whole chromosomal assembly of a reference genome of the species is available. For example, we have applied this method to poplar aneuploids (*see* **Note 1**). Please *see* Refs. [10, 20] for additional descriptions of the *GFP-tailswap* haploid induction cross and the DNA extraction procedure.

1. *GFP-tailswap* haploid inducer (Line CS66982: http://www.arabidopsis.org/servlet/TairObject?id=4502003353&type=stock)
2. Wild-type *Arabidopsis thaliana*
3. MS growth media
4. Illustra nucleon phytopure kit (GE Healthcare) or any suitable protocol that can yield high-quality DNA

2.2 Genomic Library Construction

Any suitable PCR-free method for Illumina sequencing (*see* **Note 2**) methods can be used for genomic library construction. Please *see* Ref. [20] for additional descriptions using the KAPA Hyper Prep kit.

1. KAPA Hyper Prep kit (KAPA Biosystems)
2. NextFlex-96 adapters (Bioo Scientific) (*see* **Note 3**)

2.3 Dosage Plotting and Visualization

Please *see* Refs. [20, 21] for additional descriptions of dosage plotting and visualization using Illumina sequencing data.
Software:

1. Linux- or MacOS- or Unix-based computer system
2. Python2.6 or Python2.7
3. AllPrep (if necessary, https://github.com/Comai-Lab/allprep)

4. BWA version 0.5.7 (http://bio-bwa.sourceforge.net/)
5. c1.18-dev or 1.19 (http://SAMtools.sourceforge.net/)
6. BWA-DoAll (http://comailab.genomecenter.ucdavis.edu/index.php/Bwa-doall)
7. Bin-by-sam (http://comailab.genomecenter.ucdavis.edu/index.php/Bin-by-sam)

Data:

1. High-quality demultiplexed FASTQ read sequence files (".fq" files, as output by the sequencing facility or the demultiplexing program of choice)
2. Reference genomic file in FASTA format, in this case the TAIR10 *Arabidopsis* genome (ftp://ftp.arabidopsis.org/home/tair/home/tair/Sequences/whole_chromosomes/)

2.4 Chromothripsis Reconstruction Using SMuFin

Software:

1. Software from 2.3
2. SMuFin (http://cg.bsc.es/smufin/)
3. Circos (http://www.circos.ca/documentation/tutorials/configuration/installation/)
4. circos-links-fromSMuFin.py (https://github.com/haploidgenomics/SMuFinForChromothripsis/blob/master/circos-links-fromSMuFin.py)

Data:

1. Uninterleaved, paired-end, high-quality demultiplexed FASTQ read sequence files (".fq" files) of "tumor" (individual with chromothripsis) and "normal" (diploid sibling) samples
2. Reference genomic file in FASTA format, in this case the TAIR10 *Arabidopsis* genome (ftp://ftp.arabidopsis.org/home/tair/home/tair/Sequences/whole_chromosomes/)

2.5 Verification of SMuFin Analysis

Software:

1. Software from 2.3
2. PRICE (http://derisilab.ucsf.edu/software/price/)
3. Blast interface (https://blast.ncbi.nlm.nih.gov/Blast.cgi)
4. junction-bins-fromSMuFin.py (https://github.com/haploidgenomics/SMuFinForChromothripsis/junction-bins-fromSMuFin.py)
5. batch-specific-junction-bin-search.py (https://github.com/haploidgenomics/SMuFinForChromothripsis/blob/master/batch-specific-junction-bin-search.py)
6. batch-uninterleaver-PRICE-Assembler.py (https://github.com/haploidgenomics/SMuFinForChromothripsis/blob/master/batch-uninterleaver-PRICE-Assembler.py)

Data:

1. Interleaved FASTQ ".fq" file and the corresponding SAM ".sam" file

3 Methods

The steps involved in the production and analysis of dosage variation are summarized in Fig. 1 and explained in detail below.

3.1 Genome Elimination Cross and DNA Extraction

1. Identify *GFP-tailswap* haploid inducer (HI) homozygous for the *cenh3-1* mutation with the genotype: *cenh3-1/cenh3-1 GFP-tailswap/GFP-tailswap*. We identify *cenh3-1/-1* homozygous plants using a dCAPS genotyping protocol [10, 22] from a segregating population from a selfing of a line with this genotype: *cenh3-1/CENH3 GFP-tailswap/GFP-tailswap*.
2. The primary and early inflorescences are crossed with pollen from wild-type (CENH3 +/+) plants without the need for emasculation of the *GFP-tailswap* HI because it is almost completely male sterile.

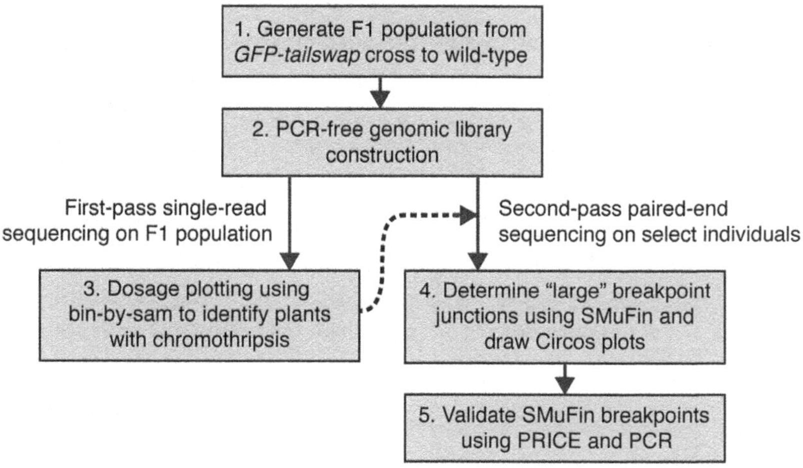

Fig. 1 Suggested workflow for the detection and characterization of chromothripsis in *Arabidopsis thaliana*. An F1 population is first obtained (**step 1**) by making haploid induction crosses. DNA is extracted from adult leaf tissue and subjected to PCR-free genomic library preparation for Illumina sequencing (**step 2**). Up to 384 individuals can be screened per lane to identify chromothriptic individuals (**step 3**—*see* **Note 3**). Libraries containing chromothriptic individuals are pooled along with a suitable diploid sibling from the same cross for a second round of sequencing at higher depth and with a minimum of 100PE sequencing. Using the SMuFin tool, "tumor" (individuals that harbor chromothripsis) and "normal" (diploid sibling) samples are compared pairwise to identify discordant breakpoint junctions, which are then plotted for visualization on Circos (**step 4**). Finally, breakpoints are validated using an independent de novo assembler called PRICE to assist in the design of breakpoint-specific PCRs primers (**step 5**)

3. Collect mature fruit and sterilize F1 seeds using suitable method.
4. F1 seeds are plated on MS media, stratified at 4 °C for at least 2 days, and germinated.
5. Germinated seedlings are then transplanted on soil and grown in long day conditions.
6. Collect ~1 mg of rosette leaf tissue per plant for DNA extraction using Illustra phytopure kit or an equivalent DNA extraction protocol. To ensure library-quality DNA, run 0.5 μg of the DNA prep on a 1% TAE gel to confirm the presence of high molecular weight DNA bands with minimal smearing.

3.2 Genomic Library Construction

1. Use 1 μg of input DNA for each library preparation.
2. Shear the DNA using a Covaris E220 Focused-ultrasonicator and the following shearing settings: peak incident power (175), duty factor (5%), cycles per burst (200), and treatment time (60 s).
3. Perform library construction as recommended by the manufacturer (*see* Ref. [20]).
4. Pool samples as recommended by the sequencing facility, using 50 SR sequencing (*see* **Note 3**).

3.3 Dosage Plotting and Visualization

1. Perform basic read demultiplexing and quality control, for example, using the Allprep.py script.
2. For each individual, align the reads to the reference genome using bwa to generate .sam alignment files. If performing the sequence alignments using the bwa-doall.py script, the typical sam files can be found in the/sam folder.
3. Record read coverage across the reference sequence and compare samples to each other to identify variation in coverage specific to each sample. In this initial screen for chromothripsis from an F1 population with low coverage, we recommend using 500 kb bins using the bin-by-sam.py script.

   The plots in Fig. 2 were generated using these parameters:

   ```
   bin-by-sam -s 500000 -o 500kb_dosageplots.txt
   ```

4. Plot the Sample/control. Visualize the dosage plots to identify normal and chromothriptic samples, as shown in Fig. 2.

3.4 Chromothripsis Reconstruction Using SMuFin

1. Resubmit samples that exhibit chromothripsis, considered the "tumor" sample, along with a matched "normal" sample, which in this case is a diploid sibling from the same sequencing run. We highly recommend sequencing chromothriptic "tumor"

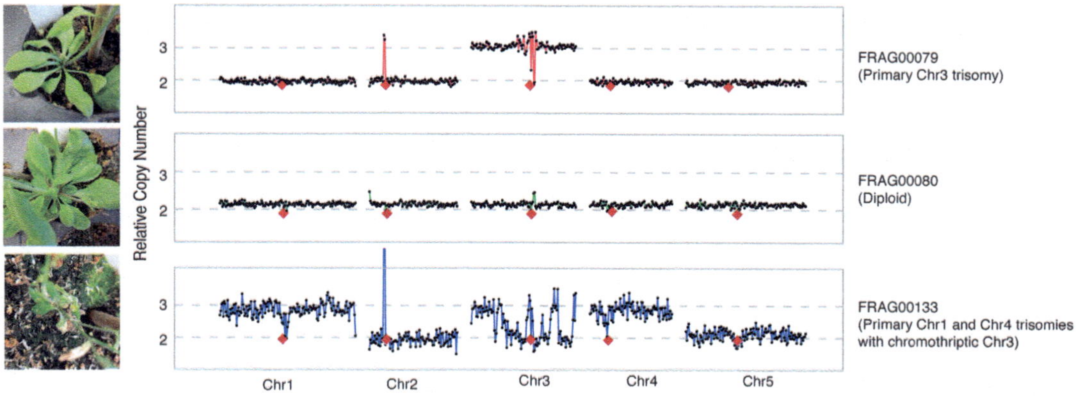

Fig. 2 Dosage plots of three F1 individuals from a *GFP-tailswap* haploid induction cross. Three individuals are presented here, using a bin size of 500 kb. Centromere positions are indicated with red diamonds. From these plots, we ascertained that FRAG00079 is a primary Chr3 trisomic, FRAG00080 is a diploid, and FRAG00133 is trisomic for Chr1 and Chr4 but also carries an extra copy of Chr3 that has undergone chromothripsis, based on the rapid oscillating copy number changes along Chr3

samples and control "normal" samples together in a pooled fashion. Much like designing an RNAseq expression analysis, the precision and accuracy of using the SMuFin tool is greatly improved with matched samples derived from the same sequencing run (*see* **Note 4**).

2. Perform basic read demultiplexing and quality control, for example, using the Allprep.py script. To generate a high-density dosage plot, as well as for SMuFin analysis, the output paired-end reads should be uninterleaved, with R1 and R2 reads in separate files.

3. Create high-density dosage plots of the chromothriptic individual by mapping one of the paired-end files (this prevents double counting) on all samples using bwa (e.g., use bwa-do-all). Use the matched diploid sample as the control during bin-by-sam.py analysis using a bin size of 5 kb (–s 5000).

   For example, the plot in Fig. 3 was generated using these parameters:

```
bin-by-sam.py -s 5000 -c FRAG00080_aln.sam -o 5kb_
dosageplot.txt
```

4. Extract from averaged dosage values from this high-density dosage plot to use as the scatter plot file for Circos and name it plot.txt (**step 9**). The plot file requires a tab-separated file: chr \t start \t end \t dosage without any intervening lines.

   Example of the first three lines from the plot file for Circos:

chr1	1	5000	2.812
chr1	5001	10000	3.047

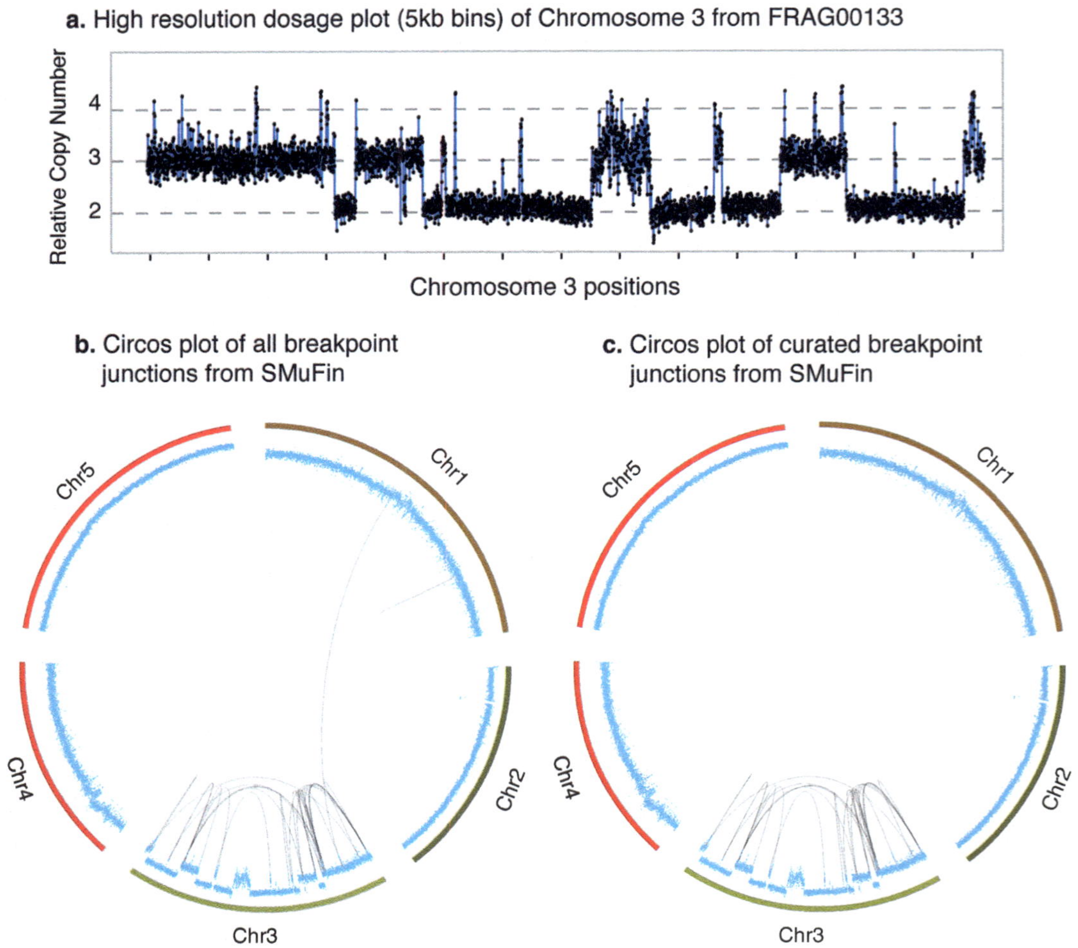

Fig. 3 Detailed dosage plot and Circos plots of SMuFin output from chromothriptic individual FRAG00133. (**a**) A high-density (5 kb bin) dosage plot zoomed in on the chromothriptic chromosome 3 of FRAG00133 shows the major duplicated (copy number of three) and triplicated (copy number of four) blocks. (**b**) Circos plot representing all the breakpoint junctions identified using SMuFin for FRAG00133. (**c**) Curated Circos plot of FRAG00133, obtained after removing breakpoint junctions unsupported by the dosage plot of chromothriptic chromosome 3

chr1 10001 15000 2.596

5. Run SMuFin using the .fq files. Detailed instructions for how to run SMuFin can be found in the support section of: http://cg.bsc.es/smufin/. In this case, SMuFin analysis was performed on a standalone Unix server with 194 Gb of RAM.

The SMuFin analysis for Fig. 3 was generated using these parameters:

```
mpirun --np 12 ~usr/bin/SMuFin \
--ref ~usr/genomes/TAIR10_all.fas \
--normal_fastq_1 FRAG00080_1.txt \
```

```
--normal_fastq_2 FRAG00080_2.txt \
--tumor_fastq_1 FRAG00133_1.txt \
--tumor_fastq_2 FRAG00133_2.txt \
--patient_id FRAG00133 --cpus_per_node 1
```

6. The breakpoint junctions from chromothriptic events are found within the "somatic_large_SV" .txt file from the SMuFin output and can be used for Circos plotting (Fig. 3).
7. To plot these breakpoints, a link file is extracted using the circos-links-fromSMuFin.py script from "somatic_large_SV" file.

   The link file for Fig. 3 was generated using these parameters:

   ```
   circos-links-fromSMuFin.py somatic_large_SVs.
   FRAG00133 link_FRAG00133.txt
   ```

   The tab-separated link file output:

chr3	5849924	5850024	chr3	5854392	5854492
chr3	194890	194990	chr3	1156019	1156119
chr3	202278	202378	chr3	8256931	8257031

7. Plot Circos using plot.txt from **step 5** and link.txt from **step 7**.

   For the configuration (conf) used, please refer to https://github.com/haploidgenomics/SMuFinForChromothripsis/blob/master/circos_FRAG00133.conf
8. Manually curate the Circos plot, if appropriate. In the case of chromothriptic chromosomes in *Arabidopsis*, the evidence so far suggests that the entire chromosome is inherited as a unit in a Mendelian fashion (Tan et al. 2015). Therefore, we are able to manually curate breakpoint junctions related just to the chromothriptic chromosome (chromosome 3 in the case of FRAG00133) by removing links between unrelated chromosomes.
9. Re-plot Circos from the manually curated link files.

3.5 Verification of SMuFin Analysis

1. Reconstruct the breakpoints identified from SMuFin with the de novo PRICE assembler and using extracted reads that surround the breakpoint junctions from SMuFin's "large_somatic_SV" output. Comparison of SMuFin generated breakpoint junctions to previously manually curated, and de novo assembled breakpoints for individual FRAG00062 [8] show the concordance for the majority of junctions (Fig. 4).
2. Align the paired-end interleaved reads to the reference genome using *bwa* (e.g., use bwa-do-all).
3. Create a bin file to extract a specific fastq read file associated with the breakpoint junctions from the "somatic_large_SV" file using the junction-bins-fromSMuFin.py script.

   The bin file was created using these parameters:

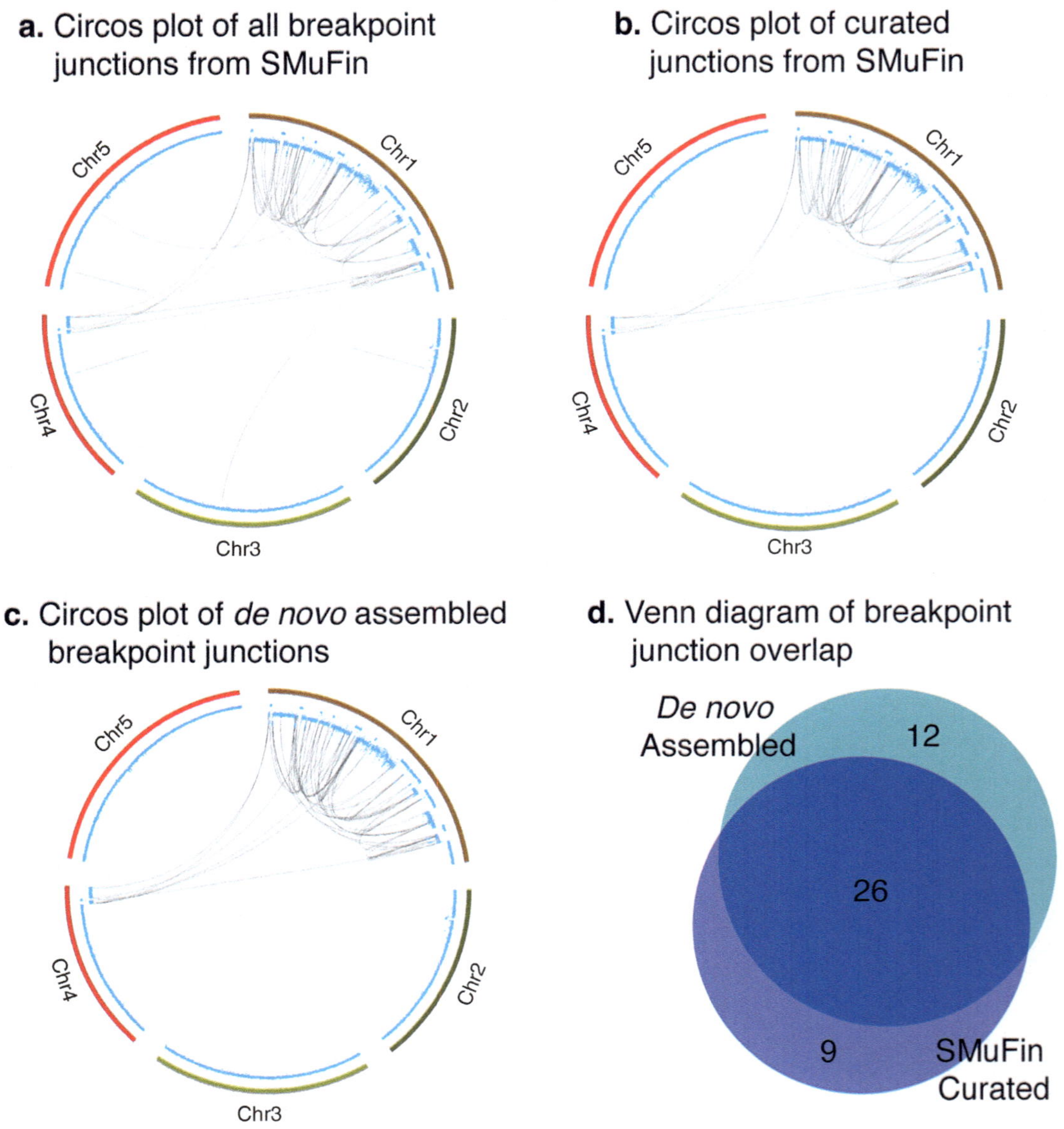

Fig. 4 Comparison of breakpoint junctions using de novo junction finding methods and SMuFin for chromothriptic individual FRAG00062. (**a**) Circos plot of all breakpoint junctions identified by SMuFin for FRAG00062. (**b**) Curated Circos plot of FRAG00062, obtained after removal of breakpoint junctions unrelated to the dosage plot of chromothriptic chromosome 1. (**c**) Circos plot of de novo assembled FRAG00062 breakpoint junctions from Tan et al. (2015). (**d**) Venn diagram showing the overlap between de novo assembled breakpoint junctions and breakpoint junctions identified using SMuFin

```
junction-bins-fromSMuFin.py somatic_large_SVs.
FRAG00133 bin_FRAG00133.txt
```

The first three lines from the link file are:

Chrom	BinStart	BinEnd
Chr3	5849424	5850424

Chr3 194390 195390

4. Extract paired-end reads using the associated .fq and _aln.sam files.

   The bin-specific .fq files were created using these parameters:

```
batch-specific-junction-bin-search.py -b bin_
FRAG00133_07.txt -Q
```

5. Reconstruct the junctions using the batch-uninterleaver-PRICE-Assembler.py in a folder containing all the junction specific .fq files extracted from **step 4**.
6. Perform blast searches from each .fa file and search for contigs that span breakpoint junctions in the /Assemblies folder. Contigs created by PRICE that are spanning breakpoint junctions have a tell-tale signature of discordance mapped to the chromothriptic chromosome, as exemplified in Fig. 5.
7. Verify the breakpoint junctions using PCR and primers designed to amplify a DNA contig that include the junction (*See* **Note 5**). PCR primers can be designed using the Primer3 program. Additional methods that can be used for the verification of breakpoint junctions are also discussed in **Note 6**.

4 Notes

1. The approach described here has led to the identification of chromothriptic individuals in a population of gamma-mutagenized poplar individuals (Fig. 6), consistent with the idea that chromothripsis might be more common than anticipated.
2. A recent study assessed different methods for detecting chromothriptic events, using various software. In our experience de novo reconstruction from a junction to junction approach [8] has the best concordance with the breakpoints identified by SMuFin. Part of our success is likely related to the use of PCR-free methods of sequencing [23] which reduces biases.
3. The number of samples that can be pooled in an Illumina sequencing lane varies depending on genome size and expected output of the sequencer. For *A. thaliana*, which has a 125 megabase genome, we recommend 500,000 reads per sample for the initial detection of potential chromothriptic individuals. If more than 96 individuals are pooled in a lane, we recommend using dual-indexed adaptors, which allow for higher degrees of multiplexing at lower cost.
4. In our experience, this is one of the most crucial steps to ensure that SMuFin runs consistently. The "tumor" samples and "normal" control samples should be pooled and sequenced together at equal depth for best results.

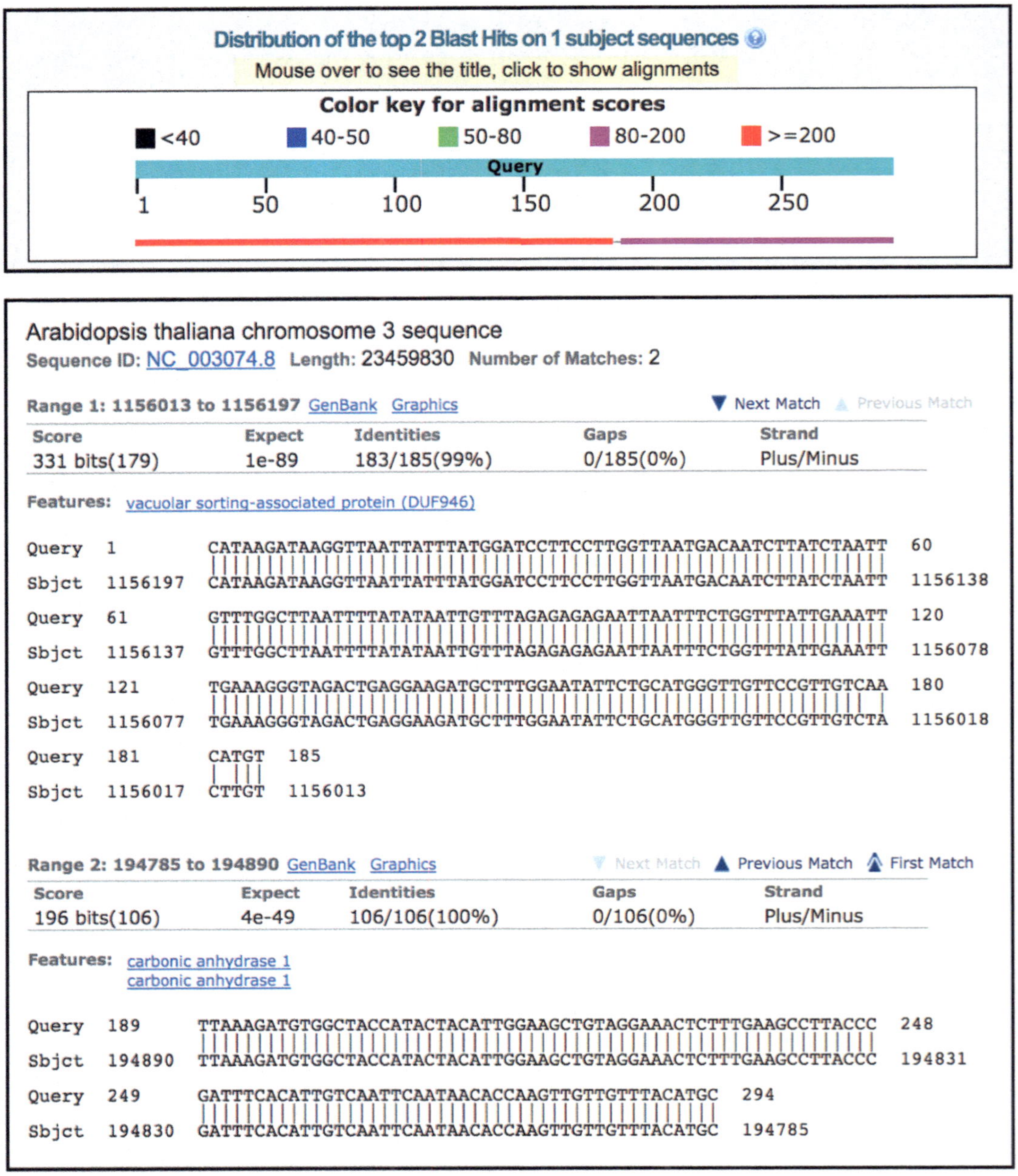

Fig. 5 Example of a PRICE assembled contig for breakpoint junctions. The top panel shows the discordant blast resulting from a PRICE assembled contig for one of the bins identified from SMuFin analysis of FRAG00133. The bottom panel shows the contiguous breakpoint junction for positions Chr3: 1,156,013 to Chr3: 194,890, in concordance with the predicted breakpoint based on SMuFin analysis

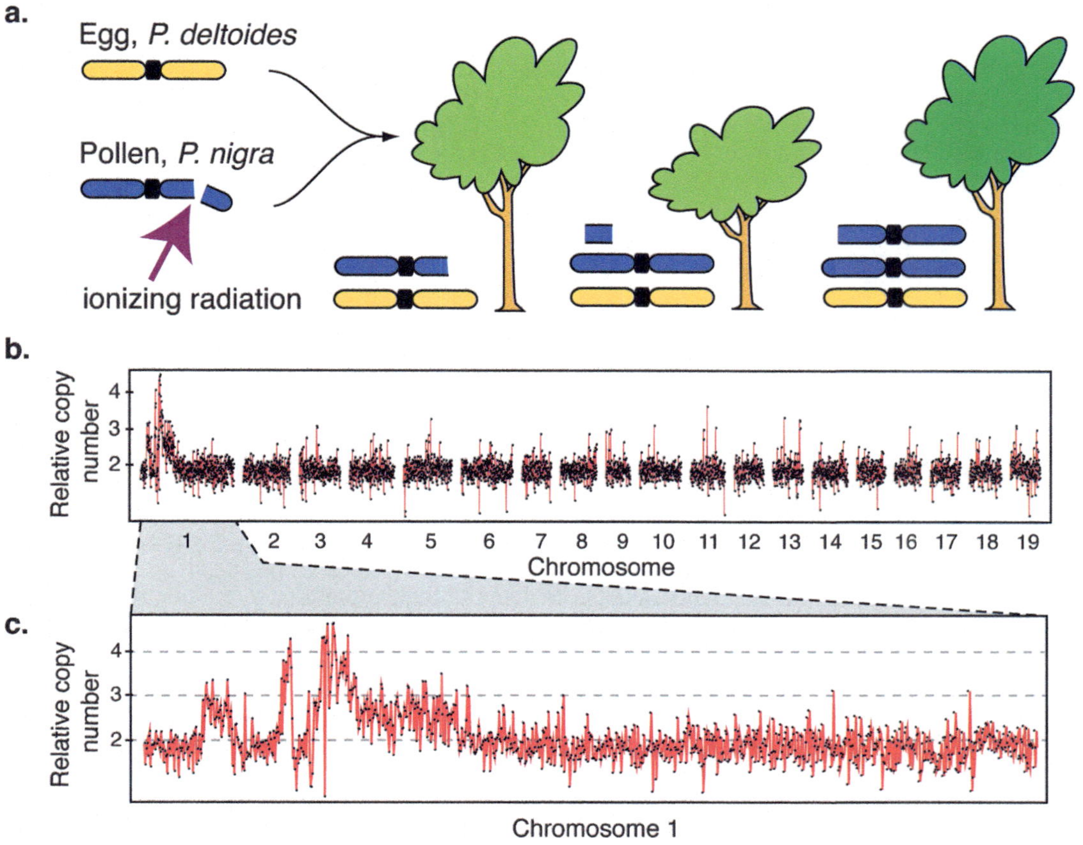

Fig. 6 Identification of chromothriptic individuals in a population of poplar gamma mutants. (**a**) A population of interspecific hybrid poplar siblings was created by crossing a *P. deltoides* female individual to pollen grains from a *P. nigra* male individual. The *P. nigra* pollen was previously gamma irradiated, which creates double-stranded breaks, resulting in large-scale insertions and deletions throughout the genome. The resulting population is a group of sibling trees with slightly variable chromosomal content and exhibiting phenotypic variation. (**b**) Each F1 sibling was characterized for the presence of insertions and deletions using low-pass Illumina sequencing and dosage plotting, as described in this protocol. Specifically, 384 individuals were pooled into a single lane of HiSeq 4000, 50 SR sequencing. Dosage plots were constructed for each, using 100 kb bins. (**c**) For individuals with potential chromothriptic karyotypes, a second round of sequencing was performed, allowing for more precise definition of the dosage variation (50 kb bins are shown here for chromosome 1)

5. Validation of each breakpoint junction remains challenging, but using both de novo reconstruction and SMuFin can potentially produce better results if a high-quality contig is required.
6. Other methods producing long reads that do not require assembly or bioinformatics reconstruction can be considered too. Until there are ways to selectively sequence the chromothriptic chromosomes only, or alternative sequencing methods are available at lowered cost, the throughput of using Illumina remains relevant.

References

1. Alioto TS, Buchhalter I, Derdak S et al (2015) A comprehensive assessment of somatic mutation detection in cancer using whole-genome sequencing. Nat Commun 6:10001. https://doi.org/10.1038/ncomms10001
2. Britt AB, Kuppu S (2016) Cenh3: an emerging player in haploid induction technology. Front Plant Sci 7:357. https://doi.org/10.3389/fpls.2016.00357
3. Crasta K, Ganem NJ, Dagher R et al (2012) DNA breaks and chromosome pulverization from errors in mitosis. Nature 482:53–58
4. Gernand D, Rutten T, Varshney A et al (2005) Uniparental chromosome elimination at mitosis and interphase in wheat and pearl millet crosses involves micronucleus formation, progressive heterochromatinization, and DNA fragmentation. Plant Cell 17:2431–2438
5. Henry IM, Zinkgraf MS, Groover AT et al (2015) A system for dosage-based functional genomics in poplar. Plant Cell 27:2370–2383
6. Ishii T, Karimi-Ashtiyani R, Houben A (2016) Haploidization via chromosome elimination: means and mechanisms. Annu Rev Plant Biol 6:421–438
7. Kloosterman WP, Tavakoli-Yaraki M, van Roosmalen MJ et al (2012) Constitutional chromothripsis rearrangements involve clustered double-stranded DNA breaks and nonhomologous repair mechanisms. Cell Rep 1:648–655
8. Korbel JO, Campbell PJ (2013) Criteria for inference of chromothripsis in cancer genomes. Cell 152:1226–1236
9. Leibowitz ML, Zhang CZ, Pellman D (2015) Chromothripsis: a new mechanism for rapid karyotype evolution. Annu Rev Genet 49:183–211
10. Liu P, Erez A, Nagamani SC et al (2011) Chromosome catastrophes involve replication mechanisms generating complex genomic rearrangements. Cell 146:889–903
11. Maheshwari S, Tan EH, West A et al (2015) Naturally occurring differences in CENH3 affect chromosome segregation in zygotic mitosis of hybrids. PLoS Genet 11(1):e1004970. https://doi.org/10.1371/journal.pgen.1004970
12. McDermott DH, Gao JL, Liu Q et al (2015) Chromothriptic cure of WHIM syndrome. Cell 160:686–699
13. Neff MM, Neff JD, Chory J et al (1998) dCAPS, a simple technique for the genetic analysis of single nucleotide polymorphisms: experimental applications in Arabidopsis Thaliana genetics. Plant J 14:387–392
14. Pellestor F, Gatinois V, Puechberty J et al (2014) Chromothripsis: potential origin in gametogenesis and preimplantation cell divisions. A review. Fertil Steril 102:1785–1796
15. Ravi M, Chan SW (2010) Haploid plants produced by centromere-mediated genome elimination. Nature 464:615–618
16. Ravi M, Marimuthu MP, Tan EH et al (2014) A haploid genetics toolbox for Arabidopsis Thaliana. Nat Commun 5:5334. https://doi.org/10.1038/ncomms6334
17. Stephens PJ, Greenman CD, Fu B et al (2011) Massive genomic rearrangement acquired in a single catastrophic event during cancer development. Cell 144:27–40
18. Subrahmanyam N, Kasha K (1973) Selective chromosomal elimination during haploid formation in barley following interspecific hybridization. Chromosoma 42:111–125
19. Tan EH, Comai L, Henry IM (2016) Chromosome dosage analysis in plants using whole genome sequencing. Bio-protocol 6(13): e1854. https://doi.org/10.21769/BioProtoc.1854
20. Tan EH, Henry IM, Ravi M et al (2015) Catastrophic chromosomal restructuring during genome elimination in plants. elife 4:e06516. https://doi.org/10.7554/eLife.06516
21. Zhang CZ, Leibowitz ML, Pellman D (2013) Chromothripsis and beyond: rapid genome evolution from complex chromosomal rearrangements. Genes Dev 27:2513–2530
22. Zhang CZ, Spektor A, Cornils H et al (2015) Chromothripsis from DNA damage in micronuclei. Nature 522:179–184
23. Zinkgraf M, Haiby K, Lieberman MC et al (2016) Creation and genomic analysis of irradiation hybrids in Populus. Curr Protoc. Plant Biol 1:431–450. https://doi.org/10.1002/cppb.20025

Chapter 9

RNA-Seq Analysis to Detect Abnormal Fusion Transcripts Linked to Chromothripsis

Anne-Laure Bougé, Florence Rufflé, Sébastien Riquier, Benoit Guibert, Jérôme Audoux, and Thérèse Commes

Abstract

RNA-Seq approach enables the detection and characterization of *fusion or chimeric* transcript associated to complex genome rearrangement. Until now, these events are classically identified at DNA level.

Here we describe a complete procedure including a novel way of analyzing reads that combines genomic locations and local coverage to directly infer chimeric junctions with a high sensitivity and specificity, allowing identification of different classes of chimeric RNA events. We also recommend the best practices for the bioinformatics analysis and describe the experimental process for RNA validation using real-time PCR and sequencing.

Key words RNA-Seq, Chimeras, Genomic rearrangement, Chromothripsis, Transcription, Chimeric RNAs, Fusion genes, NGS

1 Introduction

Complex genome rearrangements involving the simultaneous generation of tens to hundreds of genomic rearrangements constituted a new phenomenon called chromothripsis [1]. The use of transcriptome sequencing has not yet been extensively used for the discovery of abnormal transcripts expressed from genomic rearrangements in cases of chromothripsis associated breakpoints [2, 3]. RNA-Seq approaches are increasingly acquiring a key role not only as means of understanding the mechanisms responsible for complex diseases but also to efficiently identify novel transcriptional events including fusion transcripts also called chimeric RNA (chRNA) generated by genomic breakpoints [4, 5]. As chromothripsis could induce complex chromosome rearrangements, the resulting abnormalities at the transcriptional level are also complex. However, coupling DNA and RNA analysis could be an efficient way for the combined identification of the subset of abnormal transcripts expressed from genomic rearrangements.

Franck Pellestor (ed.), *Chromothripsis: Methods and Protocols*, Methods in Molecular Biology, vol. 1769, https://doi.org/10.1007/978-1-4939-7780-2_9,

Once identified, the impact of these transcriptional modifications at protein level can be also predicted.

Though RNA-Seq may enable new chRNA profiling, it remains a methodological challenging issue to extract, with high confidence, biological candidates from artifacts [6]. We propose a novel way of analyzing reads that combines genomic locations and local coverage to classify the biological events, to directly infer splice and chimeric junctions within a single read [7, 8]. The procedure is based on the CRAC algorithm that allows extraction of transcriptional events, irrespective of annotation. Its main advantage is its high sensitivity and specificity, allowing identification of rare chRNA events. We present a complementary ChimCT module to aggregate, annotate chRNA reads. The procedure classifies reads into four classes, depending on the exon organization. The first class corresponds to fusion transcripts arising from two different chromosomes; the second class includes parts of two genes belonging to the same chromosome strand. The third and fourth classes involve non-collinear transcription on the same chromosome.

Moreover in addition to classes and scoreValue given by ChimCT, we recommend appropriated filters and in silico validations in order to select the more confident chRNA candidates. We also recommend the best practices for RNA-Seq and describe the experimental process for chRNA validation using real-time PCR and sequencing (Fig. 1a).

2 Materials

2.1 Total RNA Extraction, Quantification, and Quality Assessment

1. RNAlater (Ambion).
2. RNeasy Qiagen kit (Qiagen).
3. β-Mercaptoethanol: 14.3 M.
4. RNAse-free DNase set (Qiagen).
5. Ethanol: 100%.
6. RNAse-free water.
7. Microcentrifuge.
8. Filter tips.
9. Micropipettes.
10. Microtubes: 1.5 mL.
11. Disposable gloves.
12. Vortex, rotor-stator homogenizer, bead homogenizer, or syringe and needles (*see* **Note 1**).
13. NanoDrop® ND-1000 Spectrophotometer (NanoDrop ND-Thermo Fisher Scientific).
14. 2100 Bioanalyzer (Agilent Technologies).
15. RNA 6000 Nano Kit (Agilent Technologies).

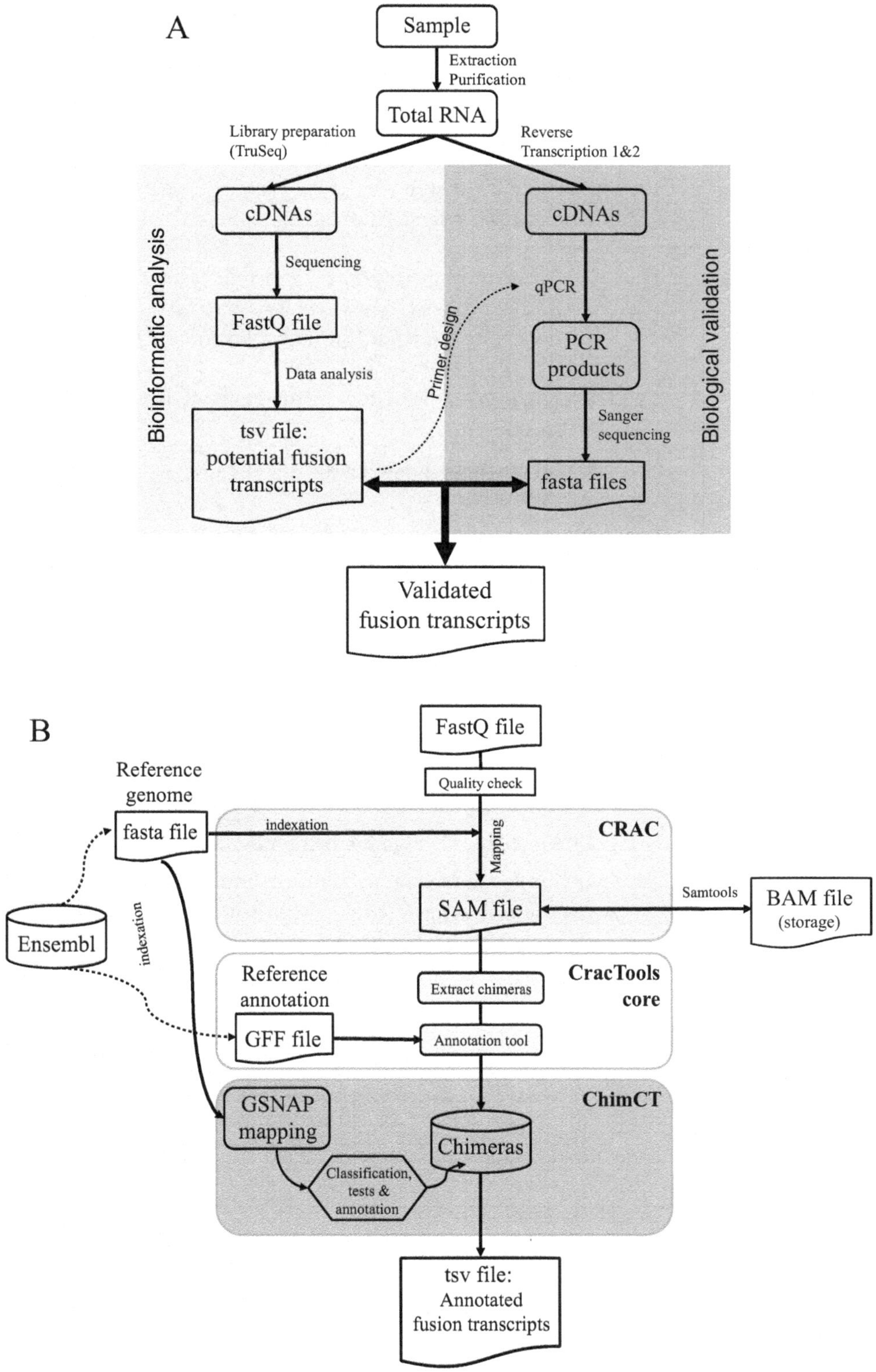

Fig. 1 Workflow for the detection of gene fusions in chromothripsis using RNA-Seq. (**a**) Global representation of the different steps required to identify chimeric events from the sample preparation to the biological validation of chRNA candidates. (**b**) Schematic representation of the RNA-Seq data analysis pipeline dedicated to gene fusion identification

2.2 Library Preparation and RNA Sequencing

1. TruSeq Stranded mRNA Sample Preparation Kit (Illumina).
2. Illumina NextSeq platform.

2.3 DATA Analysis

See **Note 2** for recommended requirements.

2.3.1 Fastq Quality Check

Download and install FastQC (https://www.bioinformatics.babraham.ac.uk/projects/fastqc/).

2.3.2 Reference Genome

Download the human reference genome fasta file from Ensembl (ftp://ftp.ensembl.org/pub/release-89/fasta/homo_sapiens/dna/Homo_sapiens.GRCh38.dna.toplevel.fa.gz).

2.3.3 Alignment Step: CRAC

(a) Download your fastq files (output from Illumina sequencing platform).
(b) Download and install CRAC version 2.5.0 (http://crac.gforge.inria.fr/).
(c) Samtools: SAM file compression.
(d) Download and install Samtools v1.5 (http://www.htslib.org/download/).

2.3.4 ChimCT: Post-Processing of CRAC's Output

Download and install the following programs and files:

- CracTools-core v1.2 (http://cractools.gforge.inria.fr/softwares/cractools/).
- ChimCT v0.14 (https://github.com/Bio2M/cractools-chimct).
- GSNAP (http://research-pub.gene.com/gmap/).
- Our home-made human genome annotation files: ensembl.gff and ensembl-est.gff (https://github.com/Bio2M/methods_mol_biol_2017). These files are built from the Ensembl Perl API (https://github.com/Ensembl version 89). *See* **Note 3** if you need to rebuild it.

2.3.5 Filtering ChimCT Output

Download and install R (https://cran.r-project.org) and Rstudio (https://www.rstudio.com/products/rstudio/download/).

2.3.6 Data Visualization

Download and install Circos (http://circos.ca/).

2.4 Manual Annotations

Tsv file (ChimCT output).

BLAT from UCSC genome browser http://genome.ucsc.edu/cgi-bin/hgBlat?command=start

ENSEMBL genome browser: http://www.ensembl.org/Homo_sapiens/Info/Index

2.5 Primers Design

Sequences of reconstructed chimeric reads given in chimCT output tsv file (column 14: Primers) and validated through the manual annotation step (gap or overlap correctly identified).

Primer3Plus web interface http://www.bioinformatics.nl/cgi-bin/primer3plus/primer3plus.cgi

2.6 Reverse Transcription

1. Total RNA: 1 μg.
2. Rnase-free water.
3. Rnase-free tips.
4. Micropipettes.
5. MultiScribe® Reverse Transcriptase: High-capacity cDNA Archive Kit (Applied Biosystems).
6. ImProm-II™ Reverse Transcriptase: Improm-II Reverse Transcription System (Promega).
7. PCR microtubes.
8. Thermal cycler.
9. Ice.

2.7 Quantitative PCR

1. 384-well PCR plate.
2. cDNA sample: 1 ng/μL.
3. Forward and reverse primers (Eurofins Genomics, Germany): High Purity Salt Free, 100 μM stock solution.
4. epMotion 5070 automated pipetting system (Eppendorf).
5. LightCycler®480 SYBR Green I Master (Roche Diagnostics): 2×.
6. LightCycler®480 Real-Time PCR System (Roche Diagnostics).

2.8 PCR Product Purification and Sanger Sequencing

1. Amplification products in PCR 384 plate.
2. Scalpels.
3. MinElute PCR Purification Kit (Qiagen).
4. NanoDrop® ND-1000 Spectrophotometer (NanoDrop ND-Thermo Fisher Scientific).
5. DNA sequencing company (e.g., Eurofins Genomics, TubeSeq Service).

3 Methods

3.1 RNA Extraction, Quantification, and Quality Assessment

1. Store cells/tissue extracts in RNAlater following manufacturer's instructions to avoid RNA decay.
2. Before extraction, centrifuge cell extracts for 5 min at 5000 × g and remove RNAlater (upper phase).

3. Extract total RNA with the RNeasy Qiagen kit according to manufacturer's instructions (*see* **Note 4**) and elute RNA with 33 μL RNase-free water. Apply the following options:
   - 10 μL β-mercaptoethanol per mL of RLT lysis buffer.
   - After the first wash (RW1 buffer), add 10 μL of DNAse stock solution (keep at −20 °C until use) to 70 μL RDD buffer.
   - Perform DNase treatment to remove residual DNA as indicated in the RNeasy kit.
4. Quantify total RNA using 1.5 μL of sample, using NanoDrop® ND-1000 Spectrophotometer.
5. Dilute RNA to operating range (25–500 ng/μL), and assess RNA quality on RNA 6000 Nano chip using the Agilent 2100 Bioanalyzer.
6. Select only samples with RNA integrity score above 8 (RIN > 8) (*see* **Note 5**) and total amount greater than 3 μg to perform RNA sequencing and further biological qPCR validations.
7. Hold RNA samples at −80 °C.
8. Send your RNA samples to the sequencing platform according to its recommendations.

3.2 Library Preparation and RNA Sequencing

We recommend to choose the stranded and pair-end protocol with a read length of 100 bp (*see* **Note 6**). This protocol could generate good quality libraries adapted to the bioinformatics analysis described here.

The amount of total RNA required for the TruSeq® Stranded mRNA Sample varies from 250 ng to 1 μg.

The library preparation process includes eight steps with the TruSeq Stranded mRNA Sample Preparation Kit from Illumina followed by a sequencing step (https://support.illumina.com/content/dam/illumina-support/documents/documentation/chemistry_documentation/samplepreps_truseq/truseqstrandedmrna/truseq-stranded-mrna-sample-prep-guide-15031047-e.pdf):

1. mRNA purification and fragmentation.
2. First-strand cDNA synthesis.
3. Second-strand synthesis.
4. 3′end adenylation.
5. Adapter ligation.
6. PCR amplification.
7. Library validation (*see* **Note 7**).
8. Library normalization and pooling (multiplex sequencing option).

3.3 Data Analysis: Chimeric RNA Identification

An overview of the full analysis pipeline is given in Fig. 1b.

For this section, command lines are given as example. This example is based on the analysis of one regular human sample using stranded paired-end RNA-Seq (Illumina technology, 120 million reads, 2 × 100 bp).

3.3.1 Fastq Quality Check

During the initial sequencing process, different types of errors can possibly occur, including contaminations, error of sequencing, and PCR artifacts, and can critically impair the accuracy of chimera extraction. We recommend to use FastQC to check your sequencing files quality-related parameters.

1. FastQC works with a graphical interface, open your fastq or fastq.gz file with FastQC.
2. For all the analyzed quality-related parameters, FastQC makes a quick evaluation of the results and shows corresponding icons: green tick for entirely normal, orange triangle for slightly abnormal, or red cross for very unusual. Check that all the icons are green before proceeding to the data analysis (*see* **Note 8** for troubleshooting guide).

3.3.2 Reference Genome Indexation: Preprocessing Step

CRAC and GSNAP use methods to compress the size of the genome and make alignment on the genome faster. Note that indexes build by CRAC and GSNAP are different and both are required for the procedure. You only need to do it once, and when done you can save some space by keeping the index and deleting the fasta genome.

1. Build indexes with crac (http://crac.gforge.inria.fr/getting-started/)

```
zcat Homo_sapiens.GRCh38.fa.gz > Homo_sapiens.GRCh38.fa
```

```
crac-index index myCRAC_Index Homo_sapiens.GRCh38.fa
```

2. Build indexes with GSNAP (http://research-pub.gene.com/gmap/src/README)

```
gmap_build -g -d myGSNAP_Index Homo_sapiens.GRCh38.fa
```

3.3.3 Alignment Step: CRAC

As input, CRAC takes fasta or FASTQ files either in raw format or compressed using gzip.

CRAC can output a SAM/BAM file (containing extra information on the biological predictions).

You can run the crac command with different options depending on your data and the information you need to retrieve. Here are the main options that are used to extract chimeric reads.

Run CRAC with the following options:
Mandatory options:

- i/path/to/myCRAC_Index (without extension) is the name of the index previously built with the crac-index binary file.
- r <reads_file1.fastq> [reads_file2.fastq] is FASTQ file(s) containing the reads. Note that the number of files depends if single or paired reads.
- k <int> is the length of the k-mer to be used to map the reads on the reference <index_file>. It must be chosen to ensure (as much as possible) that a k-mer has a very high probability to occur a single time on the genome. We recommend to use k = 22 for human and mammal genomes.
- o, --sam <crac_output.sam> is the output file in SAM format or print on STDOUT with "-o -" argument to directly go the next command (compression of the SAM file to BAM) without saving the SAM file (reduce file storage space).

Usual facultative options:

- detailed-sam to add more information in SAM output file (required for ChimCT analysis).
- summary <crac_summary.txt> save some statistics about mapping and classification.
- stranded must be specified if reads are produced by a stranded protocol of RNA-Seq (not stranded by default).

See **Note 9** for other optional arguments.

Example:

```
crac -i /path/to/myCRAC_Index -k 22 --stranded
--summary Sample_crac_summary.txt --detailed-sam -r
Sample_read1.fastq.gz Sample_read2.fastq.gz -o - |
<next cmd>
```

<next cmd>=command to run samtools view (*see* Subheading 3.3.4).

3.3.4 Samtools: SAM File Compression and Indexation

1. Compression

```
<previous cmd> | samtools view -bS - | samtools
sort - -o Sample_crac_sorted.bam
```

<previous cmd> = command to run CRAC (*see* Subheading 3.3.3).

2. Indexation (facultative)

```
samtools index Sample_sorted.bam > Sample_crac_
sorted.bai
```

Samtools index produce a BAI file that is useful to visualize your alignment results with a tool like IGV (software.broadinstitute.org/software/igv/).

3.3.5 ChimCT: Post-Processing of CRAC's Chimeric RNAs

As input, ChimCT takes BAM files. ChimCT outputs a tsv file (containing many information about chimeric reads and depending on the following options).

Run ChimCT with the following options:

1. Mandatory options
   - n <SampleName>.
   - g /path/to/annotations_file.gff.
   - s <sorted_output.bam>.
2. Usual facultative options
   - Stranded. Make use of stranded information to output chimeras in the right orientation. Stranded PE reads are interpreted as reverse-forward reads.
   - Verify-splice. Verify splices found by CRAC using the GFF3 annotation and convert them into chimera if needed.
3. GSNAP use (ChimCT calls GSNAP to perform a parallel annotation of splicing events)
   - gsnap-exe gsnap.
   - gsnap-genome-name Genome.
   - gsnap-genome-directory/path/to/myGSNAP_Index (without extension) is the name of the index previously built with the gmap_build binary file.
4. Output: files and formats
   - Detailed-sam. Print the support and location profiles for the selected reads.
   - Primers. Print the sequence of concatenated paired-end reads with junction position highlighted for easier primers design.
   - Summary <chimct_summary.txt>. Specify file to print summary stats about the chimera pipeline in a separate output file.

See **Note 10** for other optional arguments.

Example:

```
chimCT --stranded --detailed-sam --primers --summary
Sample_chimct_summary.txt --verify-splice --gsnap-
genome-name GRCh38 --gsnap-genome-directory /indexes/
gsnap/GRCh38 --gsnap-exe gsnap -n SampleName -g /anno-
tations/ensembl.gff -s Sample_cracOutput_sorted.bam >
Sample_chimct.tsv
```

ChimCT tsv file: for each analyzed chimera, the pipeline gives related information including a unique read identifier annotated by the pipeline, class, number of reads, etc.

The following paragraph details the content of ChimCT tsv file columns:

1. Id: A unique read Id for each chimera. This Id is formed by specifying the sample name and the chimeric junction Id (given by CRAC) separated by a colon.
2. Name: Names of the left and right genes involved in the fusion transcript separated by "---." When reads were located on a non-annotated, transcribed region, the corresponding "N/A" annotation was mentioned.
3. Chr (1): Chromosome number of the 5′ part of the chimera (left).
4. Pos1: Genomic position of the 5′ part of the chimeric junction.
5. Strand1: Genomic strand of the 5′ part of the chimera (+1 or −1).
6. Chr (2): Chr number of the 3′ part of the chimera (right).
7. Pos2: Genomic position of the 3′ part of the chimeric junction.
8. Strand2: Genomic strand of the 3′ part of the chimera.
9. Chim value: The chim value takes into account methodological parameters and ambiguities including the read mapping (location profile) and the read coverage (support profil). *See* column 13 for more details about location and support profiles.
10. Spanning junction: Chimeric junction read coverage (normalized per billion of reads). A junction-spanning read is a paired-end read that covers the chimeric junction (Fig. 2).
11. Spanning PE: Coverage of paired-end reads (normalized per billion of reads). Spanning paired-end reads include junction-spanning reads and their corresponding paired reads and paired reads that spans the chimeric junction without sequencing it (Fig. 2).
12. Class: Chimera classification from 1 to 4. ChimCT procedure includes classifying chRNA into four categories depending on exon organization. This classification is depicted in Fig. 3a, b, and can be summarized as follows:
    - Class 1: the exons are located on different chromosomes.
    - Class 2: the exons are collinear but most likely belong to different genes, to be verified using annotation.
    - Class 3: the exons are on the same chromosome and same strand, but not in the order in which they are found on DNA.
    - Class 4: the exons are on the same chromosome but on different strands.

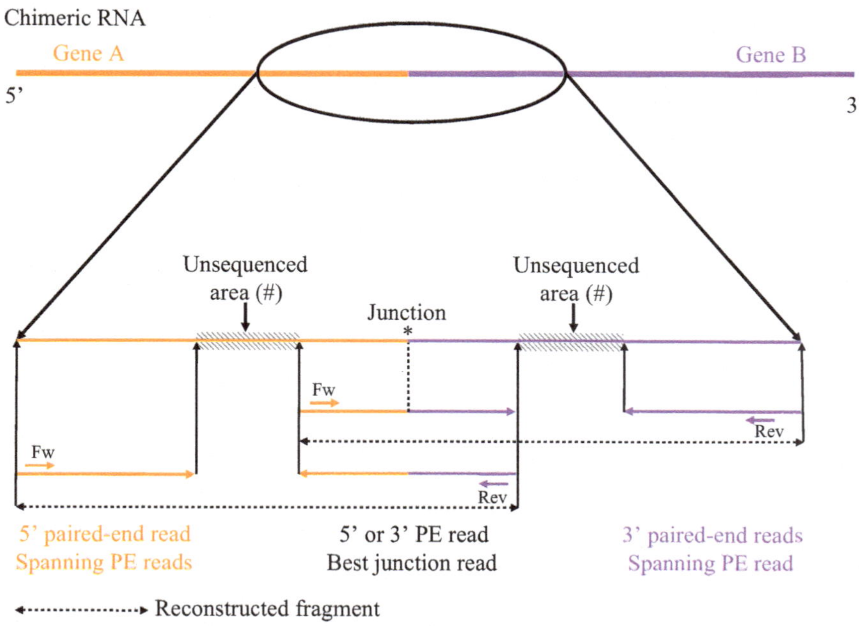

Fig. 2 PCR primers are designed to target each side of the junction point (***asterisk***). We use the ChimCT reconstructed read that corresponds to the selection of the best junction-spanning read (centered on the chimeric junction), and its corresponding paired-end read (that can be in 3′ or 5′) separated by an unsequenced gap. Fw and Rev indicate, respectively, forward and reverse primers designed to validate the chRNA by qPCR

13. Comments: Chimeric reads specific characteristics (*see* **Note 11**); these information are useful for the next step (chimeras filtering).
14. Others: Chimera annotations (*see* **Note 12**).

3.3.6 Filtering ChimCT Output (tsv File)

We recommend the following filtering steps:

1. Discard candidate fusions (chRNA) with less than four raw reads covering the fusion breakpoint (column 14 or "Others": field "Nb_spanning_reads = number").
2. Set the Chim Value minimum threshold at 60 to remove most of false positive corresponding to pseudogenes and splicing events detected by GSNAP.
3. Delete remaining splice events flagged by GSNAP (GSNAP Mapping comment indicates that all the reads are mapped by GSNAP).
4. Because chimeras involving adjacent genes could reflect annotation issues [9], we discard Class2 chimeras (genes on the same chromosome and same strand) involving adjacent genes within a 3 Kb distance region.
5. Discard chimeras flagged as "anchored" (PCR artifact).

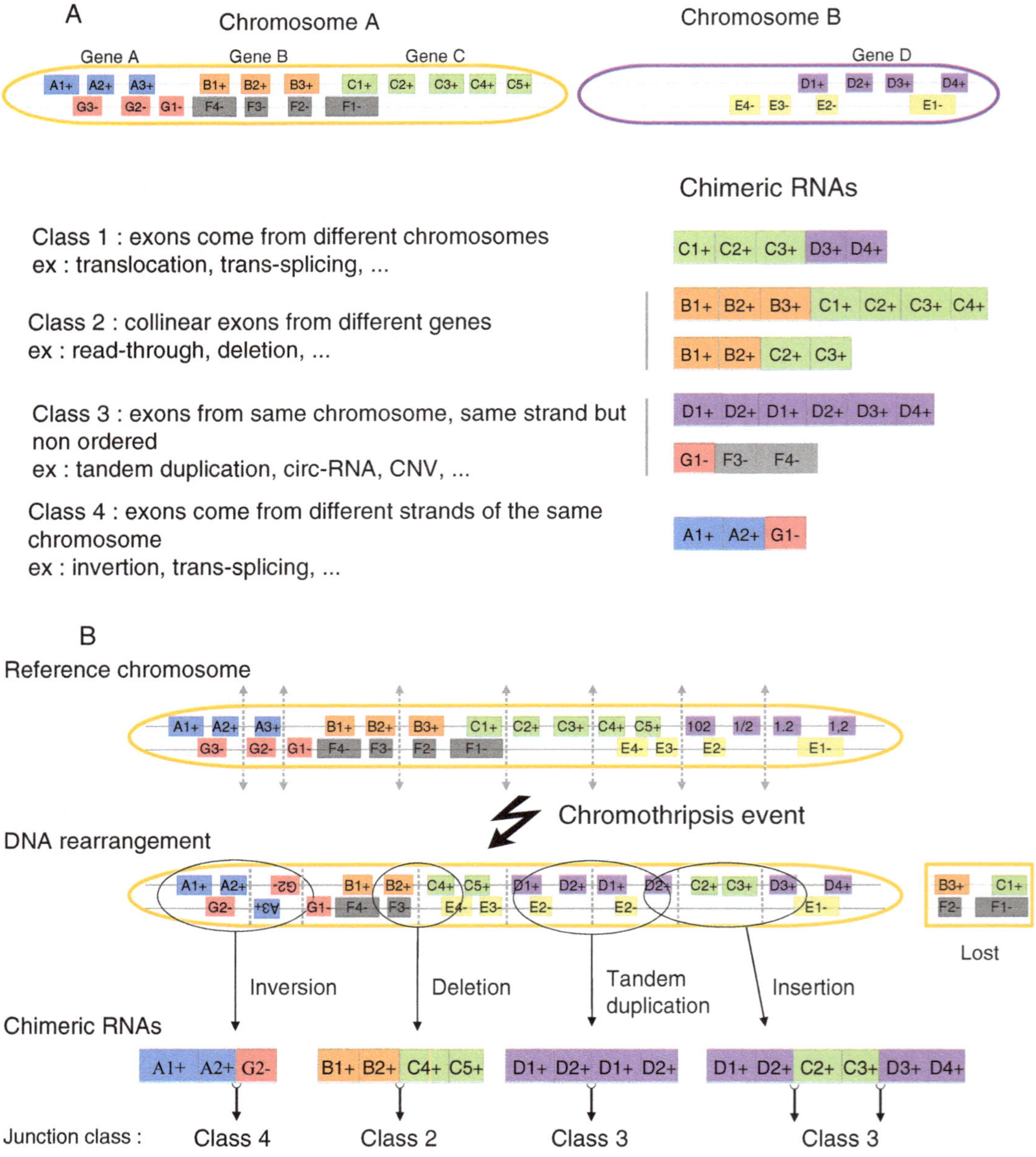

Fig. 3 Chimeric RNA classification and chromothripsis. (**a**) Schematic representation of chRNA classification, depending on exon organization. (**b**) Schematic representation of potential DNA rearrangement and corresponding chimeric RNAs due to chromothripsis. Letters indicate genes; the corresponding exons are numbered and +/− refer to the DNA strands. The random reconstructed chromosome gives rise to potential chRNA belonging to the four different classes, and all of them can be identified using our pipeline. Note that class 1 chRNAs resulting from fusion between two different chromosomes are absent from the representation

6. Discard chRNAs corresponding to fusion of two unannotated transcripts: column 2 or "Name" entry = "N/A---N/A."

Depending on your data and the types of event you are looking for, you can add other filters like:

- Use the "high/low support" mention (cf column 13) to discard in one hand low coverage fusion transcripts and, in the other hand, chimeras that are highly expressed suggesting false-positive results (pseudogenes, superfamily genes, genes with highly variable domains, etc.) (*see* **Note 11**).
- Discard chimeras which junction-spanning reads are not associated to paired-end reads, (*see* "strange paired-end support" in column 13). Warning: chimeras involving more than two genes could be flagged as "strange paired-end support."

You may also use the chimeras' subclassification to select your candidates:

- Class1 chimeras can be divided into two subclasses: "exon-exon borders" (junction point mapped at +/− 20 nt far from annotated exon borders) and "others."
- Class2 chimeras can be divided into three subclasses: "readthrough" (chimeras involving adjacent genes) and "short-" and "long-fusion distance" (chimeras involving genes really far from each other > 300 kb).
- Class3 chimeras can be divided into three subclasses: "overlapping" (5′ part of the junction-spanning read overlaps the 3′ part) or "short-" and "long-fusion distance" (cf comments column in tsv file).
- Class4 chimeras can be divided into two subclasses: "overlapping" and "others."
- This subclassification can help you to select a subgroup for specific validation process like the use of FISH techniques.

3.3.7 Data Visualization

Finally, you can visualize your results using Circos and easily identify major chromosome domains involved in chromothripsis events (Fig. 4).

3.4 Manual Annotations

1. Look at potential chimeras in tsv file and select abnormal transcripts you want to research (e.g., this can be done per classes or subclasses).
2. For each candidate chimera, you need first to check the support and location profiles and second to reconstruct and check the full cDNA fragment sequence. Then you can search for potential effects like fusion protein synthesis or other.

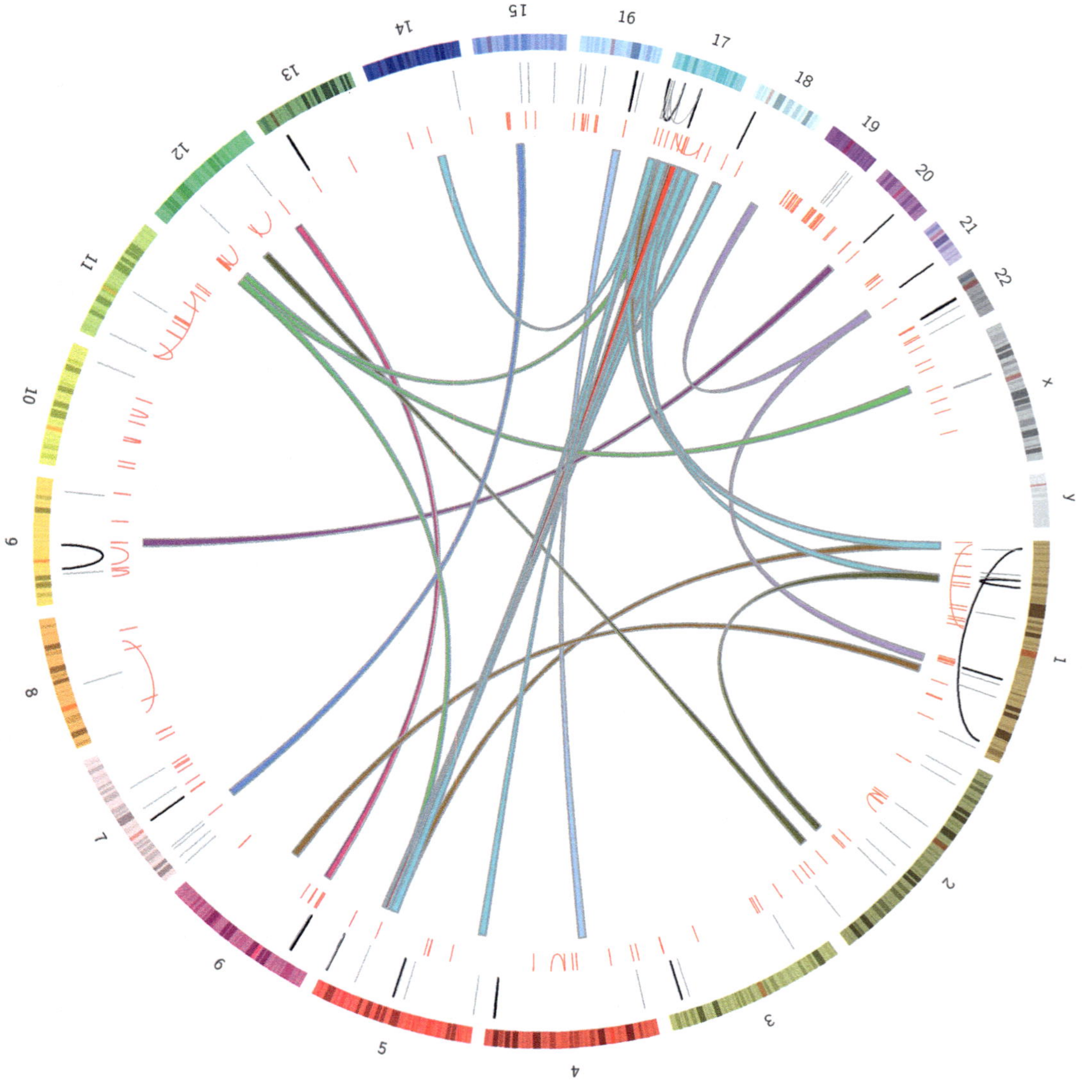

Fig. 4 Transcriptome rearrangement in one AML (acute myeloid leukemia) tumor. Chromosomes are arranged circularly end-to-end with each chromosome's cytobands marked in the outer ring [10]. The inner rings display class 2 (red), class 3 (gray), and class 4 (brown) predicted fusion transcripts. Within the circle, all predicted class 1 fusion genes are shown as arcs

3.4.1 Support and Location Profiles Check

To discard false positives, check the support and location profiles depicted in the tsv file (cf column 13).

If the support profile is very low in the break, this means that the chimera is very rare. Also if the support of k-mers inside the break is not stable, it is an indicator that the event is likely to be a false positive. The location profile can also have very valuable information. For example, if the k-mers at each side of the break have multiple locations on the genome (density > 1), this means that the chimera is ambiguous. Also a high variability of this profile can traduce a low-complexity region that could lead to the identification of a false chimera.

However, there are some exceptions to these rules, and location and support profiles should not be used as the only criteria to discard chRNA (*see* **Note 13**).

3.4.2 Reconstruction of the Full cDNA Fragment

For each chRNA of interest, you can retrieve the reconstructed sequence from the ChimCT tsv file (cf Primers column). However, RNA-Seq relies on the fragmentation and sequencing of cDNA pieces. Using pair-end sequencing, fragments are sequenced from both ends, and a good RNA fragmentation step generates pieces that are longer (~350–400 bp) than the sequencing reads (PE reads, 2 × 100 bp) leaving an unsequenced gap in the middle of the fragment. Do as follows to evaluate this gap length and reconstruct the full cDNA fragment if needed.

1. Map the reads sequences (best junction-spanning read and its corresponding paired-end read) to the human genome GRCh38 with BLAT (UCSC http://genome.ucsc.edu/cgi-bin/hgBlat?command=start).
2. Check that PE reads do not overlap. If they do, manually correct ChimCT reconstructed sequence by deleting the duplicated sequences (*see* **Note 14**).
3. Try to determine the length and the sequence of the gap between the two reads using following guide:
   (a) Use the fragment mean length information (*see* **Note 7**) to estimate the gap size.
   (b) If the mapping interpretation looks easy (sequences map to well-annotated regions and involve only two exons), the full reconstruction is not really required, and you can design primers using the reconstructed sequence provided by ChimCT in primers column and the library fragment mean size.
   (c) If the mapping interpretation looks complex (mapping result is not in concordance with ChimCT annotation), try to identify biological events (splicing, SNPs, insertions, deletions, repeats, polymorphisms, etc.) that could help class the chimera as a false positive (*see* **Notes 15** and **16**).
   (d) Identify complementary annotations using the ENSEMBL genome browser to determine exons and spliced variants involved in the transcript.
   (e) Use the fragment mean length to estimate the gap size.
   (f) Reconstruct the estimated full fragment sequence if possible.

3.4.3 Evaluate the Potential Impact

Use ExPASy translate tools and PIR (Protein Information Resource) to identify affected protein domain and to evaluate potential protein damage related to selected chRNA.

3.5 Primers Design

Primers are used to perform chimera validation by qPCR followed by Sanger sequencing of the PCR product.

Use Primer3Plus web interface (http://www.bioinformatics.nl/cgi-bin/primer3plus/primer3plus.cgi) to design your primers as described in the PCR strategy (Fig. 2).

Follow the next recommendations:

1. Use the full reconstructed fragment sequence and/or the estimated fragment size from the manual annotation step to determine the amplicon length, and design your primers.
2. Best Sanger sequencing results are obtained for 150–300 nt long PCR products.
3. The amplification area should be centered on the chimeric junction, and primers are designed to target each side of the junction.
4. Primers melting temperature (TM) should be around 60 °C; avoid TM variation between your primers.
5. Primers length should range from 18 to 27 nucleotides.
6. Avoid self-complementarity and secondary structures.
7. Check that primers have unique localization on reference genome performing a BLAT.
8. Order the synthesis of your primers.

3.6 Reverse Transcription

1. Use 1ug of RNA to perform reverse transcription with random primers and MultiScribe® Reverse Transcriptase.
2. Apply the following option: add RNase inhibitor as suggested in manufacturer protocol.
3. Use 1ug of RNA to perform a second reverse transcription reaction with ImProm-II™ Reverse Transcriptase (*see* **Note 17**).

   Perform both protocols following manufacturer's instructions.

3.7 Quantitative PCR

1. Dilute forward and reverse primers to 4 μM.
2. Mix 3 μL of Master Mix with 0.66 μM forward and reverse primers (4 μL final volume).
3. Add 2 μL of each cDNA sample (1 ng/μL) (*see* **Note 18** for cDNA quantity check).
4. Load samples onto the 384-well PCR plate using an automated pipetting system (*see* **Note 19**).

5. Perform amplifications on the LightCycler®480 according to the following conditions:
6. Set temperature at 95 °C for 5 min, then 55 cycles as follows: 95 °C for 10s, followed by T°C depending on TM for 10s, and 72 °C for 10s. Ultimately, perform the melting curve analysis ranging from 95 °C to 60 °C.
7. Check the amplification specificity: for each sample, the graph of the negative first derivative of the melting curve should give a specific peak corresponding to the amplified transcript.
8. Select samples to be sequenced: samples with TM value peaks different from those found in negative control are considered as potential positive targets and retained for sequencing.

3.8 PCR Products Purification for Sanger Sequencing

1. Use a scalpel to remove the optical adhesive cover on selected wells of the 384 plate.
2. Collect and transfer your PCR products with a micropipette to 0.5 mL microtubes.
3. Purify the PCR products using the MinElute PCR Purification Kit according to manufacturer's instructions.
4. Elute DNA with 12 μL of nuclease-free water.
5. Use 1 μL of DNA solution to quantify with the NanoDrop Spectrophotometer.
6. Dilute the sample to 2 ng/μL (150–300 nt long PCR product).
7. Dilute sequencing primers to 10pM/μL.
8. Send 15 μL of sample and 15 μL of sequencing primers to Eurofins Genomics (TubeSeq Service) to perform Sanger sequencing.

   Note that you can choose your own DNA sequencing company. You should then dilute the sample and sequencing primers as recommended, and send the required amount of materials.
9. Check the sequencing results.

   The company provides two fasta files of clipped or not-clipped sequences, one pdf file including the sequencing chromatogram, and one scf file.

   Align the fasta sequence using BLAT (https://www.bioinformatics.org/sms/rev_comp.html) to the reference genome, and compare the sequenced amplicon with the initial chimeric read.

4 Notes

1. Note that RNA extraction requires a very efficient step of sample homogenization and cell lysis step. Therefore, you need to adapt the material used to your sample type.
2. Minimum hardware requirements and recommendations:

   Computer with at least 4 CPUs or 8–12 threads and at least 32 Go of memory (RAM).

   Files storage:

   - Reference genome fasta file (human genome fasta file, 15Go; fasta.gz, 1Go; can be deleted after the indexation step).
   - Reference genome indexes (CRAC human genome index, 2.3Go; GSNAP human genome index, 6Go).
   - Input fastq files (fastq.gz: 2× 5Go*).
   - CRAC output: SAM (267Go*, facultative storage) and BAM files (21Go* per sample).
   - *Indicative numbers related to the analysis of one regular fastq file (120 millions of paired-end or PE reads: 2 × 100 bp).
   - Total runtime of all programs: ~16 h (28Go RAM, 12 threads); depending on your installation and material, multiple jobs can be run in parallel.
3. How to build ChimCT annotation files (using CracTools): Run the following commands:

```
buildGFF3FromEnsembl Name_of_your_species > ensem-
bl.gff

buildGFF3FromEnsembl --est Name_of_your_species >
ensembl-est.gff
```

   Name_of-your_species = "Homo sapiens", "Mus musculus", … (cf http://www.ensembl.org/info/about/species.html).
4. All reagents (except DNAse) must be stored at room temperature. At this step special precautions should be taken when working with RNA to maintain an RNase-free environment (use gloves and RNase-free material and solution).

   Take care to completely remove ethanol before elution step (do not hesitate to add an extra centrifugation step or increase the air drying time).
5. We recommend not to use low-quality or degraded RNA as this can lead to the biased sequencing results (short fragments, overrepresentation of 3′ends, etc.) and analysis issues.

6. We recommend to take into account the following parameters to choose your RNA-Seq strategy:

   Read length: CRAC precision and sensitivity increase with read length; we recommend a minimum read length of 100 bp.

   Fragment length: take care that the fragments are long enough for the paired-end reads not to overlap (>200 bp, at least 300 bp).

   Single vs. paired-end reads: CRAC is more precise with paired-end compared to single reads.

   Coverage: to detect low-level chimeras, we recommend at least 40 millions of sequenced paired reads.

   Stranded: more precise information is used by ChimCT to annotate chimeric reads.

7. Take care that you select libraries with DNA fragment ranging from 200 to 400 bp. Indeed short fragments (<2 × 100 bp) induce sequencing analysis issues as paired reads will overlap.

8. FastQC analyze the following parameters:

   - Per base and per sequence quality scores: A quality too low is the sign of a high sequence error rate and should impair accuracy of the mapping.
   - GC content of reads: A normal distribution centered on the overall GC content of the human genome indicates the absence of contamination or other kinds of biases.
   - Per base N content: A proportion of N bases exceeding few percents indicate that the sequencing was unable to make decent base calling.
   - Read length: All the reads should have the same length; otherwise, there is an anomaly in the sequencing.
   - Overrepresented k-mers and sequences: Useful to detect varied biases and contaminations.
   - Adapter content: A high content of known Illumina adapter sequence indicates that they should be trimmed.
   - To correctly interpret parameters status, you must take into account the context and conditions of the library preparation. However, we recommend to reconsider the initial data if the quality scores are too low or if adapter sequences are present. Then filter quality and trim reads using FASTX-Toolkit (http://hannonlab.cshl.edu/fastx_toolkit/) or trimmomatic (http://www.usadellab.org/cms/?page=trimmomatic) before running the data analysis pipeline.

9. CRAC supplementary options:

   - no-ambiguity <none> discard biological events (splice, svn, indel, chimera) which have several matches on the reference index. Note that if CRAC has not identified a biological cause for the multimapped read, it is classified as a biological undetermined event and discarded.

- fr/--rf/--ff set the mates alignment orientation (--rf by default). If Illumina data are usually --rf, you should be aware that data can sometimes be provided in fr orientation.
- nb de threads <number>. To improve the full running time, it is highly recommended to use multithreading (> = 10, depending on reference genome and data size).

10. ChimCT supplementary options:

- gsnap-nb-threads = Integer. Number of threads used to run GSNAP (default 4). To improve the full running time, it is highly recommended to use multithreading (> = 10, depending on reference genome and data size) for GSNAP Mapping.
- keepIg Ig is a superfamily gene. Because such genes have highly variable domains, by default CRAC considers chimeric reads corresponding to Ig as false positives and discards them. This option allows you to save them.
- spanning-reads = file Specify basename to output chimeric spanning reads (PE and single) in fasta format.

11. Detailed ChimCT comments:

Annotation = pseudogene. If one of the two genes implicated in the chimeric junction is characterized as a "pseudogene" in the annotations (GFF file from Ensembl).

Stringent tests = (low/high) support. The normalized support is calculated as the number of read assigned to the chimera (junction-spanning reads) per billion of sequenced reads.

If the chimeras has a normalized support >5000 or <20, it will be flagged as "high support" and "low support," respectively.

chimera_anchored. For a given chimera, if all the junction-spanning reads are similar (i.e., same position of break in the read), the chimera is flagged as "anchored" (potential PCR artifact).

PairedEndChimeras = 'strange_paired_end_support'. If there is less spanning PE reads than junction-spanning reads.

GSNAP Mapping = '(number) of read mapped out of (number)'. The number of reads mapped by GSNAP out of the total number of reads supporting the chimeric junction (junction-spanning reads) is reported. If all the chimeric reads are mapped by GSNAP, the chimera is considered as a false positive (splice event).

Fusiondistance = short(number)/overlap (class3). For class3 chimeras, the fusion distance which is the distance between the two ends of the fusion (junction-spanning read) is calculated. If the two fragments are overlapping, the chimera is flagged as "overlap." If the distance is less than 1000 bp, the chimera is flagged as "short," and the fusion distance is given.

12. Others: chimeras annotations

mRNA_IDs: Ensembl mRNA ID.

Annotation_type: Ensembl transcript type.

Description_genes: Ensembl gene type.

Exon_IDs: Ensembl ID of the closest exons. (Reads are annotated according to Ensembl annotations file by giving priority to location in exons of the annotated genes.)

Exon_end_distances: distances between the chimeric junction and the closest annotated exons.

Exon_ranks: Ensembl exon ranking related to the full transcript.

Read_id: ID of the most representative chimeric read (selection of the longest read centered at the junction point) = "best chimeric read."

Pos_junction: location of the junction point in the best chimeric read.

Read_seq: sequence of the best chimeric read.

Profil_support: given only if option --detailed sam is used; the support profile is an array that indicates for each k-mers (22 nt) of a read its number of occurrence in the whole read collection (Fig. 5).

Profil_location: given only if option --detailed sam is used; the location profile is an array that indicates for each k-mer (22 nt) of a read its number of occurrence in the reference genome. A break is an interruption of k-mers mapping and is visualized by a stretch of 0 in location profile. The end of the break traduces the localization of the chimeric junction point (Fig. 5).

Nb_spanning_reads: number of junction-spanning reads in the raw file.

Nb_spanning_PE: number of spanning PE reads in the raw file.

Primers: given only if option --primers is used; for each fusion transcript, the CRAC software provided a reconstructed sequence comprising, on one hand, the chimeric junction sequence based on the most representative read (best read) and, on the other hand, the corresponding paired read sequence. The symbol (#) marks the gap (fragment that was not sequenced), and the (*) symbol marks the junction point. This information is useful to design qPCR primers (*see* Subheading 3.5).

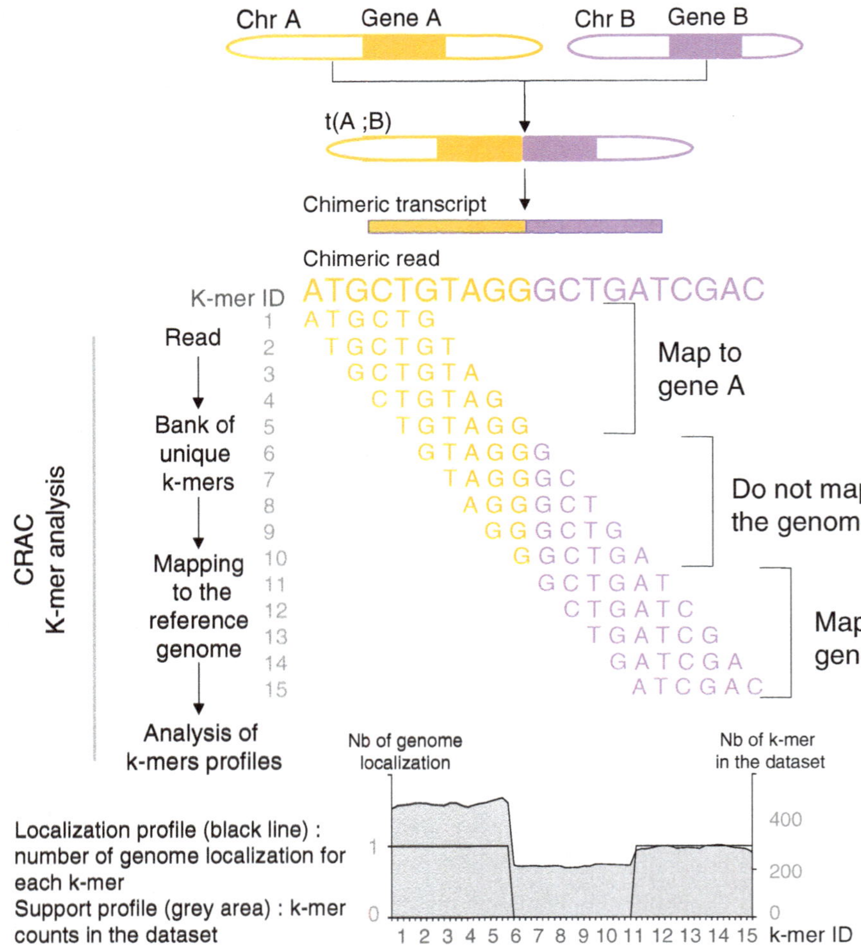

Fig. 5 Anatomy of a chimeric junction identified with CRAC algorithm

Example:
ATTTAATACTACGACAGCCACTTTGGGCTTTGGAGCCCCCCAGGCCCCAG*TAGCTGTGCGGTCAGAGAAGAAACGCCTTAGGAAGCCAAGCAAGTGGCTT#CAAGTGGCTTTTGGAATATACAGAAGAATATGATCAGATATTTGCTCCTAAGAAAAAACAAAAGAAGGTACAGGAGCAGGTGCACAAGGTAAGTTCCCGC

BEST CHIMERIC READ (PE READ1) *Junction point #Gap PE READ2

13. As the high/low support thresholds strongly depend on your sequencing depth, you may need to adapt them to your data.
14. When RNA fragmentation generates short fragments (less than 200 bp), the sequencing procedure may generate overlapping PE reads, and the reconstructed sequence given by ChimCT is wrong (duplicated sequence in the middle of the

fragment). It is then necessary to manually rebuild the right fragment before designing primers by deleting the duplicated sequence.

15. Example of complex event leading to chimera misclassification. Careful examination of a potential chimera annotated as a class1 chRNA involving the gene A and an unannotated region of the genome revealed a false-positive result. Indeed, the mapping of the best junction-spanning read with BLAT showed that both side of the fusion point mapped to different exons from the same gene. If this event should have been classified as a splicing event by GSNAP, the presence of a point mutation in the first base of the second exon (splice site) leads to the misclassification.

16. Class 3 chRNA with the mention "fusion distance = overlap" in the comment column requires almost systematic reconstruction of the fusion transcript for the design of primers. Such chRNA often corresponds to repeated sequences. Repeated sequences in the cDNA fragment (e.g., FLT3-Internal Tandem Duplication) raise problems because the first location and support profiles are unusual and could lead to misclassification of the chimera as false positive at Subheading 3.4.1 (Fig. 6a, b) and second primers design needs special attention.

    Primers designed using chimCT reconstructed sequence could anneal in multiple places in this repeat containing fragment. When annotation is well known, a way to solve this is to manually rebuild the transcript to design new primers beyond the repeated sequences.

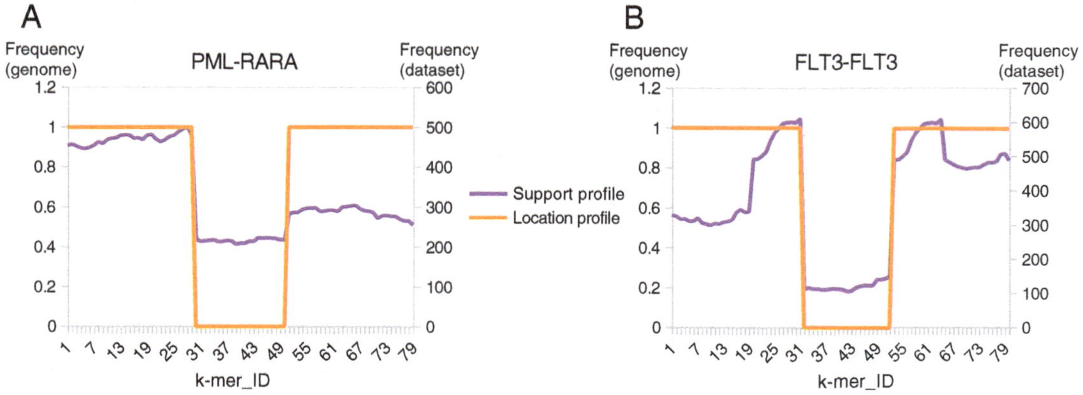

Fig. 6 Graphical representation of support and location profiles. The horizontal axis shows the k-mer identification number. The left vertical axis shows the number of time each k-mer is found on the reference genome. The right vertical axis shows the number of time each k-mer is found in the dataset. Graph A shows the example of a classical gene fusion between genes located in different chromosomes and showing usual profiles. Graph B shows the example of a chimera showing unusual profiles resulting from an internal tandem duplication

FLT3-ITD

Primers = CTGGAATAGAAAGGCTAACAGAAAAGTGTTT GG**ACAGTGGGTGTCGAGCAGTA**CTC TAAACATGAGTGAAGCCATAAAAGGG TTCCTGGTCAAGTGCTGT[G#ATAATGAGTACTTCTA CGTTGATTTCAGAGAATATGAATATGAT CTCAAATG*T]GATTTCAGAGAATATGAATAT GATC**TCAAATGGGAGTTTCCAAGAG**AA

[sequence to amplify]

Repeated sequences separated by the breakpoint (*)

Primer forward: **ACAGTGGGTGTCGAGCAGTA**

Primer reverse: **CTCTTGGAAACTCCCATTTGA**

unsequenced region, gap (estimated to be ~210 nt).

17. To check for possible chRNA formation induced by transcriptional artifacts, each sample is reverse transcribed with two different enzymes.
18. qPCR check:

    In order to test the reverse transcription step, a first PCR with housekeeping genes (ex. RPS9) is carried out on all experimental samples. This check gives an overview of the cDNA concentration per sample. A Cycle Threshold $\leq$22 indicates a sufficient amount of cDNA and samples are retained for chRNA detection.
19. For each sample, qPCR reactions are performed in triplicate with an automate process to assure best reproducibility of amplification.

References

1. Pellestor F, Gatinois V, Puechberty J et al (2014) Chromothripsis, an unexpected novel form of complexity for chromosomal rearrangements. Med Sci (Paris) 30(3):266–273
2. Wu C, Wyatt AW, McPherson A et al (2012) Poly-gene fusion transcripts and chromothripsis in prostate cancer. Genes Chromosom Cancer 51:1144–1153
3. Boeva V, Jouannet S, Daveau R et al (2013) Breakpoint features of genomic rearrangements in neuroblastoma with unbalanced translocations and chromothripsis. PLoS One 8(8):e72182. https://doi.org/10.1371/journal.pone.0072182
4. Byron SA, Van Keuren-Jensen KR, Engelthaler DM et al (2016) Translating RNA sequencing into clinical diagnostics: opportunities and challenges. Nat Rev Genet 17(5):257–271
5. Mertens F, Johansson B, Fioretos T et al (2015) The emerging complexity of gene fusions in cancer. Nat Rev Cancer 15:371–381
6. Beaumeunier S, Audoux J, Boureux A et al (2016) On the evaluation of the fidelity of supervised classifiers in the prediction of chimeric RNAs. BioData Min 9:34
7. Philippe N, Salson M, Commes T et al (2013) CRAC: an integrated approach to the analysis of RNA-seq reads. Genome Biol 14:R30. https://doi.org/10.1186/gb-2013-14-3-r30
8. Rufflé F, Audoux J, Boureux A et al (2017) New chimeric RNAs in acute myeloid leukemia. F1000Res 6:1302
9. Philippe N, Bou Samra E, Boureux A et al (2014) Combining DGE and RNA-sequencing data to identify new polyA+ non-coding transcripts in the human genome. Nucleic Acids Res 42(5):2820–2832
10. Krzywinski MI, Schein JE, Birol I et al (2009) Circos: an information aesthetic for comparative genomics. Genome Res 19(9):1639–1645

Part III

The Causal Mechanisms of Chromothripsis

Chapter 10

Experimental Determination of Checkpoint Adaptation by Mitotic Shake-Off and Microscopy

Lucy H. Swift and Roy M. Golsteyn

Abstract

Cells that undergo checkpoint adaptation arrest at and then abrogate the G2/M cell cycle checkpoint to enter mitosis with damaged DNA. Cells surviving this process frequently contain micronuclei, which can lead to genomic change and chromothripsis. In this chapter we describe how to induce checkpoint adaptation and detect it by time-lapse video and immunofluorescence microscopy and how to isolate cells undergoing checkpoint adaptation from a total cell population.

Key words Checkpoint adaptation, Checkpoint kinase 1, Cyclin-dependent kinase 1, DNA damage, HT-29 cells, Micronuclei, Microscopy, Mitosis, Mitotic shake-off

1 Introduction

Checkpoint adaptation is a process in which eukaryotic cells, including human cancer cells, may undergo upon exposure to genotoxic agents such as radiation or chemotherapy. This process involves entry into mitosis with damaged DNA and is characterized by three sequential steps: first, DNA damage must induce a G2/M cell cycle arrest; second, cells must overcome this arrest; and third, cells must enter mitosis with damaged DNA [1]. Human cancer cells are able to undergo checkpoint adaptation when treated with genotoxic agents commonly used in the clinic, such as ionizing radiation [2, 3], camptothecin [4], etoposide [4], and cisplatin [5]. Checkpoint adaptation might contribute to genomic instability in treated cancer cells since cells that survive this process display an increased number of micronuclei [6]. Micronuclei are structures in cells where DNA is prone to continuous damage [7, 6] and the chromosomes inside them are more likely to undergo chromothripsis than chromosomes in the main nucleus [7].

The precise biochemical pathways that regulate checkpoint adaptation are not yet known [8], although the enzyme checkpoint kinase 1 (Chk1) must either be dephosphorylated [2, 4] or

Franck Pellestor (ed.), *Chromothripsis: Methods and Protocols*, Methods in Molecular Biology, vol. 1769,
https://doi.org/10.1007/978-1-4939-7780-2_10,

degraded to permit mitosis [9]. The dephosphorylation step is associated with pharmacological amounts of genotoxic cancer treatments. Furthermore, inhibitors of Chk1, when applied to cells treated by genotoxic agents, induce a checkpoint adaptation-like state [10, 11]. Given the scientific and medical interest in genotoxicity and its effects upon genomic instability, it is therefore necessary to have a reliable and reproducible method to investigate checkpoint adaptation in human cells.

We describe a series of experimental techniques that can be used to determine whether cells are undergoing checkpoint adaptation. A key feature of the biology of checkpoint adaptation is the capacity of cells to acquire a rounded morphology as they enter mitosis [12]. This physical change can be easily observed and permits the isolation of mitotic cells from interphase cells; and it has been previously demonstrated that rounded cells are indeed in mitosis and not in apoptosis [4, 5]. Following observation of rounded cells, it is important to determine if these cells are undergoing the three steps of checkpoint adaptation. A convenient method to do this is by the collection of cells by mitotic shake-off, also known as mechanical shake-off, where rounded, weakly adherent mitotic cells are separated from flattened, strongly adherent interphase cells. We use HT-29 human colon carcinoma cells to study checkpoint adaptation because these cells display a pronounced rounded morphology when in mitosis [4, 5]. HT-29 cells also spend more time in mitosis following treatment as compared to other cell lines [13], and this enables them to be collected by mitotic shake-off. The distinctive rounded morphology of mitotic HT-29 cells that have undergone checkpoint adaptation makes it simple to observe them by light microscopy and confirm their isolation from the adherent interphase cells (Fig. 1).

Immunofluorescence microscopy is also a valuable method for studying checkpoint adaptation. It can be used to investigate which cells in a population are undergoing checkpoint adaptation and if cells that survive this process have micronuclei. Cells that undergo checkpoint adaptation have damaged DNA that can be detected by staining for the presence of histone γH2AX, a molecular marker of damaged DNA [14]. To investigate if the rounded cells observed by time-lapse video or light microscopy are mitotic, condensed chromosomes and phosphorylated Ser10 histone H3, a molecular marker of mitotic chromosome condensation [15], can be detected. The methods of Western blotting and enzyme activity assays can also be used to complement microscopy studies. The measure of cyclin B protein levels can be used to confirm that cells are competent to enter mitosis. Cyclin B is required for cells to enter mitosis, and cyclin B accumulates in the G2 phase of the cell cycle. In addition, Cdk1 enzyme is active in mitotic cells and not active in interphase cells, and these activities can be compared by enzyme assays; this assay has been previously described in this series [16]. Finally,

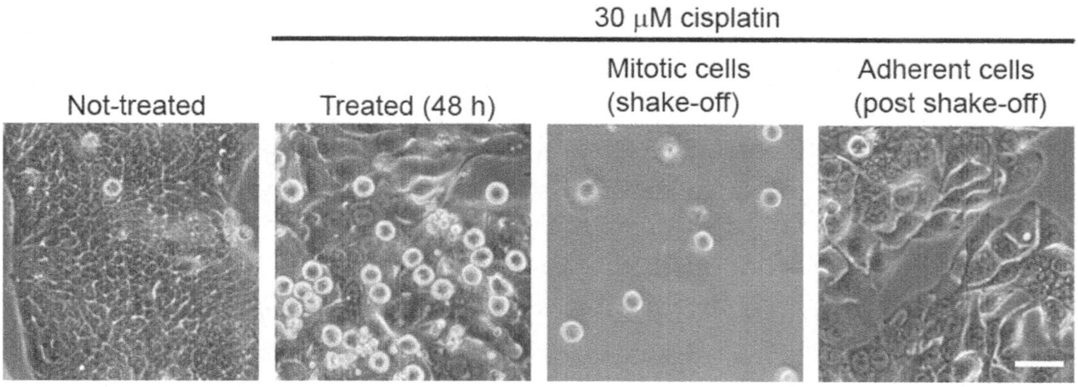

Fig. 1 HT-29 cells were either treated with dimethyl sulfoxide (DMSO), a solvent for cisplatin (non-treated) or treated with 30 μM cisplatin. At 48 h posttreatment (treated 48 h), images of live cells were taken by phase-contrast microscopy. The rounded cells are undergoing checkpoint adaptation. The rounded cells were collected by mitotic shake-off and transferred to a new flask (mitotic cells shake-off). The cells remaining after shake-off in the treated 48 h flask consisted predominantly of adherent, interphasic cells (adherent post shake-off). The white bar represents 25 μm

once it has been confirmed that cells are undergoing checkpoint adaptation, survival cells can be investigated for the presence of micronuclei, by observation of DNA and the nuclear lamina [6].

2 Materials

2.1 Culturing HT-29 Cells

1. HT-29 (ATCC HTB-38) human colon carcinoma cells.
2. RPMI 1640 media supplemented with 10% heat-inactivated fetal bovine serum (FBS).
3. Corning tissue culture-treated flasks.

2.2 Time-Lapse Video Microscopy

1. Dilution series of genotoxic agent.
2. Corning tissue culture plates suitable for imaging.
3. Microscope capable of running nonfluorescent time-lapse video imaging for 96 h.

2.3 Mitotic Shake-Off

1. HT-29 cells.
2. Corning tissue culture-treated T75 cm^2 flasks.
3. Phosphate-buffered saline (PBS): 137 mM NaCl, 3 mM KCl, 100 mM Na_2HPO_4.
4. RPMI 1640 media supplemented with 10% heat-inactivated fetal bovine serum (FBS).
5. Heavy but thin metal object, such as a metal wrench.

2.4 Immunofluorescence Microscopy

1. Sterile glass coverslips.
2. If using poly-L-lysine-coated coverslips, 0.1 mg/mL filter sterilized poly-L-lysine solution diluted in water. To coat sterile coverslips with sterile poly-L-lysine: In a flow hood, cover surface layer of each coverslip in 0.1 mg/mL poly-L-lysine solution, and let it stand for 10 min, remove excess liquid, and air-dry in flow hood.
3. Fine-tipped tweezers.
4. 6-well plates or 35 mm tissue culture dishes.
5. Phosphate-buffered saline (PBS): 137 mM NaCl, 3 mM KCl, 100 mM Na_2HPO_4.
6. 3% paraformaldehyde diluted in PBS, pH 7.
7. 50 mM NH_4Cl in PBS.
8. PBS containing 0.2% (v/v) Triton-X-100.
9. PBS-T: PBS containing 0.1% Tween 20.
10. 3% (w/v) bovine serum albumin (BSA) in PBS-T. Store at 4 °C and make fresh for each experiment.
11. A humidified chamber, such as a shallow tray or 10 cm petri dish, containing damp paper towel on the bottom with a sheet of parafilm over the top. The tray should be covered with a lid to prevent evaporation of antibody solution.
12. Antibodies: anti-phospho-Ser10 histone H3 (Cedarlane/Millipore, 06-570(CH)), anti-cyclin B (Santa Cruz, SC-752), anti-histone γH2AX (Cedarlane/Millipore, 05-636(CH)), anti-lamin A/C (Santa Cruz, SC-6215), anti-rabbit Alexa Fluor 594 (Invitrogen, R37117), anti-mouse Alexa Fluor 488 (Invitrogen, A11059), anti-goat Alexa Fluor 488 (Invitrogen, A110055).
13. 1 μg/mL DAPI (4′6-diamidino-2-phenylindole) in PBS.
14. Mounting agent (ProLong Gold Antifade reagent—Molecular probes, P36934).

3 Methods

3.1 Culturing HT-29 Cells

1. Culture HT-29 cells in RPMI 1640 medium supplemented with 10% (v/v) heat-inactivated FBS.
2. Grow cells at 37 °C in a humidified incubator with 5% CO_2.
3. Plate at 3.0×10^5 cells/T25 cm^2 flask and 1.0×10^6 cells/T75 cm^2 flask and incubate for 72 h prior to treatment (*see* **Note 1**).
4. Treat cells when they reach 50–60% confluency (*see* **Note 2**).

3.2 Time-Lapse Video Microscopy

1. Seed HT-29 cells in a 12-well plate (*see* **Note 3**) and incubate for 72 h.
2. Treat cells with a dilution series of genotoxic agent, beginning with a sub-cytotoxic concentration of agent and ending with a supra-cytotoxic concentration of agent. Include a vehicle only treatment as a negative control. Note that checkpoint adaptation occurs in cells treated with pharmacological concentrations of genotoxic agents. At supra-cytotoxic concentrations, cells may die by apoptosis [5].
3. Collect light images every 20 min for 96 h using a time-lapse video microscope with a 20× magnification objective.
4. Observe time-lapse video images to determine the concentration of genotoxic agent and treatment time that induces the maximum number of rounded (mitotic) cells (*see* **Note 4**).
5. Individual cells can be observed and manually scored for a rounded morphology, indicative of mitosis, between 0 and 96 h to quantify the percentage of the treated population undergoing checkpoint adaptation.

3.3 Mitotic Shake-Off

1. Seed HT-29 cells at 1.0×10^6 cells/T75 cm^2 flask 72 h prior to treatment. Multiple flasks may be needed per treatment if a large number of mitotic cells are required for further study. For most experiments two or three T75 flasks per treatment condition are sufficient.
2. Treat cells with the desired concentration of genotoxic agent for the appropriate length of time, as determined by observing cells using either time-lapse video or light microscopy. We use either 50 nM camptothecin or 30 μM cisplatin.
3. After treatment, gently remove media, wash cells gently with PBS, and add fresh media at 1 mL/25 cm^2.
4. Place flask on benchtop, tap all sides of the flask, along the bottom edge, with medium force using heavy metal object (such as the handle of a metal wrench) (*see* **Note 5**).
5. Observe cells under a microscope, and when mitotic cells have detached, gently collect the media from the flask.
6. It may be necessary to add fresh media and repeat this procedure to collect all of the mitotic cells.
7. Collected mitotic cells can be either fixed onto poly-L-lysine-coated coverslips and stained for immunofluorescence microscopy, lysed for use in Western blotting [4] or to detect Cdk1 kinase activity [17, 16], or cultured in new flasks with fresh media, to investigate cells that survive checkpoint adaptation.

3.4 Immunofluorescence Microscopy

1. For cells being treated on coverslips, plate 1×10^5 cells on sterile glass coverslips in 6-well plates and incubate for 72 h prior to treatment. For analysis of cells from mitotic shake-off, plate cells onto sterile coverslips coated with poly-L-lysine immediately after shake-off (*see* **Note 6**).
2. All steps should be at room temperature unless otherwise indicated. All wash steps should be for 5 min with 2 mL/well volume.
3. Gently remove media from wells and wash once with PBS.
4. Fix cells at room temperature for 20 min in 1 mL/well 3% paraformaldehyde diluted in PBS.
5. Wash once with PBS, and quench fixative with 1 mL/well 50 mM NH_4Cl in PBS for 10 min.
6. Wash twice with PBS and permeabilize cells with 1 mL/well 0.2% (v/v) Triton-X-100 in PBS for 5 min.
7. Wash three times with PBS-T (0.1% Tween 20).
8. Block cells with 2 mL/well 3% (w/v) BSA in PBS-T for 30 min.
9. Prepare a humidified chamber, as described in Subheading 2.4, **item 11**.
10. Prepare primary antibody in 3% BSA diluted in PBS-T (100 μL needed per 18 × 18 mm coverslip). For our work we used the conditions shown in Table 1.
11. Using fine-tipped tweezers, place coverslips, cell side facing up, onto parafilm in humidified chamber. Add primary antibody to cover the surface of the coverslip. Cover with a lid to prevent evaporation and incubate. Antibody dilution and incubation conditions will depend on the antibody in use.

Table 1
Primary antibodies used in immunofluorescence microscopy

Use to detect	Antibody	Dilution	Incubation instructions
Mitotic cells	Anti-phospho-Ser10 histone H3	1:1000	18 h 4 °C
Cells in G2/early M	Anti-cyclin B1	1:100	2 h Room temperature
Damaged DNA (*see* **Note 6**)	Anti-histone γH2AX	1:400	2 h Room temperature
Micronuclei	Anti-lamin A/C	1:150	2 h Room temperature

Table 2
Secondary antibodies used in immunofluorescence microscopy

Primary antibody	Secondary antibody	Dilution	Incubation instructions
Anti-phospho-Ser10 histone H3	Anti-rabbit Alexa Fluor 594	1:400	2 h Room temperature in dark
Anti-cyclin B1	Anti-rabbit Alexa Fluor 594	1:400	2 h Room temperature in dark
Anti-histone γH2AX (*see* **Note 7**)	Anti-mouse Alexa Fluor 488	1:400	2 h Room temperature in dark
Anti-lamin A/C	Anti-goat Alexa Fluor 488	1:150	2 h Room temperature in dark

12. Put coverslips back in wells of plate and wash three times with PBS-T.
13. Prepare secondary antibody in 3% BSA diluted in PBS-T (100 μL needed per 18 × 18 mm coverslip). For our work we used the conditions shown in Table 2. Keep prepared secondary antibody in the dark.
14. Using fine-tipped tweezers place coverslips, cell side facing up, onto parafilm in humidified chamber. Add secondary antibody, as per step 11. From this step on coverslips should be kept in the dark. We use aluminum foil to keep coverslips in the dark and limit the time coverslips are exposed to the light when performing washes.
15. Put coverslips back in wells of plate and wash once with PBS-T.
16. Add 1 mL/well 1 μg/mL DAPI in PBS and incubate for 15 min.
17. Wash three times with PBS-T.
18. Using fine-tipped tweezers, gently blot the edges of the coverslip with a tissue (Kimwipe) to remove excess liquid and slowly mount on a clean microscope slide using one drop of mounting agent, such as ProLong Gold, per coverslip (*see* **Note 7**).
19. Observe slides using a fluorescence microscope and capture images. Cells positive for staining can be quantified using image analysis software such as ImageJ (*see* **Note 8**).

4 Notes

1. HT-29 cells need more time than some other cell lines (such as HeLa or U2OS) to adhere to culture vessels. Cells should be left to adhere for at least 48 h before treatment, and the duration should be consistent for each experiment. We grow HT-29 cells on Corning tissue culture-treated flasks for optimal adherence. Seeding numbers for all protocols may need to be optimized so that cells are at optimum confluency (50–60%) at time of treatment.
2. The 50–60% confluency of cells provides maximum yield of mitotic cells. Cells that reach a high confluency cycle slower and fewer mitotic cells can be collected.
3. Cells can be seeded in plates with a different number of wells according to the number of treatments being added and microscope capabilities.
4. The non-treated population will have cells that continuously cycle, whereas cells that undergo checkpoint adaptation will begin to accumulate in mitosis at a certain treatment time (e.g., in the case of cisplatin-treated cells, 48 or 72 h posttreatment). If a time-lapse video microscope is not available, cells may be observed using a light microscope to determine the optimal concentration of genotoxic agent and treatment time to use in further experiments.
5. Using too much force to tap the sides of the flask will result in cracking it or may detach interphase cells. We tap the flasks along the bottom of the flask, close to the benchtop, where the plastic is stronger. The handle of a metal wrench is ideal because it is thin and so will place the force mostly along the bottom of the flask. The tool we use is 1 × 25 cm and weighs 175 g.
6. Treated cells do not undergo checkpoint adaptation synchronously, although we have observed 30% of a cell population in mitosis at one time. It is not known how prior synchronization would affect checkpoint adaptation, but it is noteworthy that some synchronization methods activate the DNA damage checkpoint. Furthermore, the majority of cells that undergo checkpoint adaptation will die at some point after collection, and so a large number of cells need to be plated to have adequate live cells. This number will depend on the treatment conditions prior to shake-off and how long the cells need to be cultured for after seeding and so should be determined experimentally. We suggest starting with 5.0×10^5 cells collected by mitotic shake-off. Cells collected by shake-off and seeded onto coverslips should be incubated until they adhere to or settle on the coverslips before fixing and staining (minimum 20 min).

7. It is important to minimize bubbles when adding mounting agent to the slide. If a large number of bubbles are introduced, wipe the slide clean and add a new drop of mounting agent. Lowering the coverslip onto the slide slowly also helps to prevent bubbles from being introduced.
8. We have observed that HT-29 cells display pan nuclear histone γH2AX staining following treatment with genotoxic agents [4, 5], as opposed to distinct histone γH2AX foci observed in some other cell lines.

Acknowledgments

Our research is funded by the Canada Foundation for Innovation (CFI) and the Alberta Innovates Sustainability Fund. We thank the members of the Cancer Cell Laboratory, the University of Lethbridge, and Dr. Aru Narendran (University of Calgary) for the support.

References

1. Toczyski DP, Galgoczy DJ, Hartwell LH (1997) CDC5 and CKII control adaptation to the yeast DNA damage checkpoint. Cell 90:1097–1106
2. Syljuåsen RG, Jensen S, Bartek J et al (2006) Adaptation to the ionizing radiation-induced G2 checkpoint occurs in human cells and depends on checkpoint kinase 1 and Polo-like kinase 1 kinases. Cancer Res 66:10253–10257
3. Rezacova M, Rudolfova G, Tichy A et al (2011) Accumulation of DNA damage and cell death after fractionated irradiation. Radiat Res 175:708–718
4. Kubara PM, Kernéis S, Studeny A et al (2012) Human cells enter mitosis with damaged DNA after treatment with pharmacological concentrations of genotoxic agents. Biochem J 446:373–381
5. Swift LH, Golsteyn RM (2016) Cytotoxic amounts of cisplatin induce either checkpoint adaptation or apoptosis in a concentration dependent manner in cancer cells. Biol Cell 108:127–148
6. Lewis CW, Golsteyn RM (2016) Cancer cells that survive checkpoint adaptation contain micronuclei that harbour damaged DNA. Cell Cycle 15:3131–3145
7. Crasta K, Ganem NJ, Dagher R et al (2012) DNA breaks and chromosome pulverization from errors in mitosis. Nature 482:53–58
8. Swift LH, Golsteyn RM (2014) Genotoxic anti-cancer agents and their relationship to DNA damage, mitosis, and checkpoint adaptation in proliferating cancer cells. Int J Mol Sci 15:3403–3431
9. Zhang YW, Otterness DM, Chiang GG et al (2005) Genotoxic stress targets human Chk1 for degradation by the ubiquitin-proteasome pathway. Mol Cell 19:607–618
10. Ferry G, Studény A, Bossard C et al (2011) Characterization of novel checkpoint kinase 1 inhibitors by in vitro assays and in human cancer cells treated with topoisomerase inhibitors. Life Sci 89:259–268
11. On KF, Chen Y, Ma HT et al (2011) Determinants of mitotic catastrophe on abrogation of the G2 DNA damage checkpoint by UCN-01. Mol Cancer Ther 10:784–794
12. Terasima T, Tolmach L (1963) Growth and nucleic acid synthesis in synchronously dividing populations of HeLa cells. Exp Cell Res 30:344–362
13. Gascoigne KE, Taylor SS (2008) Cancer cells display profound intra- and interline variation following prolonged exposure to antimitotic drugs. Cancer Cell 12:111–122
14. Mah LJ, El-Osta A, Karagiannis TC (2010) gammaH2AX: a sensitive molecular marker of DNA damage and repair. Leukemia 24:679–686

15. Hendzel MJ, Wei Y, Mancini MA et al (1997) Mitosis-specific phosphorylation of histone H3 initiates primarily within pericentromeric heterochromatin during G2 and spreads in an ordered fashion coincident with mitotic chromosome condensation. Chromosoma 106:348–360
16. Lewis CW, Taylor RG, Golsteyn RM (2016) Measurement of Cdk1/Cyclin B kinase activity by specific antibodies and western blotting. Methods Mol Biol 1342:337–348
17. Lewis CW, Taylor RG, Kubara PM et al (2013) A western blot assay to measure cyclin dependent kinase activity in cells or in vitro without the use of radioisotopes. FEBS Lett 587:3089–3095

Chapter 11

A Role for Retrotransposons in Chromothripsis

Dustin C. Hancks

Abstract

Chromothripsis is a mutational event driven by tens to hundreds of double-stranded DNA breaks which occur in a single event between a limited number of chromosomes. Following chromosomal shattering, DNA fragments are stitched together in a seemingly random manner resulting in complex genomic rearrangements including sequence shuffling, deletions, and inversions of varying size. This genomic catastrophe has been observed in cancer genomes and the genomes of patients harboring developmental and congenital defects. The mechanisms catalyzing DNA breakage and coordinating the "random" assembly of genomic fragments are actively being investigated. Recently, retrotransposons—a type of "jumping gene"—have been implicated as one means to generate double-stranded DNA breaks during chromothripsis and as sequences which can contribute to the final configuration of the derived chromosomes. In this methods chapter, I discuss how to apply available bioinformatic tools and the hallmarks of retrotransposon mobilization to breakpoint junctions to assess the role for active and inactive retrotransposon sequences in chromothriptic events.

Key words Chromothripsis, Retrotransposon, *Alu*, L1, LINE-1, SVA, Nonallelic homologous recombination, Deletion, Inversion

1 Introduction

It is well established that genetic mutations result in human disease including the initiation and progression of cancer. Thus, significant effort has been invested in characterizing the mechanisms and nature of genetic mutations in patients. Mutational analysis has been transformed over the past decade with the implementation of high-throughput next-generation DNA sequencing (NGS). Together with advances in computational biology, whole-genome sequencing (WGS) has become feasible and commonplace. Perhaps naturally, whole-genome sequencing analysis has been rapidly applied to cancers of various origins and stages to understand genetic factors shaping the biology of these malignancies.

An important finding came when characterizing paired-end sequencing data from the genome of a patient with chronic lymphocytic leukemia [1] with 42 somatically acquired rearrangements

Franck Pellestor (ed.), *Chromothripsis: Methods and Protocols*, Methods in Molecular Biology, vol. 1769,
https://doi.org/10.1007/978-1-4939-7780-2_11,

involving four chromosomes. Additional analysis of other cancer genomes using NGS identified similar mutational profiles. Specifically, novel complex rearrangements, which included sequence shuffling, inversions, and oscillation between two copy number states involving only a few chromosomes, indicated pulverization of DNA in a single event. Subsequently, this phenomenon was termed chromothripsis—meaning "chromosome shattering" [1].

To date, chromothripsis has now been observed in many types of cancers with some types of bone cancers being the most prevalent [1–3]. This complex class of rearrangements in cancer cells often involves tens to hundreds of chromosomal fragments dispersed across only a limited number of chromosomes (e.g., 2–4). Further investigations have identified similar genomic rearrangements in patients with congenital abnormalities [4, 5]. Notably, at least one instance has been reported where a stably segregating chromothripsis event was identified in healthy individuals [6, 7].

The cellular mechanisms driving chromothripsis and mediating DNA repair of damaged chromosomes are actively being investigated. Of particular interest are defects in chromosomal segregation which can result in the formation of structures termed micronuclei [8, 9]. Chromosomes partitioned into micronuclei often become highly fragmented and reassembled in a random manner due to defects in DNA replication. Likewise, hallmarks of chromothripsis have been observed in cells where telomeres have fused—a phenomenon known as telomere crisis—to generate dicentric chromosomes [10]. Along with these studies, one report identified a role for mobile DNA such as retrotransposons, also known as "jumping genes," as factors contributing to a germline chromothripsis event [7].

Retrotransposons are genetic elements widespread in nature and major components of most genomes (Fig. 1) [11]. In the human genome, retrotransposons are responsible for ~35% of the genomic real estate [12]. Long INterspersed Element-1 (LINE-1 or L1) is the only retrotransposon in humans able to duplicate

Fig. 1 (continued) transcription requires an upstream enhancer sequence and a downstream pol III terminator (TTTT). (**c**) SINE-VNTR-*Alu* (SVA) elements are active composite hominid-specific retrotransposons [41], >2700 copies in the human genome. Starting at the 5′-end, SVA elements consist of a hexameric repeat (CCCTCT$_n$), an *Alu*-like domain derived from two antisense *Alu* elements, a GC-rich VNTR, and a SINE-R domain derived from an extinct HERV-K retrovirus. SVA transcription can occur upstream or within the element. Upstream transcriptional initiation [42, 43] or transcriptional readthrough can generate chimeric transcripts which if mobilized will produce 5′- and 3′-transductions [37], respectively. (**d**) In addition to mobilizing retrotransposon RNAs, L1 can copy and paste cellular RNAs to new genomic locations to produce processed pseudogenes [44, 45] also known as retrogenes (~8000 in human genome [46]). Typically, processed pseudogene insertions resemble cDNA sequences (e.g., lack introns), terminate in a 3′-polyA tail, and are flanked by a target-site duplication. *UTR* untranslated region, *ORF* open-reading frame, *AA* amino acid, *kB* kilobase, *VNTR* variable number of tandem repeats, *env* envelope, *LTR* long terminal repeat. Not drawn to scale

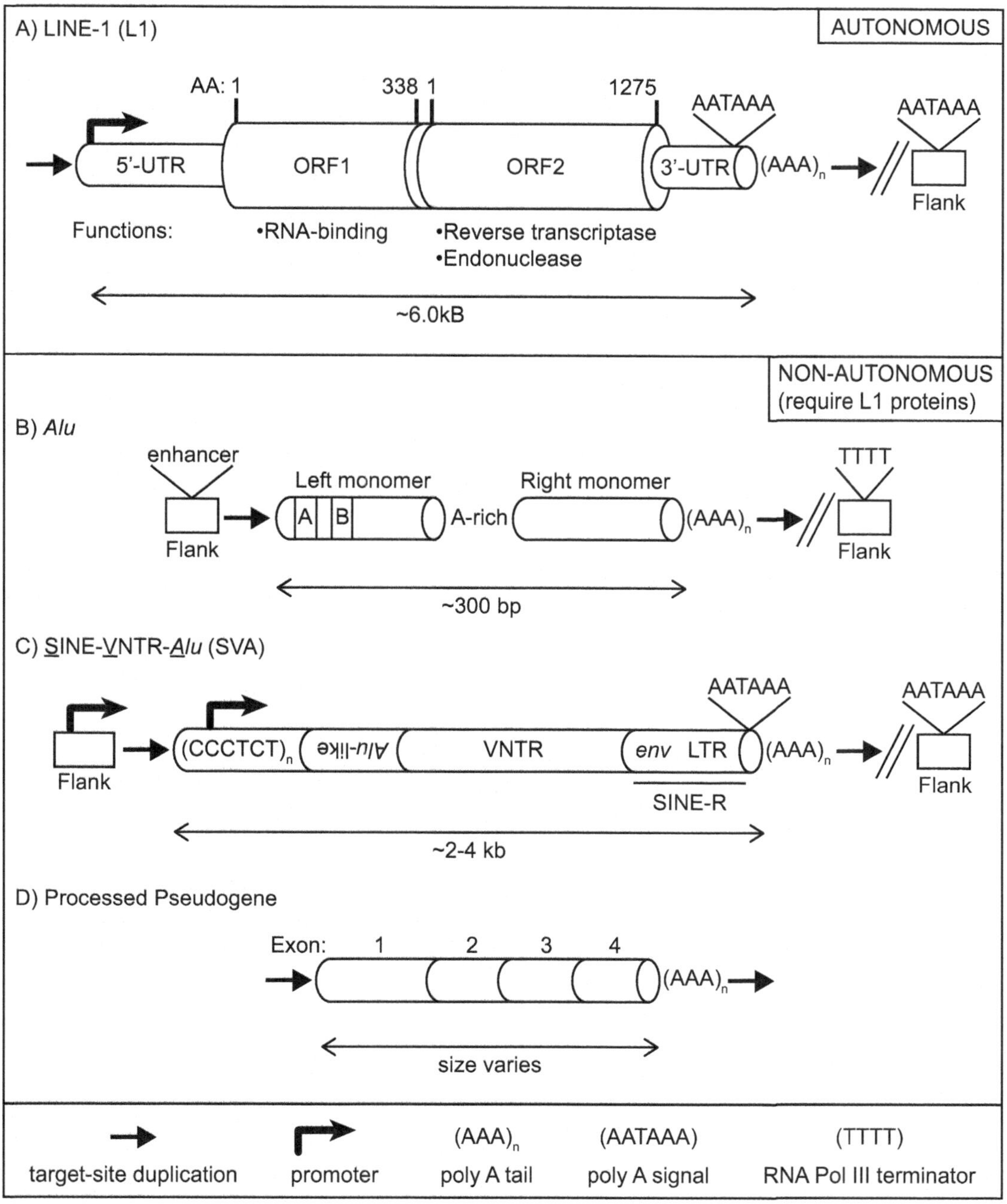

Fig. 1 Retrotransposons active in the human genome [11]. (**a**) Long INterspersed Element-*1* (LINE-1 or L1) is the only autonomous retrotransposon in the human genome (> 500,000 copies). Proteins encoded by L1 (ORF1p and ORF2p) are required for mobilization of its own RNA and other retrotransposon RNAs. L1 encodes its own promoter in the 5′-UTR (*black bent arrow*), a protein with RNA-binding activity (ORF1p), and a protein with reverse transcriptase and endonuclease activities (ORF2p). A full-length L1 terminates at its 3′-end in a polyA tail of variable length (AAA_n). Majority of L1-mediated retrotransposition events are flanked by a target-site duplication of varying length (4–20 bp, *black horizontal arrows*). Occasionally, transcriptional readthrough will occur by bypassing the weak polyA signal (AATAA) and terminating transcription at a downstream sequence in the flank generating a chimeric transcript. If mobilized, these chimeric transcripts generate 3′-transductions [36]. (**b**) *Alu* elements are ~300 bp active primate-specific Short INterspersed Elements (SINEs), > 1.2 million in the human genome [40]. *Alus* are RNA pol III transcripts which encode an (**a**) and (**b**) box. Efficient *Alu*

itself (autonomous) (Fig. 1a). Mobilization of L1 sequences occurs via an RNA intermediate—a process termed retrotransposition. A full-length active L1 encodes two proteins, both of which are required for retrotransposition *in cis* [13]. ORF1p displays RNA-binding activity, while ORF2p encodes reverse transcriptase [14] and DNA endonuclease (EN) [15] activities. Along with its own RNA, L1 ORF2p is able to copy other RNAs, such as *Alu* elements (Fig. 1b), hominid-specific SVA elements (Fig. 1c), and even protein-coding RNAs (e.g., processed pseudogenes) (Fig. 1d), to new genomic locations by a coupled integration reverse-transcription mechanism [16] (Fig. 2a). Briefly, DNA is cleaved at an L1 EN target sequence (5′-TTTT/AA-3′) to produce a free 3′-OH which is subsequently used by ORF2p to prime first-strand cDNA synthesis. Along with contributing to genome expansion through RNA-based duplication, it is well established that due to their high copy numbers, retrotransposons can result in nonallelic homologous recombination (NAHR) to generate indels or genetic inversions [17, 18].

Analysis of a stably segregating germline chromothripsis event [6, 7] identified the insertion of a 5′-truncated SVA element at one of the breakpoint junctions (BPJs) into a sequence resembling the L1 EN consensus cleavage site (5′-TTTT/AA-3′). This insertion was associated with a 110 kb deletion in the derived chromosome 5′- of the SVA and attached to a distal DNA fragment located ~4.7 megabases away relative to the reference genome. Notably, the 5′-end of the SVA insertion displayed microhomology with sequence at the BPJ of the distal fragment. Previous studies have indicated that reverse transcription by ORF2p of RNAs like SVA may function as "molecular Band-Aids" at double-stranded DNA (dsDNA) breaks [19] and that L1-mediated insertions are often associated with large genomic deletions (Fig. 2b) [20, 21]. These data suggest a model where ORF2p EN activity can be a source of DNA breaks and may bridge distal DNA fragments brought in close proximity presumably through chromatin looping during chromothripsis [7].

Further characterization of this mutational event identified two pairs of *Alu* elements flanking four breakpoint junctions, one of which was associated with the SVA insertion. Specifically, in the reference genome, there were two *Alu* elements in the same orientation flanking the 110 kb sequence deleted on the derived chromosome. In the second instance, there were two *Alu* elements on the same chromosome in opposite orientation relative to each other downstream of the 110 kb deletion flanking a 620 kb sequence inverted during chromothripsis. Both *Alu*-mediated deletions (Fig. 3a) and inversions (Fig. 3b) have been

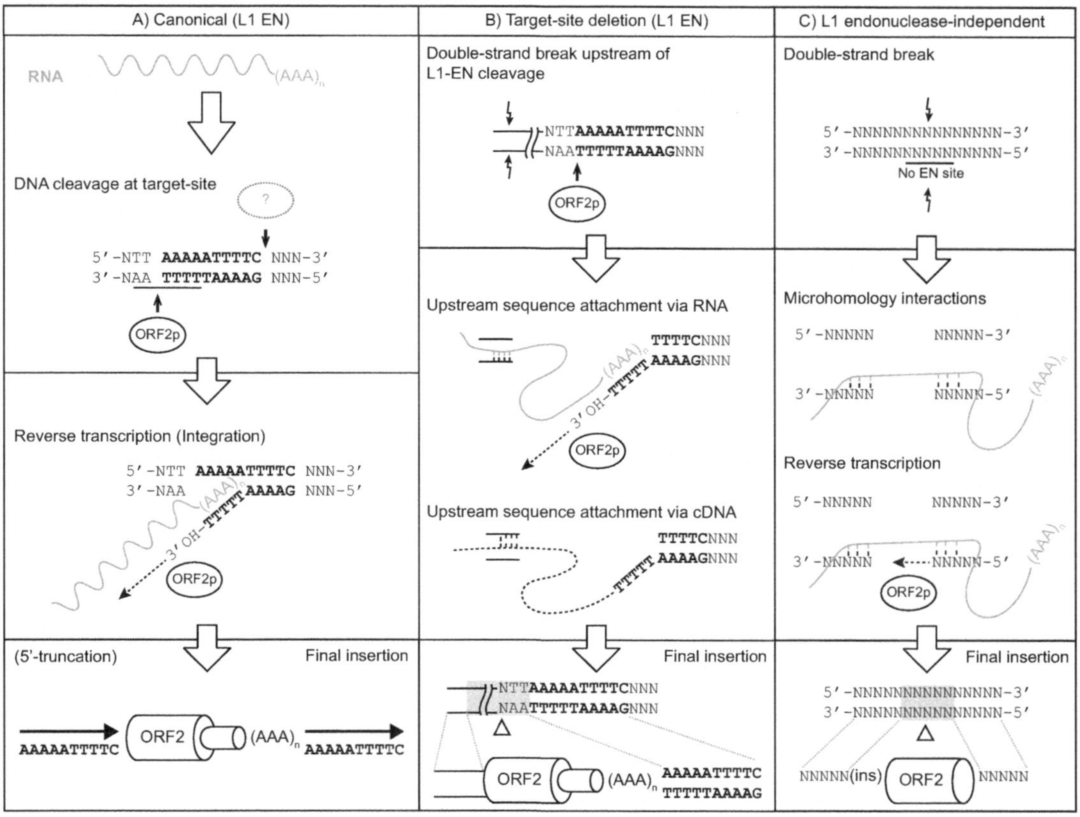

Fig. 2 Modes of L1-mediated retrotransposition. (**a**) Majority of insertions occur by a coupled integration and reverse-transcription mechanism termed target-primed reverse transcription (TPRT) [16]. Specifically, L1 ORF2p cleaves the bottom strand of its target sequence (5′-TTTT/AA-3′) followed by reverse transcription of L1 RNA (*gray line*) using the liberated 3′-OH. The mechanism of top-strand cleavage and resolution of the insertion are incomplete but likely involve cellular factors including DNA repair proteins. Briefly, the insertion is characterized by a target-site duplication (*horizontal black arrows*), 3′-polyA tail, and in this instance 5′-truncation. Most genomic L1 insertions (99%) are highly truncated at their 5′-end. (**b**) L1 EN-dependent target-site deletions. Although this pathway also requires L1 EN activity, the major difference is that a double-strand break occurs upstream of L1 EN bottom-strand cleavage. This DSB ultimately results in the loss of DNA sequence 5′- of the insertion. It is hypothesized that microhomology interactions either by the RNA or nascent first-strand cDNA stabilize this intermediate. These insertions are characterized by deletions 5′- of the target-site (gray box), L1 EN cleavage site, frequent 5′-truncations, and a 3′-polyA tail. (**c**) L1 EN-independent insertions are an alternative means by which retrotransposon sequences can be deposited into the genome. Specifically, it has been reported that when double-strand breaks occur in DNA, retrotransposon RNAs can function as molecular Band-Aids by bridging DNA fragments through microhomology interactions. As in in the other pathways, L1 ORF2p reverse transcribes the RNA into cDNA at the DNA break. These insertions are characterized by integration at sites sharing no resemblance to L1 EN cleavage site, target-site deletions, 5′- and 3′-truncations of the retrotransposon sequence including lack of a 3′-polyA tail, and untemplated nucleotides (ins) at the junctions of the insertion. cDNA—complementary DNA. Not drawn to scale

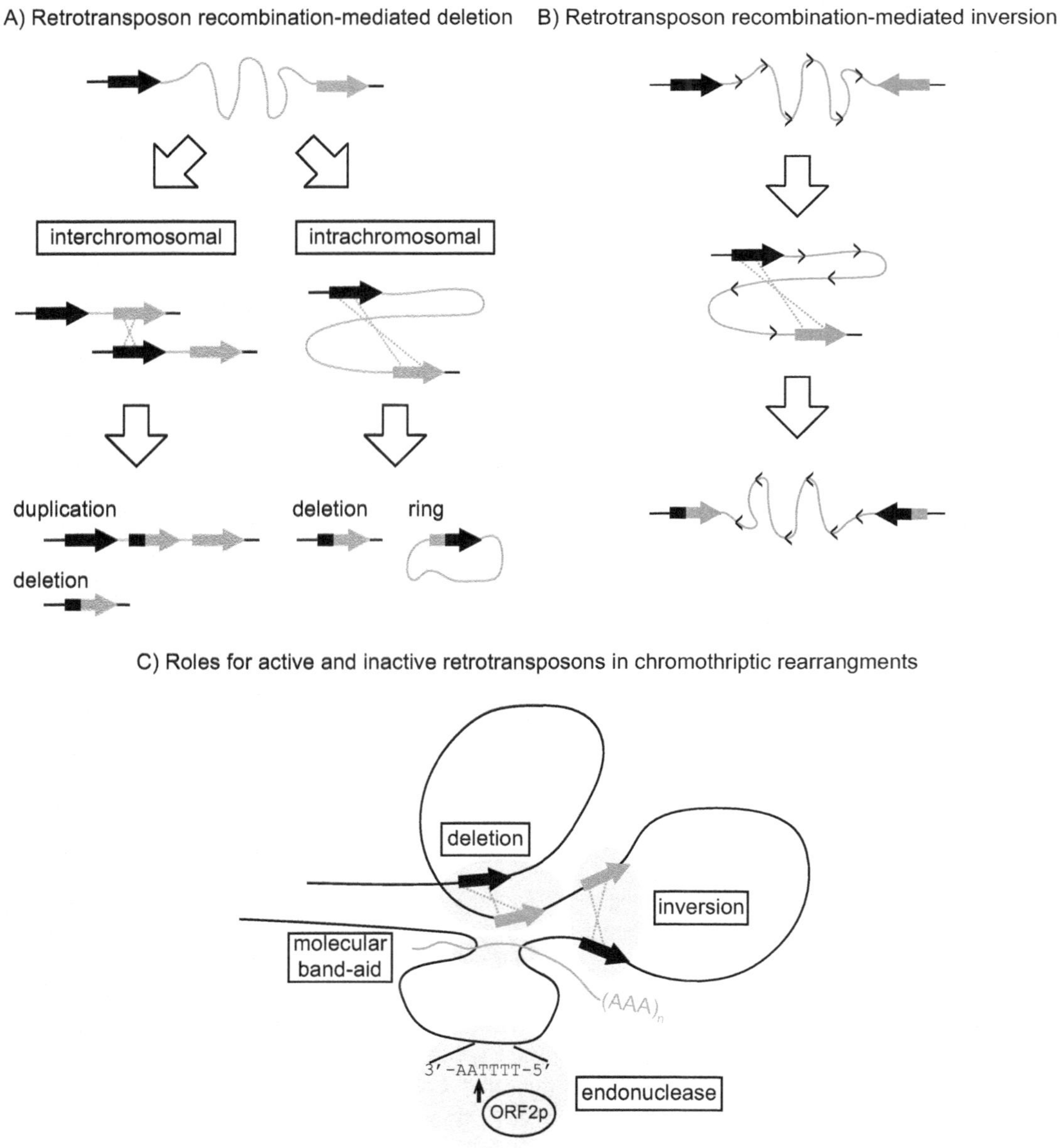

Fig. 3 Genomic rearrangements mediated by genomic retrotransposons. The rearrangements described here can be produced, in principle, by any genomic repeat. (**a**) *R*etrotransposon *r*ecombination-*m*ediated *d*eletions (RRMDs) can transpire due to nonallelic homologous recombination. Mispairing of nonhomologous repetitive elements in the same orientation, such as *Alus* (*black and gray arrows*), between homologous chromosomes (interchromosomal, *left*), can generate a duplication on one chromosome and a deletion on the other, while mispairing between repetitive elements on the same chromosome (intrachromosomal, *right*) can result in a genomic deletion and a ring fragment. RRMDs have resulted in human genetic disease [17] and contributed significantly to genome evolution [22]. A hallmark of recombination between repetitive elements is the production of a chimeric element (*half-black, half-gray arrow*) [40]. (**b**) Retrotransposon recombination-mediated inversions can occur when repetitive elements such as *Alus* in opposite orientation mispair followed by recombination [23]. Recombination results in the generation of chimeric elements (*half-black, half-gray arrows*) and an inversion of the intervening sequencing (reversal of *black arrowheads* on *gray line*). (**c**) Model for how retrotransposons may contribute to chromothripsis. NAHR between retrotransposons (*black and gray arrows*) or other repetitive elements can produce genomic deletions or inversions. Retrotransposon RNAs can serve as a means to bridge distal fragments via microhomology interactions. L1 ORF2p is a source of DNA endonuclease activity. Not drawn to scale

characterized as mechanisms contributing to primate genome evolution [22, 23], and in some cases, *Alu*-mediated rearrangements have resulted in single-gene disease [17]. These data highlight how *Alu* elements in the genome can contribute to rearrangements in germline chromothripsis events [7].

Perhaps the most striking about this case of chromothripsis was that at some of the BPJs, including the distal BPJ which the 5′-end of the SVA element was ultimately attached to, sequences resembling the L1 EN cleavage site were identified. Together these data indicate a model by which active and inactive retrotransposons can contribute to chromothriptic rearrangements (Fig. 3c): (1) generation of dsDNA breaks, (2) orientation of DNA fragments to impact derived chromosome structure, (3) function as molecular Band-Aids, and (4) bridge distal DNA fragments. Interestingly, several studies indicate that retrotransposons, such as L1, are active where chromothripsis is known to occur, namely, cancer and cells responsible for producing germ cells (e.g., embryonic stem cells) [24, 25]. Based on these observations, methods are described below to aid in the characterization of retrotransposon sequences in chromothriptic events.

2 Materials

The methods here begin making the assumption that the breakpoint junctions from the DNA sequence analysis have been identified. Briefly, the UCSC genome browser (genome.ucsc.edu) [26] is a powerful tool to analyze breakpoint junctions. In addition, the analysis outlined below involves manual inspection of breakpoint sequences complemented by the UCSC genome browser for visualization of genomic features such as repetitive elements like *Alu* sequences.

2.1 Breakpoint Junction Sequences

These data should be obtained by aligning sequencing data to the reference genome sequence (e.g., hg19). Several computational tools have been developed to aid in the characterization of retrotransposon insertions (reviewed in [27]).

2.2 RepeatMasker (www.repeatmasker.org)

RepeatMasker [28] is an analysis program freely available online. It contains libraries and annotations for known repetitive elements including retrotransposon sequences. RepeatMasker is frequently used by transposon researchers and in the annotation of transposable elements during genome analysis. RepeatMasker is able to identify different types of retrotransposons along with the subfamily which they belong to. Notably, only certain retrotransposon sequences, such as *Alu*Ya5 and L1-Hs, are active (e.g., retrotransposition competent) in the human genome; however, the ability to mitigate nonallelic homologous recombination by a retrotransposon is not known to be restricted to specific subfamilies.

2.3 UCSC Genome Browser (genome.ucsc.edu)

The UCSC genome browser [26] is a powerful tool to easily visualize a variety of genomic features including repetitive sequences like retrotransposons.

1. To display annotations for repetitive DNA, from the browser page, scroll below the browser window to the section labeled "Repeats" and select "full" from the drop-down heading below the RepeatMasker heading. This selection will add a "track" on the browser displaying different repeat families (e.g., LINE, SINE). The color of each annotated repeat (gray-scaled box) corresponds to the similarity of this repeat to the known RepeatMasker consensus.
2. To obtain information pertaining to specific repetitive elements, select the repeat (gray box). This will display a subsequent page and element specific information including subfamily, sequence length of element, and element orientation (e.g., positive strand). The orientation of the element is useful in determining whether sequences such as a pair of *Alu* elements may have contributed to the rearrangement resulting in either an indel or genetic inversion.

3 Methods

3.1 Characterization of Retrotransposon Element Insertions

1. If insertions of notable size are identified at breakpoint junctions in a chromothriptic genome, these sequences should be screened using RepeatMasker to determine their identity including type and subfamily. Indeed, not all insertions identified at breakpoint junctions may be derived from retrotransposon sequences. While insertion of a small number of nucleotides is likely due to cellular DNA repair machinery, some retrotransposon sequences can be very short in length (e.g., 46 bp) or may only consist of a homopolymer of adenines derived from the polyA tail from the mRNA (*see* **Note 1**).
2. Manually characterize insertions for hallmarks of L1-mediated retrotransposition (Fig. 2). Retrotransposon insertions can be broadly grouped into two classes [11, 29]: (1) L1 endonuclease-dependent (Fig. 2a) and (2) L1 endonuclease-independent (ENi) (Fig. 2c) [19, 30] (*see* **Note 2**). The most common hallmarks of EN-dependent insertions include:
   (a) Insertion at a sequence resembling the L1 EN site, polyA tail of variable length at the 3′-end of the insertion, most insertions being 5′-truncated (e.g., lacking sequence at the 5′-end relative to the full-length retrotransposon consensus sequence), target-site duplication of varying length (4–20 bp) directly flanking the insertion, 3′-transductions (e.g., duplication of downstream sequence due to tran-

scriptional readthrough) (*see* **Note 3**), and occasionally untemplated nucleotides at the 5′-end of the insertion

3. If a retrotransposon sequence is present at a BPJ but does not display hallmarks of L1 EN-dependent insertions, the insertion may have occurred in an L1 EN-independent manner (Fig. 2c) (*see* **Notes 4** and **5**). In the context of chromothripsis, this class of insertions has the potential to serve as a bridge through microhomology-mediated interactions to link fragmented DNA. Common hallmarks of L1 ENi insertions include:
   (a) Truncation at both the 5′- and 3′-ends, lack of polyA tail, and insertion of untemplated nucleotides 5′- and 3′- of the insertion

3.2 Characterization of L1 Endonuclease Cleavage Sites

1. Assuming no additional mutations have occurred, the easiest means to identify the L1 EN cleavage site is to inspect the 5′-target-site duplication (Fig. 2a). For example, if the following bold sequence 5′-**TTAAAACTG**-3′ is the 5′-TSD in the same orientation as the L1-mediated insertion, then the L1 EN cleavage site was (relative to the bottom strand, underlined) 5′-TTTT/AA-3′ where the "/" indicates the insertion site (*see* **Notes 6** and **7**).

3.3 Characterization of the Contribution of Inactive Retrotransposons in Chromothripsis

Chromothripsis is characterized by rearrangements, e.g., inversions and deletions involving numerous DNA fragments [31]. It has been established that direct repeats such as two *Alu* elements on different chromosomes or the same one may mediate nonallelic homologous recombination resulting in sequence deletion, duplication, or inversion (Fig. 3) [17, 18].

1. To identify similar events at breakpoint junctions, go to the BPJ on UCSC genome browser with the RepeatMasker track on.
2. Look for two of the same type of repetitive elements (e.g., *Alu* elements) immediately adjacent to the BPJ (*see* **Note 8**). In the case of a deletion or inversion in the derived chromosome, one expectation if repetitive sequences were involved in NAHR is that you are able to identify two repeats of the same family flanking the BPJ either in the same or opposite orientation (*see* **Note 9** and **10**).
3. If repetitive (retrotransposon) elements are identified in the reference genome flanking a deletion, duplication, or inversion, inspect the sequence element in the derived chromosome for chimeric nature—a hallmark of recombination—using nucleotide sequence alignment (*see* **Note 8**).

4 Notes

1. The short insertion size of retrotransposons is due to their inherent nature to become truncated—primarily at their 5′-end—during integration. For example, an *Alu* insertion was identified in the *SERPINC1* gene associated with type 1 antithrombin deficiency only six base pairs long with a 40 bp polyA tail [29, 32]. Similarly, insertions containing only a homopolymer stretch of adenines (e.g., mRNA polyA tail) and lacking any known retrotransposon sequence have also been characterized [33].
2. Majority of insertions are L1 EN dependent (Fig. 2a) and integrate by a mechanism termed target-primed reverse transcription (TPRT) [16]. Briefly, ORF2p cleaves the bottom strand of DNA at a sequence resembling its target site (5′-YYYY/RR-3′, where Y = pyrimidine and R = purine with 5′-TTTT/AA-3′ being most frequent) [15, 34, 35]. Cleavage liberates a 3′-OH which is used to prime first-strand cDNA synthesis by ORF2p. Currently, the mechanism of top-strand DNA cleavage and 5′-attachment of the cDNA to upstream DNA are incomplete. The resolution of the insertion presumably involves cellular DNA repair machinery.
3. Weak polyA signals encoded by L1 and SVA often promote transcriptional readthrough by RNA polymerase in turn generating "chimeric" RNAs consisting of the retrotransposon sequence, a polyA tail, 3′-transduced downstream sequence, and another polyA tail (Fig. 1) [36, 37].
4. L1 ENi insertions (Fig. 2c) are less common and were first identified in cell lines lacking key DNA repair proteins using engineered L1 elements containing an inactivating mutation in the L1 endonuclease domain of ORF2p [19]. L1 ENi insertions have since been identified and characterized in detail in the human genome [30].
5. Another sequence feature occasionally observed for both L1 EN and ENi insertions is deletion of genomic sequence at the insertion site also known as target-site deletions (Fig. 2b). For example, as noted above, a large deletion of 110 kb occurred 5′- of the inserted SVA element in the case of Nazaryan et al. [7]. For L1 EN insertions associated with deletions, a sequence resembling the L1 EN consensus cleavage site (5′-TTTT/AA-3′) should be identifiable at the pre-insertion locus, also commonly referred to as the empty site.
6. It has long been hypothesized that if left unrestrained, ORF2p EN can be deleterious to the cell [38]. Although the L1 EN is thought to cleave only its target sequence on the bottom strand

during classical TPRT, evidence from insertions resulting in single-gene disease lacking TSDs but inserting at L1 EN sites and Nazaryan et al. [7] suggest that ORF2p can cleave the top-strand following bottom-strand cleavage at sequences resembling the L1 EN cleavage consensus sequence.

7. The canonical L1 cleavage site is 5′-TTTT/AA-3′ with almost all variations displaying similarity to 5′-YYYY/RR-3′. To date, several datasets including those characterizing L1-mediated insertions using next-generation sequencing have been published (reviewed in [27]). Gilbert et al. (2005) [34] contains a list of 100 characterized insertions produced by engineered L1s in cell culture. Hancks and Kazazian [29] contains an up-to-date list of over 120 published insertions resulting in single-gene disease in humans and their hallmarks of retrotransposition.
8. Retrotransposon sequences including *Alu* elements are very common in the human genome. Repeat elements involved in a NAHR event must be directly at the breakpoint and not "nearby." Due to their abundance, many repeats by chance will likely be close to a breakpoint. Further indication that a repetitive (retrotransposon) element was involved in a NAHR event is the presence of either a solo chimeric element in the case of a deletion or duplication or two chimeric elements in the case of an inversion in the derived chromosomes.
9. Along with L1s, *Alus*, and SVA elements, endogenous retroviruses (ERVs), which comprise ~8% of the human genome, are also frequent mediators of NAHR [39]. For instance, most ERVs in the genome are present as solo long terminal repeats (LTRs). Considering that the majority of ERV insertions initially are full length (LTR-gag-pol-env-LTR), these solo LTRs are evidence that ERVs frequently undergo NAHR presumably to reduce the burden of these selfish genetic elements.
10. While high-sequence identity is important to facilitate NAHR between two repetitive sequences, it is not absolutely required. For example, in Nazaryan-Petersen et al. [7], the inversion was mediated by two *Alu* elements from two distantly related subfamilies *AluSq* and *AluJb*, which share ~77% nucleotide identity by BLASTn analysis.

Acknowledgments

D.C.H. is funded by a K99/R00 Pathway to Independence Award from the National Institutes of Health (USA, NIGMS) and the Cancer Prevention & Research Institute of Texas (CPRIT).

References

1. Stephens PJ, Greenman CD, Fu B et al (2011) Massive genomic rearrangement acquired in a single catastrophic event during cancer development. Cell 144:27–40. https://doi.org/10.1016/j.cell.2010.11.055
2. Forment JV, Kaidi A, Jackson SP (2012) Chromothripsis and cancer: causes and consequences of chromosome shattering. Nat Rev Cancer 12:663–670. https://doi.org/10.1038/nrc3352
3. Storchová Z, Kloosterman WP (2016) The genomic characteristics and cellular origin of chromothripsis. Curr Opin Cell Biol 40:106–113. https://doi.org/10.1016/j.ceb.2016.03.003
4. Kloosterman WP, Guryev V, van Roosmalen M et al (2011) Chromothripsis as a mechanism driving complex de novo structural rearrangements in the germline. Hum Mol Genet 20:1916–1924
5. Kloosterman WP, Cuppen E (2013) Chromothripsis in congenital disorders and cancer: similarities and differences. Curr Opin Cell Biol 25:341–348. https://doi.org/10.1016/j.ceb.2013.02.008
6. Bertelsen B, Nazaryan-Petersen L, Sun W et al (2016) A germline chromothripsis event stably segregating in 11 individuals through three generations. Genet Med 18:494–500. https://doi.org/10.1038/gim.2015.112
7. Nazaryan-Petersen L, Bertelsen B, Bak M et al (2016) Germline chromothripsis driven by L1-mediated retrotransposition and Alu/Alu homologous recombination. Hum Mutat 37:385–395. https://doi.org/10.1002/humu.22953
8. Crasta K, Ganem NJ, Dagher R et al (2012) DNA breaks and chromosome pulverization from errors in mitosis. Nature 482:53–58. https://doi.org/10.1038/nature10802
9. Zhang C-Z, Spektor A, Cornils H et al (2015) Chromothripsis from DNA damage in micronuclei. Nature 522:179–184. https://doi.org/10.1038/nature14493
10. Maciejowski J, Li Y, Bosco N et al (2015) Chromothripsis and kataegis induced by telomere crisis. Cell 163:1641–1654. https://doi.org/10.1016/j.cell.2015.11.054
11. Richardson SR, Doucet AJ, Kopera HC et al (2015) The influence of LINE-1 and SINE retrotransposons on mammalian genomes. Microbiol Spectr 3:MDNA3–0061–2014. https://doi.org/10.1128/microbiolspec.MDNA3-0061-2014
12. Kazazian HH Jr, Moran JV (2017) Mobile DNA in health and disease. N Engl J Med 377:361–370. https://doi.org/10.1056/NEJMra1510092
13. Moran JV, Holmes SE, Naas TP et al (1996) High frequency retrotransposition in cultured mammalian cells. Cell 87:917–927. https://doi.org/10.1016/S0092-8674(00)81998-4
14. Mathias SL, Scott AF, Kazazian HH et al (1991) Reverse transcriptase encoded by a human transposable element. Science 254:1808–1810
15. Feng Q, Moran JV, Kazazian HH Jr et al (1996) Human L1 retrotransposon encodes a conserved endonuclease required for retrotransposition. Cell 87:905–916. https://doi.org/10.1016/S0092-8674(00)81997-2
16. Luan DD, Korman MH, Jakubczak JL et al (1993) Reverse transcription of R2Bm RNA is primed by a nick at the chromosomal target site: a mechanism for non-LTR retrotransposition. Cell 72:595–605. https://doi.org/10.1016/0092-8674(93)90078-5
17. Deininger PL, Batzer MA (1999) Alu repeats and human disease. Mol Genet Metab 67:183–193
18. Cordaux R, Batzer MA (2009) The impact of retrotransposons on human genome evolution. Nat Rev Genet 10:691–703. https://doi.org/10.1038/nrg2640
19. Morrish TA, Gilbert N, Myers JS et al (2002) DNA repair mediated by endonuclease-independent LINE-1 retrotransposition. Nat Genet 31:159–165. https://doi.org/10.1038/ng898
20. Symer DE, Connelly C, Szak ST et al (2002) Human L1 retrotransposition is associated with genetic instability in vivo. Cell 110:327–338. https://doi.org/10.1016/S0092-8674(02)00839-5
21. Gilbert N, Lutz-Prigge S, Moran JV (2002) Genomic deletions created upon LINE-1 retrotransposition. Cell 110:315–325. https://doi.org/10.1016/S0092-8674(02)00828-0
22. Han K, Lee J, Meyer TJ et al (2007) Alu recombination-mediated structural deletions in the chimpanzee genome. PLoS Genet 3:1939–1949. https://doi.org/10.1371/journal.pgen.0030184
23. Lee J, Han K, Meyer TJ et al (2008) Chromosomal inversions between human and chimpanzee lineages caused by retrotransposons. PLoS ONE 3:e4047. https://doi.org/10.1371/journal.pone.0004047
24. Burns KH (2017) Transposable elements in cancer. Nat Rev Cancer 17:415–424. https://doi.org/10.1038/nrc.2017.35

25. Faulkner GJ, Garcia-Perez JL (2017) L1 Mosaicism in mammals: extent, effects, and evolution. Trends Genet. https://doi.org/10.1016/j.tig.2017.07.004
26. Kent WJ, Sugnet CW, Furey TS et al (2002) The human genome browser at UCSC. Genome Res 12:996–1006
27. Ewing AD (2015) Transposable element detection from whole genome sequence data. BMC Mobile DNA 6:24. https://doi.org/10.1186/s13100-015-0055-3
28. Smit, AFA, Hubley, R & Green, P. RepeatMasker Open-4.0.2013-2015. http://www.repeatmasker.org
29. Hancks DC, Kazazian HH (2016) Roles for retrotransposon insertions in human disease. Mob DNA 7:9. https://doi.org/10.1186/s13100-016-0065-9
30. Sen SK, Huang CT, Han K et al (2007) Endonuclease-independent insertion provides an alternative pathway for L1 retrotransposition in the human genome. Nucleic Acids Res 35:3741–3751. https://doi.org/10.1093/nar/gkm317
31. Zhang CZ, Leibowitz ML, Pellman D (2013) Chromothripsis and beyond: rapid genome evolution from complex chromosomal rearrangements. Genes Dev 27:2513–2530. https://doi.org/10.1101/gad.229559.113
32. Beauchamp NJ, Makris M, Preston FE et al (2000) Major structural defects in the antithrombin gene in four families with type I antithrombin deficiency—partial/complete deletions and rearrangement of the antithrombin gene. Thromb Haemost 83:715–721
33. Chen J-M, Stenson PD, Cooper DN et al (2005) A systematic analysis of LINE-1 endonuclease-dependent retrotranspositional events causing human genetic disease. Hum Genet 117:411–427. https://doi.org/10.1007/s00439-005-1321-0
34. Gilbert N, Lutz S, Morrish TA et al (2005) Multiple fates of L1 retrotransposition intermediates in cultured human cells. Mol Cell Biol 25:7780–7795. https://doi.org/10.1128/MCB.25.17.7780-7795.2005
35. Hancks DC, Kazazian HH Jr (2012) Active human retrotransposons: variation and disease. Curr Opin Genet Dev 22:191–203. https://doi.org/10.1016/j.gde.2012.02.006
36. Moran JV, DeBerardinis RJ, Kazazian HH (1999) Exon shuffling by L1 retrotransposition. Science 283:1530–1534
37. Xing J, Wang H, Belancio VP et al (2006) Emergence of primate genes by retrotransposonmediated sequence transduction. Proc Natl Acad Sci U S A 103:17608–17613. https://doi.org/10.1073/pnas.0603224103
38. Gasior SL, Wakeman TP, Xu B et al (2006) The human LINE-1 retrotransposon creates DNA double-strand breaks. J Mol Biol 357:1383–1393. https://doi.org/10.1016/j.jmb.2006.01.089
39. Mager DL, Stoye JP (2015) Mammalian endogenous retroviruses. Microbiol Spectr 3:MDNA3–0009-2014. https://doi.org/10.1128/microbiolspec.MDNA3-0009-2014
40. Batzer MA, Deininger PL (2002) Alu repeats and human genomic diversity. Nat Rev Genet 3:370–379. https://doi.org/10.1038/nrg798
41. Hancks DC, Kazazian HH (2010) SVA retrotransposons: evolution and genetic instability. Semin Cancer Biol 20:234–245. https://doi.org/10.1016/j.semcancer.2010.04.001
42. Hancks DC, Ewing AD, Chen JE et al (2009) Exon-trapping mediated by the human retrotransposon SVA. Genome Res 19:1983–1991. https://doi.org/10.1101/gr.093153.109
43. Damert A, Raiz J, Horn AV et al (2009) 5′-transducing SVA retrotransposon groups spread efficiently throughout the human genome. Genome Res 19:1992–2008. https://doi.org/10.1101/gr.093435.109
44. Wei W, Gilbert N, Ooi SL et al (2001) Human L1 retrotransposition: cis preference versus trans complementation. Mol Cell Biol 21:1429–1439. https://doi.org/10.1128/MCB.21.4.1429-1439.2001
45. Esnault C, Maestre J, Heidmann T (2000) Human LINE retrotransposons generate processed pseudogenes. Nat Genet 24:363–367. https://doi.org/10.1038/74184
46. Zhang Z, Harrison PM, Liu Y, Gerstein M (2003) Millions of years of evolution preserved: a comprehensive catalog of the processed pseudogenes in the human genome. Genome Res 13:2541–2558

Chapter 12

Generation of Micronuclei and Detection of Chromosome Pulverization

Monique N. H. Luijten, Jeannie X. T. Lee, Sixun Chen, and Karen C. Crasta

Abstract

Lagging chromosomes that arise after chromosome mis-segregation during cell division can be encapsulated within small structures known as micronuclei. A link between whole-chromosome mis-segregation and chromothripsis has been demonstrated via micronuclear chromosome pulverization. Here, we describe methods to efficiently generate micronuclei and examine downstream cell fates, specifically with regard to DNA damage and chromosome pulverization.

Key words Micronuclei, Chromosome mis-segregation, Aneuploidy, Mitosis, DNA breaks, Chromothripsis

1 Introduction

Cancer genomes are characterized by extensive alterations, including point mutations and chromosome rearrangements, which are generally considered to be the result of several independent or progressive events [1]. However, in recent years, cancer genome analyses have revealed an additional phenomenon whereby massive genomic rearrangement occurs in one single event, termed chromothripsis (from the Greek "chromo" from chromosomes and "thripsis" for shattering into pieces). Chromothriptic genomic profiles show extensive genomic rearrangements and small-scale DNA copy number changes, all restricted to a single chromosome or chromosome arm [2, 3]. Chromothripsis has been detected in (advanced) tumors and could potentially contribute to tumorigenesis through loss of tumor suppressor function, production of oncogenic chromosomal fusions, and/or formation of high copy double minute chromosomes containing oncogenes [4]. We have shown that chromothripsis can potentially be mediated through the physical isolation of single or few chromosomes into aberrant nuclear structures known as micronuclei [5].

Franck Pellestor (ed.), *Chromothripsis: Methods and Protocols*, Methods in Molecular Biology, vol. 1769, https://doi.org/10.1007/978-1-4939-7780-2_12,

Micronuclei are widely recognized as sensitive biomarkers for genotoxicity and chromosome instability in peripheral blood lymphocytes and buccal mucosa cells [6]. They are frequently detected in preneoplastic lesions and cancers and are indicative of genetic damage events that could increase the risk of developmental and degenerative diseases. Micronuclei arise from whole chromosomes or chromatid fragments [7]. Spindle defects are a major cause of whole-chromosome-containing micronuclei [8, 9]. When the microtubules of the mitotic spindle do not properly attach to the chromosomal kinetochore, chromosomes can lag behind during anaphase and subsequently get encapsulated into micronuclei at telophase during nuclear reformation. Alternatively, micronuclei can also originate from chromosome fragments as a consequence of DNA breakage [10]. Examples of micronuclei containing fragments are those derived from broken anaphase bridges [11]. Here, we focus on the generation of whole-chromosome micronuclei.

Previous studies demonstrated that whole chromosomes in micronuclei first show defective DNA damage during S phase where they undergo defective DNA replication [5]. The replication timing of the micronuclear DNA is delayed compared to the main nucleus. When micronucleated cells entered mitosis where nuclear envelope breakdown occurs, premature chromosome compaction occurred due to exposure of replicating micronuclear DNA to the cytosolic compartment. Chromosome spreads prepared from the micronucleated cells displayed this "chromosome shattering" as DNA breaks and pulverized chromosomes. Reassembly of the resulting fragments by non-homologous end joining can result in a spectrum of genomic rearrangements reminiscent of chromothripsis (Fig. 1). Live-cell imaging analyses revealed that the micronuclear chromosomes can be reincorporated into the daughter nuclei. Thus, chromosome rearrangements arising within the micronuclei have the potential to be integrated

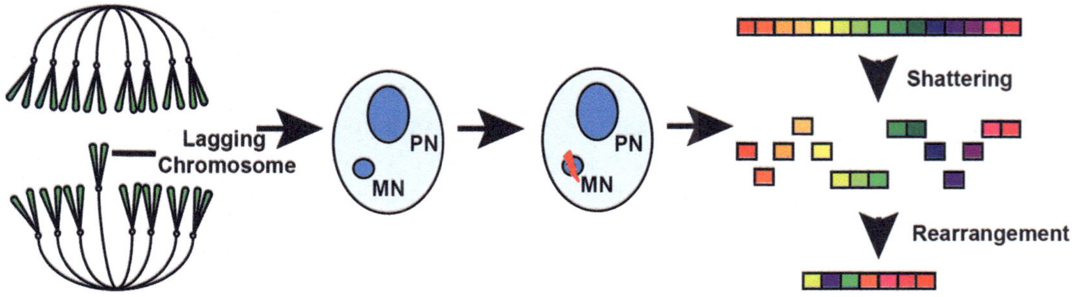

Fig. 1 Micronuclei as a mechanism of chromothripsis. Whole-chromosome segregation errors produce a lagging chromosome, which ends up in a micronucleus. Through aberrant DNA replication, the chromosome in the micronucleus can undergo shattering and rearrangement via chromothripsis. If reincorporated into the main nucleus, the rearranged chromosomes have the potential to promote tumorigenesis. *PN* primary nucleus, *MN* micronucleus

within the main genome and can potentially contribute to genomic mutations and malignant transformation of the cell. In this way, micronuclei represent an important source of genetic variation.

In this chapter, we describe protocols to generate micronuclei and study the fate of chromosomes, particularly chromosome pulverization, within the micronuclei. To study the consequences of micronuclei and resulting DNA damage, micronuclei are induced in vitro using the rapidly reversible inhibitor of microtubule polymerization, nocodazole. Treatment with nocodazole induces cells to arrest in mitosis. Upon washout from nocodazole, increased frequencies of abnormal merotelic kinetochore attachments can lead to lagging mis-segregated chromosomes [12]. We describe protocols for immunofluorescence with markers of DNA damage and centrosomes for detection of DNA damage in whole-chromosome micronuclei. Protocols for flow cytometry are added for assessment of cell cycle phase. Protocols for metaphase spreads and spectral karyotyping will be described for detection of chromosome pulverization.

2 Materials

2.1 Cell Culture and Micronuclei Induction

1. 10 cm^2 polystyrene tissue culture plates.
2. 15 cm^2 polystyrene tissue culture plates.
3. 12-well polystyrene tissue culture plates.
4. 14 mm borosilicate glass circular coverslip.
5. Microscope slides.
6. Dulbecco's Modified Eagle Medium (DMEM) or DMEM/F12, supplemented with 10% fetal bovine serum (FBS), 100 IU/mL penicillin, and 100 μg/mL streptomycin (*see* **Note 1**).
7. DMEM supplemented with 20% FBS, 100 IU/mL penicillin, and 100 μg/mL streptomycin.
8. Sterile phosphate-buffered saline (PBS).
9. 0.25% trypsin-EDTA.
10. Cell line of interest (hTERT-RPE-1 and U2OS cells were used in our study, *see* **Note 2**).
11. Nocodazole (1 mg/mL stock in DMSO).
12. p53 siRNA (*see* **Note 3**).
13. Lipofectamine RNAi Max.
14. Trypan blue stain for cell counting.
15. Automated cell counter/hemocytometer for cell counting.
16. Tissue culture hood.

17. Tissue culture incubator at 37 °C with 5% humidified CO_2.
18. Water bath.
19. Phase-contrast microscope equipped with ×100 objective for visualization of micronuclei.

2.2 Flow Cytometry

1. 6-well polystyrene tissue culture plates.
2. 0.25% trypsin-EDTA.
3. Ice-cold PBS.
4. Ice-cold absolute ethanol.
5. Propidium iodide (1 mg/mL in water).
6. RNAse (10 mg/mL).
7. 5 mL round-bottom polystyrene test tube with cap.
8. 1.5 mL Eppendorf tube.
9. Aluminum foil.
10. High-speed centrifuge.
11. Vortex.
12. Flow cytometer.

2.3 Immunofluorescence

1. PBS.
2. 16% paraformaldehyde.
3. 0.25% Triton X-100 in PBS.
4. Blocking buffer (5% FBS, 2% BSA, 0.1% Triton X-100 in PBS).
5. Rabbit γ-H2AX-Ser 139 antibody (marker for DNA damage) (Cell Signaling Technology).
6. Mouse γ-H2AX (marker for DNA damage) (clone JBW301, Upstate).
7. ACA antibody (marker for centromere) (Antibodies Inc.).
8. Fluorescence-conjugated secondary antibodies (Thermo Fisher Scientific).
9. Hoechst 333342 (10 mg/mL in water).
10. Mounting medium.
11. Standard epifluorescence microscope with excitation and emission filters for visualization of DAPI/FITC/TRITC.

2.4 Preparation of Metaphase Spreads

1. Colcemid (100 ng/mL in complete medium).
2. 75 mM KCl.
3. Carnoy's solution (3:1 methanol/glacial acetic acid).
4. Gurr buffer.
5. Giemsa stain/DAPI/Hoechst DNA stain.
6. Slides for staining or hybridization.

7. 22 × 32 mm coverslips.
8. Hotplate.
9. Phase-contrast microscope with ×100 objective.

2.5 Spectral Karyotyping

1. SkyPaint probes kit (Applied Spectral Imaging).
2. 60 μg/mL pepsin in 0.01 N HCl.
3. PBS.
4. 0.05 M $MgCl_2$ in PBS.
5. 1% formaldehyde in 0.05 M $MgCl_2$ in PBS.
6. Saline sodium citrate buffer (SSC 2× and 4×).
7. 50% formamide in SSC 2×.
8. 70% formamide in SSC 4×.
9. 0.1% Tween 20 in SSC 4×.
10. 70%, 80%, and 100% ethanol in water.
11. Coplin jar.
12. 43 °C incubator (humidified atmosphere).
13. Spectral imaging software (SkyView).
14. Microscope slides and coverslips.

3 Methods

The following protocols describe the methods to generate micronuclei in vitro using nocodazole.

3.1 Cell Culture

1. Use freshly thawed and early passage cells for all experiments. The cells should be maintained on sterile 10 cm^2 tissue culture plates in Dulbecco's Modified Eagle Medium (DMEM) supplemented with 10% fetal bovine serum (FBS), 100 IU/mL penicillin, and 100 μg/mL streptomycin at 37 °C in 5% CO_2 (humidified atmosphere). Change the culture medium every 2–3 days. At 80% confluence, split the cells using trypsin-EDTA. Cells should be passaged/maintained only for 1–2 months.
2. To set up a micronuclei generation experiment, cells should be plated from the 10 cm^2 "maintenance" dish into a sterile 15 cm^2 dish. Aspirate the culture medium from the 10 cm^2 dish, and wash the cells twice with sterile PBS, followed by incubation in 1 mL of 0.25% trypsin-EDTA for 5 min at 37 °C, until most of the cells have detached.
3. Inactivate the trypsin by addition of 9 mL 10% FBS-containing DMEM, and pellet the cells at 300 × g for 5 min. Remove the supernatant, and thoroughly resuspend the cells in 10 mL of complete medium.

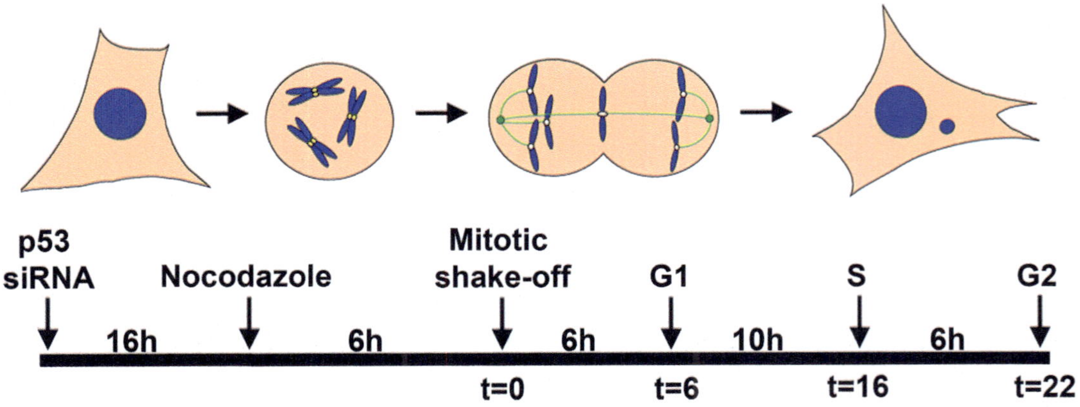

Fig. 2 Schematic of experimental procedure to induce micronuclei in vitro. After p53 knockdown to prevent aneuploidy-induced G1 arrest, cells are blocked in prometaphase for a short (6 h) interval with 100 ng/mL nocodazole. Blocked cells are collected by mitotic shake-off and released from nocodazole to complete cell division. After nocodazole release, spindles reassemble abnormally, producing merotelic kinetochore attachments, lagging chromosomes, and cells with micronuclei. At indicated time points, samples are collected for further analysis (adapted from ref. [5] with permission from Nature Publishing Group)

4. Count the cell concentration using an automated cell counter/hemocytometer, and transfer 3×10^6 cells to a sterile 15 cm^2 dish. Cellular seeding density may require optimization for each individual cell line. Top up the volume to 20 mL complete medium and grow the cells at 37 °C.

3.2 Generation of Whole-Chromosome Micronuclei

Here, we outline how to generate whole-chromosome micronuclei using nocodazole washout (Fig. 2).

1. Dilute nocodazole to 100 ng/mL (from a 1 mg/mL stock) in 20 mL complete growth medium.
2. One day after seeding, treat the cells with 100 ng/mL nocodazole for 6 h (*see* **Note 4**) to synchronize cells in mitosis. In certain cell lines, aneuploidy induced by nocodazole washout experiments can produce a p53-dependent G1 cell cycle arrest [13]. To enable cell cycle progression of the subsequent aneuploid daughter cells with micronuclei, transiently transfect the cells with p53 siRNA (50 nmol) 12 h before the nocodazole treatment (*see* **Note 3**).
3. Mechanically collect the mitotic cells using mitotic shake-off by gentle shaking or tapping of the plate or by gently washing with the medium the cells are grown in overnight (*see* **Note 5**).
4. Pellet the mitotic cells at 300 × *g* for 5 min, remove the supernatant, and wash the cells twice with prewarmed (*see* **Note 6**) DMEM to wash out nocodazole (*see* **Note 7**). Replate the mitotic cells in nocodazole-free DMEM containing 20% FBS (*see* **Note 8**), so that they can complete cell division.

3.3 Flow Cytometry to Determine Cell Cycle Progression

Flow cytometry analyses are used to determine when the cells are at the desired stage of the cell cycle (*see* **Note 9**).

1. Following mitotic shake-off, count mitotic cells, and plate 2.5×10^5 cells per 6 well in two wells with DMEM containing 20% FBS.
2. Harvest cells at various intervals by trypsin, and pellet at $300 \times g$ for 5 min at room temperature (RT). Aspirate the supernatant, resuspend the pellet in 1 mL of PBS, and transfer to a 1.5 mL Eppendorf tube.
3. Pellet the cells at $300 \times g$ for 5 min at RT, aspirate the supernatant, and resuspend the cells in 300 μL of ice-cold PBS (*see* **Note 10**). To fix cells, add dropwise 700 μL of ice-cold absolute ethanol, while vortexing at mid speed.
4. Store cells at −20 °C for at least 12 h. Cells can remain in fixative for up to 3 weeks before staining.
5. Invert/vortex the cells before centrifuging at $300 \times g$ for 5 min at RT, and aspirate the ethanol.
6. Following one wash with PBS, pellet the cells at $300 \times g$ for 5 min at RT. Resuspend the pellet in 400 μL of staining solution containing 250 μg/mL RNase A and 10 μg/mL propidium iodide in PBS at 37 °C for 30 min.
7. Analyze the samples on the flow cytometer. If samples are not run immediately, they can be stored covered in foil at 4 °C for up to a week.

3.4 Immunofluorescence Detection of DNA Damage in Whole-Chromosome Micronuclei

To detect DNA damage in micronuclei, the DNA damage marker γ-H2AX-Ser 139 and centromeric marker ACA are used.

1. Following shake-off, count the mitotic cells, and plate 1×10^5 cells on a 14 mm glass coverslip in a 12-well plate for immunofluorescence studies in nocodazole-free DMEM containing 20% FBS.
2. When the cells are at the desired stage of the cell cycle, as determined by flow cytometry, aspirate the nocodazole-free medium and wash the coverslips twice in PBS. Fix the cells in PBS containing 4% paraformaldehyde for 15 min at RT.
3. Permeabilize the cells in 0.25% Triton X-100 in PBS for 10 min at RT.
4. Block non-specific binding sites in blocking buffer for 30 min at RT and then incubate with primary antibodies diluted in blocking buffer at RT for 1 h or overnight at 4 °C. We used the antibodies at the following dilutions: rabbit γ-H2AX-Ser 139 (1:500), mouse γ-H2AX (1:500), and human ACA (1:1000).

5. Wash the coverslips twice in PBS, and incubate with fluorescence-conjugated secondary antibodies diluted 1:1000 in blocking buffer at RT for 1 h. After two washes with PBS, stain the nuclei and micronuclei with 200 ng/mL Hoechst 33342 in PBS for 10 min at RT.
6. Following three washes with PBS, mount the coverslips on a microscope slide in mounting solution.
7. Obtain *Z*-stacks with a 0.3 μm step size using a confocal microscope equipped with 488, 561, and 640 nm laser light and ×100 objective (*see* **Note 11**).

3.5 Metaphase Spreads

1. To determine the pulverization of the chromosome(s) in the micronucleus, prepare chromosome spreads from cells in the first cell cycle after release from nocodazole or from DMSO-treated controls.
2. After shake-off, count the mitotic cells, and plate 1×10^6 cells on a 10 cm^2 plate in nocodazole-free DMEM containing 20% FBS.
3. Use flow cytometry, as outlined above, to determine when the daughter cells progress to S/G2 phase after release, and at this point, add 100 ng/mL colcemid for 6 h.
4. Following 6 h incubation, harvest the cells using trypsin, pellet at 300 × *g* for 5 min at RT, and resuspend in a hypotonic solution of 75 mM KCl prewarmed to 37 °C for 18 min.
5. Fix the cells by dropwise addition of 1 mL freshly made Carnoy's solution, and pellet at 300 × *g* for 5 min at RT. Wash the samples three more times with 2 mL Carnoy's solution. Depending on the pellet size, resuspend in 200–500 μL Carnoy's solution. Cells can be stored overnight in Carnoy's solution at 4 °C or at −20 °C for months.
6. Add 30 μL of cell suspension onto a slide by holding up the slide at a 30°–45° angle, and drip the cell suspension onto the top region of the slide. Alternatively, one could also drop cells on the glass slide from a height of 0.5–1 cm.
7. Keep the slide vertically until it is ready to be completely dried on the hotplate at 37 °C for approximately 15–60 min. The slides can also be dried overnight at RT.
8. Next, add three to four drops of mounting medium containing DNA stain (DAPI or Hoechst). Place a long coverslip across and let the slide dry for 1 h at room temperature. Seal the slide with nail varnish. Alternatively, the DNA on the metaphase spread slides can be stained using Giemsa stain, by addition of 1.5 mL Giemsa stain in Gurr buffer for 3 min. Next, rinse the slide with Gurr buffer, and allow to dry on the hotplate at 37 °C for 1 h, before placing a long coverslip across.

9. Chromosomes can then be counted using a phase-contrast microscope under ×100 magnification.

3.6 Spectral Karyotyping

Spectral karyotyping can be used to check whether pulverized DNA fragments originate from a single chromosome (Fig. 3). The following steps encompass pretreatment, denaturation, hybridization, and image capture using spectral karyotyping (SKY) microscopy.

1. This protocol follows **Step 7** in Protocol 3.5 Metaphase Spreads. After preparation, the dried slide is pretreated for hybridization by equilibration in a Coplin jar containing 2× SSC for 30 min at 37°C.
2. Next, remove cytoplasmic debris by treating the slide with 60 μg/mL pepsin in 0.01 N HCl for 5 min at 37 °C (optional) (*see* **Note 12**).
3. Wash the slide two times with 5 min washes in 1×PBS.
4. Wash the slide with 1×PBS containing 0.05 M $MgCl_2$ for 5 min and fix in fresh 1% formaldehyde in PBS containing 0.05 M $MgCl_2$ for 10 min at RT.
5. Wash the slide in PBS for 5 min at RT with shaking.
6. Dehydrate using an ethanol series of 70%, 80%, and 100% ethanol in water for 3 min at RT each, without shaking.
7. Air-dry the slide until the ethanol has evaporated (1–2 min).
8. Analyze under a phase-contrast microscope to determine whether the chromosomal morphology has been preserved (*see* **Note 13**).
9. To denature the chromosomes on the slide, treat with 2× SSC containing 70% formamide for 5 min at 75°C. Dehydrate the slides in cold 70%, 80%, and 100% ethanol for 2 min at each concentration, and then air-dry.
10. Denature 10 μL SkyPaint probes per slide for 7 min at 80 °C, followed by incubation for 30 min at 37 °C.
11. For hybridization, apply 3 μL denatured probe to a 12 mm circle or 6 μL of probe for a 18 mm circle containing denatured chromosomes. Place a coverslip on the slide, seal the edges with rubber cement, and incubate the slide upside down for 2 days at 37 °C with sufficient moisture in humidified chamber. The chamber/box can be kept moist by adding 1–2 mL of water or placing a damp towel inside, simultaneously ensuring that the slide does not come into contact with water.
12. Carefully remove the rubber cement seal with tweezers from the slide, ensuring coverslip is not dragged across slide as this will scratch the metaphase chromosomes.

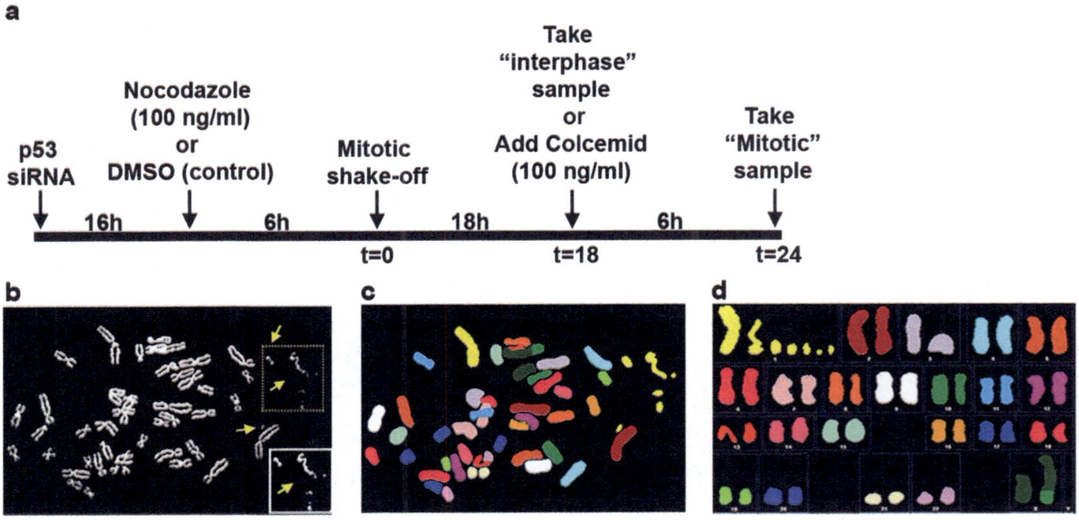

Fig. 3 Detection of chromosome pulverization in cells with micronuclei. (**a**) Schematic representation of the experimental design to determine the fate of the chromosomes in the micronuclei. After knockdown of p53 and prometaphase block in 100 ng/mL nocodazole, cells are collected by mitotic shake-off and released for resumption of cell division. When the daughter cells progress to S/G2 phase, cells are incubated with 100 ng/mL colcemid for 6 h and samples then collected for metaphase spreads. (**b**–**d**) Representative metaphase spreads showing pulverization of chromosome 1, visible as fragments of stained material clustered into a discrete locus on the spread, as demonstrated by (**b**) DNA stain (*yellow arrows* indicate chromosome fragments, brightened in inset), (**c**) pseudo-colored SKY probes and (**d**) aligned SKY karyotype (adapted from ref. [5] with permission from Nature Publishing Group)

13. Place the slide into a Coplin jar with 2× SSC with slight agitation at RT for 10 min.
14. Wash the slide three times in 2× SSC containing 50% formamide for 8 min at 43 °C, followed by two washes in 2× SSC for 8 min at 43 °C.
15. Wash the slide once in 4× SSC containing 0.1% Tween 20 for 20 min at RT.
16. Incubate the slide with 100 μL Cy5 staining reagent (vial 3 of the SkyPaint probes kit) for 45 min at 37 °C.
17. Wash the slide three times in 4× SSC containing 0.1% Tween 20 for 10 min, and incubate the slide in 100 μL Cy5.5 staining reagent (vial 4 of the SkyPaint probes kit) for 45 min at 37 °C.
18. Wash the slide three times in 4× SSC containing 0.1% Tween 20 for 10 min at RT. Apply anti-fade DAPI reagent (vial 5 of the SkyPaint probes kit) to the coverslip, and place on the slide.
19. Capture and analyze the karyotype image using spectral imaging software (SkyView) from Applied Spectral Imaging (*see* **Note 14**).

4 Notes

1. The choice of media for maintenance of cell cultures depends on the cell line of interest. In our study, DMEM was used for U2OS cells, while DMEM/F12 was used for hTERT-RPE-1 cells.
2. hTERT-RPE-1 is a human telomerase-immortalized retinal pigment epithelial cell line that is chromosomally stable with a modal chromosome number of 46. We will refer to hTERT-RPE-1 cells as RPE-1 cells henceforth. RPE-1 cells are female in origin and are known to have one karyotypically normal X-chromosome and one derivative X-chromosome with additional chromosomal material from chromosome 10 at the terminal end of the q-arm. U2OS is an osteosarcoma cell line. Both cells were able to generate micronuclei following nocodazole washout.
3. In certain cell lines including RPE-1, but not U2OS cells, aneuploidy induced by nocodazole washout experiments can produce a p53-dependent G1 cell cycle arrest [13]. Hence, to enable cell cycle progression of the subsequent aneuploid daughter cells with micronuclei, RPE-1 cells are transiently transfected with 50 nmol p53 siRNA 12 h before the nocodazole treatment. U2OS cells are less sensitive to aneuploidy-induced cellular senescence and hence do not require transfection.

   Sequence information of the siRNA pools used from Dharmacon are as follows:

   Human p53 ON-TARGETplus SMARTpool siRNA L-003329-00-0005 (J-003329-14) GAAAUUUGCGUG UGGAGUA,

   (J-003329-15) GUGCAGCUGUGGGUUGAUU,

   (J-003329-16) GCAGUCAGAUCCUAGCGUC,

   (J-003329-17) GGAGAAUAUUUCACCCUUC,
4. As prolonged mitotic arrest causes DNA damage [14], cells are arrested for a short interval with nocodazole. Nocodazole incubation duration may require optimization for each individual cell line (e.g., RPE-1 6 h, U2OS 6–8 h).
5. When mitotic cells are difficult to detach using "mechanical" approach, cells in the 15 cm^2 can be incubated with 0.25% trypsin-EDTA for 15–30 s at 37 °C. Use the phase-contrast microscope to confirm that detachment is restricted to mitotic cells only.
6. Slight fluctuations in temperature may affect cell cycle progression, lagging chromosomes, and resulting micronucleus formation. Therefore, it is imperative that medium is prewarmed to 37 °C to collect and wash mitotic cells.

7. Washout from nocodazole results in merotelic attachments and increased incidence of lagging chromosomes [12]. Wash cells using DMEM or other culture media used. We find that washes using PBS buffer somehow increases incidence of lagging chromosomes and therefore affects the generation of micronuclei, as compared to washes with culture media.
8. Upon washout from nocodazole, cells are replated into media containing 20% FBS instead of usual 10% FBS as cells growing in the former were observed to resume cell cycle progression at a rate closer to "cycling cells" than those grown in the latter.
9. Following nocodazole washout, the time for cells to complete one round of cell division may require optimization for each individual cell line. We harvested cells at the indicated time points where we deemed the majority of cells to be in that cell cycle phase by flow cytometry (e.g., for RPE-1 cells, samples for G1 phase, S phase, and G2 phase are collected 6 h, 16 h, and 22 h after mitotic shake-off, respectively).
10. Ensure there are no clumps, and use a 1 mL tip to pipette up and down approximately 10–20 times. It is imperative that the sample is present as a single-cell suspension.
11. Micronuclei are defined as small (1–5 nm in diameter), spherical structures that are in approximately the same focal plane as the primary nucleus of the same cell [15]. For quantification of γ-H2AX, primary nuclei were classified as positive when five or more foci were detected, with the threshold of foci intensity set against control sample with DNA damage (either irradiated at 2 Gy or treated with 200 ng/μL of doxorubicin for 5 h). The same threshold was applied to score γ-H2AX foci in micronuclei.
12. Slides should be pretreated with pepsin only if cytoplasm is present. This is observed as gray particulate matter surrounding the cytoplasm. There is no need for pretreatment with pepsin if there is no visible cytoplasm observed as a clear light margin surrounding the metaphase chromosomes, and the chromosomes show good morphology (black in color and not shiny).
13. The chromosomes should show their original morphology and not a "hollow" appearance of being over-digested by the enzymes. In case morphology is not retained, these cells cannot be used for SKY hybridization, and protocol should be started from the beginning.
14. Chromosomes with a "pulverized" appearance were often clustered into a discrete location on the spread. *See* Fig. 3 and Refs. [5, 16].

References

1. Jones S, Chen WD, Parmigiani G et al (2008) Comparative lesion sequencing provides insights into tumor evolution. Proc Natl Acad Sci U S A 105(11):4283–4288. https://doi.org/10.1073/pnas.0712345105
2. Stephens PJ, Greenman CD, Fu B et al (2011) Massive genomic rearrangement acquired in a single catastrophic event during cancer development. Cell 144(1):27–40. https://doi.org/10.1016/j.cell.2010.11.055
3. Korbel JO, Campbell PJ (2013) Criteria for inference of chromothripsis in cancer genomes. Cell 152(6):1226–1236. https://doi.org/10.1016/j.cell.2013.02.023
4. Holland AJ, Cleveland DW (2012) Chromoanagenesis and cancer: mechanisms and consequences of localized, complex chromosomal rearrangements. Nat Med 18(11):1630–1638. https://doi.org/10.1038/nm.2988
5. Crasta K, Ganem NJ, Dagher R et al (2012) DNA breaks and chromosome pulverization from errors in mitosis. Nature 482(7383):53–58. https://doi.org/10.1038/nature10802
6. Fenech M (2000) The in vitro micronucleus technique. Mutat Res 455(1–2):81–95
7. Norppa H, Falck GC (2003) What do human micronuclei contain? Mutagenesis 18(3):221–233
8. Cimini D (2008) Merotelic kinetochore orientation, aneuploidy, and cancer. Biochim Biophys Acta 1786(1):32–40. https://doi.org/10.1016/j.bbcan.2008.05.003
9. Ganem NJ, Godinho SA, Pellman D (2009) A mechanism linking extra centrosomes to chromosomal instability. Nature 460(7252):278–282. https://doi.org/10.1038/nature08136
10. Terradas M, Martin M, Tusell L et al (2010) Genetic activities in micronuclei: is the DNA entrapped in micronuclei lost for the cell? Mutat Res 705(1):60–67. https://doi.org/10.1016/j.mrrev.2010.03.004
11. Hoffelder DR, Luo L, Burke NA et al (2004) Resolution of anaphase bridges in cancer cells. Chromosoma 112(8):389–397. https://doi.org/10.1007/s00412-004-0284-6
12. Cimini D, Howell B, Maddox P et al (2001) Merotelic kinetochore orientation is a major mechanism of aneuploidy in mitotic mammalian tissue cells. J Cell Biol 153(3):517–527
13. Lanni JS, Jacks T (1998) Characterization of the p53-dependent postmitotic checkpoint following spindle disruption. Mol Cell Biol 18(2):1055–1064
14. Dalton WB, Nandan MO, Moore RT et al (2007) Human cancer cells commonly acquire DNA damage during mitotic arrest. Cancer Res 67(24):11487–11492. https://doi.org/10.1158/0008-5472.CAN-07-5162
15. Fenech M, Chang WP, Kirsch-Volders M et al (2003) HUMN project: detailed description of the scoring criteria for the cytokinesis-block micronucleus assay using isolated human lymphocyte cultures. Mutat Res 534(1–2):65–75
16. Obe G, Beek B, Vaidya VG (1975) The human leukocyte test system. III. Premature chromosome condensation from chemically and x-ray induced micronuclei. Mutat Res 27(1):89–101

Check for updates

Chapter 13

Detection of Impaired DNA Replication and Repair in Micronuclei as Indicators of Genomic Instability and Chromothripsis

Mariona Terradas, Marta Martín, and Anna Genescà

Abstract

The physical isolation of chromosomes within micronuclei offers an attractive mechanistic explanation for the local DNA fragmentation and clustered genome rearrangements that characterize chromothripsis. Localized shattering of the chromatin confined in micronuclei can be a consequence of defects in micronuclei basic general functions, such as DNA replication and repair. The detection of DNA repair and replication defects in micronuclei is described here, as well as the analysis of chromosome breakage and inaccurate reassembly of broken segments in the daughter cells, as indirect methods to detect chromothripsis.

Key words Micronuclei, Chromothripsis, DNA fragmentation, Chromosome rearrangements

1 Introduction

Any mechanism for chromothripsis needs to explain how tens to hundreds of breaks are generated and clustered into relatively small chromosomal regions, often a single chromosome or chromosome arm. What events induce the massive, localized, and synchronous wave of DNA breaks during chromothripsis? An initial idea was that localized shattering of a chromosome was induced by ionizing radiation under conditions of chromosome compaction, such as during mitosis when chromatin is condensed and chromosomes individualized [1]. However, it is unclear how a stray pulse of ionizing radiation would hit on one chromosome without similarly damaging the others.

Perhaps a more plausible explanation for chromothripsis is through the partitioning of a chromosome into a micronucleus, where it is shattered, while the rest of chromosomes remain intact [2, 3]. Micronuclei are aberrant nuclear structures that result from acentric chromosome fragments or whole chromosomes lagged behind during anaphase. According to one micronucleus-dependent

Franck Pellestor (ed.), *Chromothripsis: Methods and Protocols*, Methods in Molecular Biology, vol. 1769, https://doi.org/10.1007/978-1-4939-7780-2_13, © Springer Science+Business Media, LLC, part of Springer Nature 2018

origin of chromothripsis, duplication of chromosomes residing in micronuclei is markedly impaired, thereby inducing replication stress with stalling of replication forks and induction of DNA breaks [2]. Using an approach combining live cell imaging and single-cell DNA sequencing, Zhang et al. demonstrated that the damage in micronucleated cells was notably specific: it occurred only on the chromosome contained in these aberrant structures [3]. These experiments provided direct evidence for a micronucleus-dependent mechanism for chromothripsis.

Repair defects in micronuclei can also contribute to chromothripsis. DNA damage response (DDR) pathways for resolving DNA breaks and pyrimidine dimers are impaired in micronuclei. For instance, recruitment of 53BP1 and MRE11 to DSBs and of XPA and XPC to photo-lesions is markedly inefficient in micronuclei, compared to the primary nucleus of the same cell [4, 5]. Nevertheless, cells are not "blind" to micronuclear DNA damage. Phosphorylated histone H2AX (γH2AX) appears at sites of DNA damage in micronuclei, demonstrating the existence of at least a partial response, as histone phosphorylation requires the action of the ATM kinase [6]. However, recruitment of downstream factors is limited by trafficking defects through the micronuclear envelope. Therefore, repair defects can potentially contribute to the shattering of chromosomes sequestered in micronuclei.

Micronuclei present nuclear envelope defects that seem to underlie their impaired DNA repair and replication functions. Using fluorescent proteins fused to a nuclear localization signal, Hatch et al. [7] reported that 60% of micronuclei undergo an irreversible loss of compartmentalization due to defects in nuclear lamina assembly during interphase. A simple hypothesis for why DNA damage occurs when the micronuclear envelope loses its integrity is that stalled replication forks and DNA lesions provide a substrate for cytoplasmic nucleases that cannot attack an intact chromosome and whose fragmentation remains unresolved because repair factors cannot efficiently reach the micronuclear DNA.

Micronuclear replication stress and fragmentation are relevant because the genetic materials entrapped in micronuclei are not lost by cells. Huang et al. studied the fate of micronuclei containing GFP-labeled chromatin by analyzing time-lapse videos and did not observe micronuclei expulsion from the cell [8]. Direct experimental evidence for the reincorporation of micronuclear DNA to the primary nuclei of the daughter cells after mitosis was obtained using a histone fused to a photosensitive fluorescent protein [2]. As pulverized and rearranged micronuclear DNA can be reincorporated to one of the daughter nuclei after mitosis at a considerable frequency, mutations in the micronuclear DNA could potentially be integrated into the daughter nuclei and contribute to the genomic instability of the cell.

In summary, a synergistic combination of DNA repair defects and replication stress in micronuclei with abnormal nuclear enve-

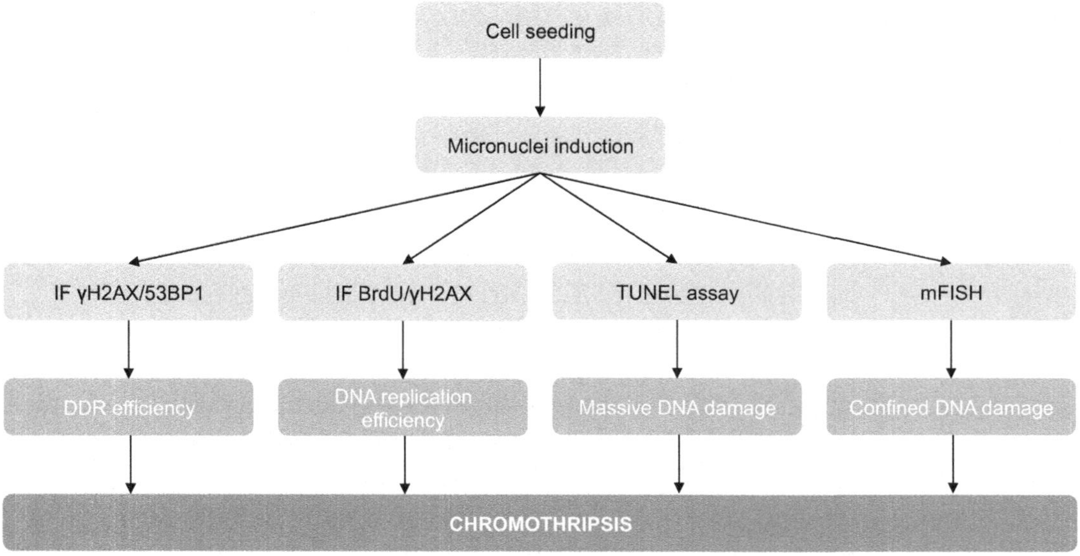

Fig. 1 Workflow summarizing the different methods described in the present chapter to detect the causes and consequences of micronuclear-dependent chromothripsis

lope might well be a cause of chromothriptic events. In this chapter, we explain several methodologies that can be used to detect causes and consequences of micronuclei-dependent chromothripsis (Fig. 1): specifically, (1) immunofluorescent detection of the DSB marker γH2AX in the micronuclei and recruitment of 53BP1 to compare DNA repair functions in micronuclei *vs* primary nuclei, (2) detection of incorporated BrdU to micronuclear DNA to ascertain asynchronous or delayed DNA replication, (3) TUNEL assay to detect massive DNA damage in the micronuclei, and (4) molecular cytogenetics (mFISH) for the detection of the rearranged products of shattering, as micronuclei provide an environment to constrain the rearrangements in a single chromosome. Together, the methodologies presented will provide an overall picture of micronucleus-dependent chromothripsis, its causes, as well as its consequences.

2 Materials

2.1 Micronuclear DNA Damage Repair Efficiency

1. PBS: phosphate-buffered saline.
2. 4% PFA: 4% paraformaldehyde (PFA) in 1×PBS; mix 1.6 g of PFA in 30 mL of distilled water and heat it to 50 °C. Add NaOH until the solution becomes transparent, and add 4 mL of 10× PBS. Cool down the solution keeping it at 4 °C for an hour. Adjust the pH to 7.4 and add distilled water up to 40 mL. PFA should be freshly made or made 1 week before

and kept at 4 °C in dark conditions. Otherwise, it could be made and kept at −20 °C and in dark conditions for 3 months.

3. Permeabilization solution 1: 0.5% Triton X-100 in 1×PBS (*see* **Note 1**).
4. PBST: 0.1% Tween 20 in 1×PBS (*see* **Note 1**).
5. Blocking solution: 3% FCS (fetal calf serum) in PBST.
6. Antibody γH2AX (05-636, Millipore, Merck) (*see* **Note 2**).
7. Antibody 53BP1 (ab21083, Abcam) (*see* **Note 2**).
8. Mounting media: antifade solution (such as Vectashield Mounting Medium for fluorescence, Vector Laboratories, Inc.) containing a chromatin dye such as DAPI (4,6-diamino-2-phenylindole) or propidium iodide (PI). If any secondary antibody used in the immunofluorescence is red, then use DAPI instead of PI.
9. Parafilm M® (Sigma Aldrich).
10. Epifluorescence microscope with appropriate filter range.

2.2 Micronuclear DNA Replication Efficiency

1. PBS: phosphate-buffered saline.
2. 4% PFA: 4% paraformaldehyde (PFA) in 1×PBS.
3. Permeabilization solution 1: 0.5% Triton X-100 in 1×PBS (*see* **Note 1**).
4. PBST: 0.1% Tween 20 in PBS (*see* **Note 1**).
5. Blocking solution: 3% FCS (fetal calf serum) in PBST.
6. Antibody γH2AX (05-636, Millipore, Merck) (*see* **Note 2**).
7. Antibody BrdU (ab8152, Abcam) (*see* **Note 2**).
8. Borate buffer saline: 10 mM borate buffer +150 mM NaCl, pH 8.2 (*see* **Note 3**).
9. Weak TX100 solution: 0.1% Triton X-100 in 1 × PBS.
10. Strong TX100 solution: 1% Triton X-100 in 1 × PBS.
11. PBS+: 0.5% BSA (bovine serum albumin) + 0.15% glycine in 1 × PBS. Add 1×PBS up to 100 mL.
12. Denaturation solution 1: 1 M HCl—0.1% Triton X-100 in 1 × PBS (*see* **Note 1**).
13. Mounting media: antifade solution (such as Vectashield Mounting Medium for fluorescence, Vector Laboratories, Inc.) containing a chromatin dye such as DAPI (4,6-diamino-2-phenylindole) or propidium iodide (PI). If any secondary antibody used in the immunofluorescence is red, then use DAPI instead of PI.
14. Parafilm M® (Sigma Aldrich).
15. Epifluorescence microscope with appropriate filter range.

2.3 Micronuclei DNA Damage: TUNEL Assay

1. PBS: phosphate-buffered saline.
2. 2% PFA: 2% paraformaldehyde in 1×PBS; mix 1×PBS and 4% PFA at 1:1 (v/v).
3. Citrate solution: 0.1% Triton X-100 in 0.1% Sodium Citrate. Add 0.1 g of sodium citrate powder to 100 mL of deionized water and mix. When diluted, add 0.1 mL of Triton X-100 to the solution, and mix to its complete dissolution (*see* **Note 5**).
4. TUNEL solution (Roche, Sigma Aldrich, or Merck).
5. Mounting media: antifade solution (such as Vectashield Mounting Medium for fluorescence, Vector Laboratories, Inc.) containing a chromatin dye such as DAPI (4,6-diamino-2-phenylindole) or propidium iodide (PI). If any secondary antibody used in the immunofluorescence is red, then use DAPI instead of PI.
6. Epifluorescence microscope with appropriate filter range.

2.4 Confined DNA Damage: mFISH

1. PBS: phosphate-buffered saline.
2. Pepsin solution: 0.05 μg/mL pepsin in deionized water. To prepare 50 mL of pepsin solution, add 500 μL of HCl 1 N in 50 mL of deionized water, mix, and heat up to 37 °C. When the solution reaches the desired temperature, add 25 μL of pepsin at 100 μg/mL to reach a final concentration of0.05 μg/mL. It is important to mix the pepsin thoroughly after adding it, as otherwise it might remain insolubilized. Use the solution immediately at 37 °C (*see* **Note 6**).
3. RNAse A solution (100 μg/mL): dilute 100 μg of RNAse A in 1 mL of water. Keep the solution frozen. Avoid multiple freezing-thawing cycles (*see* **Note 7**).
4. 20×SSC solution: To prepare 100 mL of 20×SSC solution, add 17.5 g NaCl (3 M) and 8.8 g sodium citrate (0.3 M) to 80 mL of deionized water. After complete dilution, adjust pH to 7.2. Add deionized water to a final volume of 100 mL. This solution can be autoclaved.
5. 2×SSC solution: add 10 mL of a 20×SSC solution to 90 mL of deionized water and mix.
6. 0.4×SSC solution: add 2 mL of a 20×SSC solution to 98 mL of deionized water and mix.
7. Fixing solution: 50 mM $MgCl_2$ in 1% of formaldehyde (FA). First prepare 1% of FA solution by adding 2.7 mL of 37% FA to 97.3 mL of 1×PBS. Then, mix 0.48 g of $MgCl_2$ with 1% FA solution up to a final 100 mL volume.
8. Denaturation solution 2: 26% formamide in 1×SSC, pH 7. For 100 mL of denaturing solution, add 35 mL of formamide 37%

to 15 mL of 2×SSC. Adjust pH to 7. The solution can be used more than once. Do not keep it longer than a week (*see* **Note 4**).

9. Ethanol series: 70%, 85%, 100%.
10. Rubber cement (Union Rubber Inc.).
11. Mounting media: antifade solution (such as Vectashield Mounting Medium for fluorescence, Vector Laboratories, Inc.) containing a chromatin dye such as DAPI (4,6-diamino-2-phenylindole) or propidium iodide (PI). If any secondary antibody used in the immunofluorescence is red, then use DAPI instead of PI.
12. Epifluorescence microscope with appropriate filter range.

3 Methods

The methods here described have been performed in normal human dermal fibroblasts (nHDFs) and lymphoblastoid cells, but they can be applied to any cell type with few modifications.

Carry out all procedures at room temperature (RT) unless otherwise specified.

All the incubations with antibody solutions described in the chapter should be made in a wet chamber (*see* **Note 8**) to avoid the evaporation of the solution and the drying of the sample.

3.1 Cell Seeding and Induction of Micronuclei

1. Seed the cells in a glass support that will later allow the cells to be analyzed under an epifluorescence microscope. Chamber slides or a cell culture plate with a cover glass is recommended.

   The number of cells needed depends on the experiment, but ideally the culture should be 70–80% confluent at the time of starting the experiment.

2. Micronuclei can be induced in different ways, depending on their content. In the case of micronuclei containing acentric fragments, they can be induced with ionizing radiation or radiomimetic agents, such as bleomycin. In the case of micronuclei carrying whole chromosomes, aneugenic agents such as nocodazole should be used. To induce radiation-induced micronuclei in nHDFs, an 80% confluent T25 flask is subcultured in two 1-well chamber slides at a 1:6 ratio. The day after subculture, the cells are exposed to 2 Gy of γ-rays, and 48 h later, all of the protocols described below can be applied. Alternatively, 24 h after seeding, cells are treated with 40 μg/mL of bleomycin for 24 h.

   In order to induce micronuclei carrying whole chromosomes, HDFs are seeded as previously described, and 40 h after seeding, cells are treated with 100 ng/mL of nocodazole for 8 h.

3.2 Determining the Micronuclear DNA Damage Repair Efficiency: IF γH2AX + 53BP1

1. At least 24 h after the seeding, remove medium, and rinse the cells twice with 1×PBS (phosphate-buffered saline).
2. Fix the sample with 4% PFA for 15 min at room temperature (RT). Alternatively, cells can be fixed with ice-cold methanol for 10 min at −20 °C.
3. Remove PFA and wash the cells with 1×PBS for 5 min (*see* **Note 9**).
4. Permeabilize the cells with 0.5% Triton X-100 in 1×PBS for 15 min at RT.
5. Rinse cells with 1×PBS.
6. Incubate the cells for 20 min at RT in blocking solution: 3% FCS in PBST (*see* **Note 10**).
7. Remove the blocking solution and add 30 μL/cm² of first antibody solution containing γH2AX and 53BP1 antibodies. Cover it with a piece of Parafilm M®, place it in a wet chamber, and incubate for 120 min at RT or, alternatively, incubate overnight (ON) at 4 °C. Be careful to remove any air bubble and to cover the whole area of cells to be analyzed.
8. Remove the piece of Parafilm M®, and wash the cells three times with PBST for 5 min each wash.
9. Remove the PBST, add 30 μL of the second antibody solution, cover it with Parafilm M®, and incubate in the wet chamber at RT for 90 min.
10. Repeat **step 8**.
11. Rinse the sample with distilled water and let it air-dry.
12. Apply mounting media on the slides.
13. To analyze the sample, first use the DAPI filter to look for micronucleated cells. Only those cells carrying micronuclei whose size is one third of the main nucleus—at the most—should be considered. Afterward, analyze the presence of DSBs inside the micronuclei using the adequate filter for the γH2AX labeling depending on the secondary antibody used. Once the presence of DSBs is determined, check for the presence of 53BP1. The absence of the latter could be indicative of DDR deficiency in the micronucleus. At least 100 micronucleated cells should be analyzed to determine the micronuclear DDR efficiency.

3.3 Determining the Micronuclear DNA Replication Efficiency: IF BrdU + γH2AX

In order to study the micronuclear DNA synthesis, the cells can be pulse-labeled with a base analogue, such as bromodeoxyuridine (BrdU). Afterward, the incorporation of BrdU and the presence of γH2AX are detected by immunofluorescence.

1. Seed the cells and induce the micronuclei formation as previously described.

2. Add BrdU to the culture 2 h before fixing the cells. A final 50 μM BrdU concentration should be fine; however, the time and concentration should be adjusted depending on the cell type.
3. Remove the media and rinse the cells twice in 1×PBS.
4. Fix the cells for 10 min with 4% PFA.
5. Rinse the cells with 1×PBS.
6. Permeabilize the cells with permeabilization solution 1 for 20 min.
7. Wash the sample twice with 1×PBS for 5 min each wash.
8. Block the sample with blocking solution for 30 min.
9. Incubate the cells with γH2AX antibody dissolved in blocking solution for 1 h at 37 °C. Add 30 μL/cm^2 of antibody solution, and cover the cells with a piece of Parafilm M®.
10. Remove the Parafilm M®, and wash the slides with PBST three times for 5 min each.
11. Incubate the cells for 1 h with 30 μL/cm^2 of secondary antibody solution, and cover the cells with Parafilm M®.
12. Repeat **step 10**.
13. Incubate the cells for 15 min in denaturation solution 1 to allow the BrdU antibody to enter the double helix (*see* **Note 11**).
14. Wash the sample with borate buffer saline for 5 min to restore the pH.
15. Wash the sample twice with weak TX100 solution for 10 min.
16. Wash the sample with PBS+ for 10 min.
17. Add 30 μL/cm^2 of BrdU antibody dissolved in the blocking solution. Cover the cells with a piece of Parafilm M®, and incubate the sample for 2.5 h.
18. Rinse the sample twice with strong TX100 solution.
19. Wash the sample with strong TX100 solution for 15 min.
20. Wash the sample with PBS+ for 10 min.
21. Add 30 μL/cm^2 of a secondary antibody solution that includes the second antibody of γH2AX and BrdU. Incubate the cells for 1 h at RT.
22. Rinse the sample twice with strong TX100 solution.
23. Wash the sample twice with strong TX100 solution for 10 min.
24. Wash the sample with 1×PBS for 5 min.
25. Rinse the cells in distillate water.

26. Air-dry the sample and apply mounting media.
27. Analyze the sample under an epifluorescence microscope. Using the DAPI filter, count the cells and identify the ones carrying micronuclei. Then move to the proper filters and analyze the γH2AX and the BrdU labeling. A positive signal for γH2AX and BrdU may indicate replication-associated damage likely to have occurred from collapsed replication forks and/or processing of replication intermediates by cytoplasmic nucleases.

3.4 Detecting Massive Micronuclear DNA Damage: TUNEL Assay

1. Seed the cells and induce the formation of micronuclei.
2. Remove the media and rinse the cells twice with PBS.
3. Fix the cells with 2% PFA for 40 min.
4. Wash the cells twice with 1×PBS for 5 min each wash.
5. Permeabilize the cells with freshly prepared citrate solution.
6. Wash the cells twice with 1×PBS, for 5 min each wash.
7. Incubate the sample for 1 h at 37 °C with the TUNEL solution prepared according to the manufacturer's instructions, inside a wet chamber.
8. Wash the cells twice with 1×PBS, for 5 min each wash.
9. Let the sample air-dry and apply mounting media.
10. Analyze the sample under an epifluorescence microscope. Using the DAPI filter, identify the micronucleated cells. Then, move the filter into the proper position and check for the presence of a uniform staining covering the whole micronuclear surface, indicative of massive DNA damage.

3.5 Confined DNA Damage: mFISH

Multiplex in situ hybridization (mFISH) is a 24-color karyotyping technique that allows simultaneous identification of all human chromosomes and is, therefore, a method of choice to study complex genomic rearrangements, such as chromothripsis. For our mFISH experiments, we have used the SpectraVysion™ probes and modified the denaturation times and washing steps of the Vysis protocol. Alternatively, the hybridization can also be performed using several DNA probes from different chromosomes, each labeled with different fluorochromes. This FISH must be performed on metaphase spreads that can be obtained according to standard protocols. Perform a diamond-tipped scribe marked area in the microscope slides to signal where the metaphase spreads are. The samples could be allowed to age before performing the hybridization to improve results. To age the slides, they can be kept at RT overnight, in a heater at 65 °C for 1 h or at 45 °C for 1 day. After aging, keep the samples in a dry environment, as humidity interferes with hybridization.

1. Wash the cells with 1×PBS for 5 min.
2. Digest in pepsin solution for 5–15 min at 37 °C.
3. Wash the cells with 1×PBS for 5 min.
4. Apply on the cells 20 μL of a 100 μg/mL RNAse A solution, and incubate for 30–60 min in a wet chamber at 37 °C.
5. Wash the cells twice with 2×SSC for 5 min each wash.
6. Fix the cells for 5 min with the fixing solution.
7. Wash the cells twice with 1×PBS for 5 min each wash.
8. Dehydrate the cells by washing them in a series of three ethanol solutions with increasing ethanol concentration (70%–85%–100%) for 1 min each, and allow the sample to air-dry.
9. Denature the sample DNA by placing the slide in a prewarmed denaturation solution 2 for 4 min at 72 °C. The denaturation solution must have reached the denaturing temperature of 72 °C before this step.

   Simultaneous denaturing: Alternatively, the appropriate amount of DNA probe (follow manufacturer's instructions) can be placed on the slide, and both DNAs can be simultaneously denatured at 72 °C for 5 min in an oven or using a HYBrite™ machine. In this case, proceed with **step 13**.
10. Immediately repeat **step 8**.
11. While performing **step 10**, denature the DNA probe (*see* **Note 12**) by heating the appropriate amount (follow manufacturer's instructions) up to 72 °C for 5 min.
12. Add the denatured DNA probe to the denatured and dehydrated sample DNA.
13. Cover the sample DNA and the DNA probe with a coverslip, and seal it using white rubber cement adhesive.
14. Incubate the sample for 24–48 h in a wet chamber at 37 °C.
15. Gently remove the coverslip by stretching the solidified cement using the tweezers.
16. Wash the cells with prewarmed 0.4×SSC for 1–2 min at 72 °C while stirring.
17. Wash the cells with 2×SSC-Tween20 for 1 min at RT while stirring.
18. Repeat **step 7**.
19. Apply mounting media on the slides.
20. Analyze the sample under an epifluorescence microscope equipped with appropriate filters. Look for small fragments of the same chromosome to identify massive DNA damage of the chromatin confined in micronuclei.

4 Notes

1. To add either Triton X-100 or Tween 20 to a solution, it is recommended to cut the sharp end of a tip and keep the tip inside the solution during the mix with a magnetic shaker.
2. Antibody solutions: the antibodies should be diluted in blocking solution at the concentration recommended by the manufacturer. Usually, the concentration needs to be adjusted depending on the protocol and the cell type; thus we recommend checking the literature for further information.
3. Borate buffer saline is an isotonic solution that has an alkaline buffering capacity. In these procedures, it is used after the denaturing solution treatment to restore the pH of the sample. Borate buffer saline can be purchased as tablets that are dissolved in deionized water. It can also be prepared with 10 mM sodium borate and 150 mM NaCl in deionized water, and adjust pH at 8.2. Borate concentration can vary between 10 mM and 100 mM depending on the buffering capacity needed.
4. Denaturing temperature may vary depending on the probe. Follow manufacturer's instructions.
5. Citrate solution is a permeabilizing solution that favors the TUNEL mix to reach the DNA of the sample.
6. Pepsin solution allows a certain degradation of the cell cytoplasm to favor the DNA probe to reach the target DNA in the chromosomes. When chromosome spreads are free of residual cytoplasm, this step could be omitted. Also this step results in a cleaner hybridization, with less unspecific background. Excessive pepsin time may result in target DNA with a blurry appearance due to excessive enzymatic degradation. Thus, the time of this step may vary depending on the cell type and on the quality of the chromosome extraction. Start with 10 min and check before proceeding further.
7. RNAse A solution allows degradation of the RNA present in the sample, providing a cleaner hybridization, with less unspecific background. When chromosome spreads are free of residual cytoplasm, this step could be omitted.
8. A wet chamber can easily be built with a plastic slide box. Cover the bottom of the slide box with water-soaked paper. Place the coverslip or the slide in the box, but avoid direct contact with the soaked paper. To do so, the coverslip can be placed inside a plastic petri dish that can be placed on the paper, and the slide can be placed on top of plastic caps placed on the paper. Close the box while the antibody reaction is taking place in order to preserve humidity and avoid drying of

the sample. Opaque boxes are recommended; otherwise they can be wrapped with aluminum foil.

9. Fixed cells can be kept at 4 °C in PBS ON or in 1×PBS with 0.025% (w/v) sodium azide for longer times.
10. It is recommended to prepare 25 mL of blocking solution, aliquot it in 1–2 mL volumes, and keep it at −20 °C to avoid bacterial contamination due to the presence of FCS.
11. The HCl molarity can be adjusted, starting at 0.5 M up to 2 M.
12. The DNA probe can be a single, a dual, or a poly probe. Hybridization can be performed easily using probes labeled with different fluorochromes.

References

1. Zhang W, Peng G, Lin SY et al (2011) DNA damage response is suppressed by the high cyclin-dependent kinase 1 activity in mitotic mammalian cells. J Biol Chem 286:35899–35905
2. Crasta K, Ganem NJ, Dagher R et al (2012) DNA breaks and chromosome pulverization from errors in mitosis. Nature 482:53–58
3. Zhang CZ, Spektor A, Cornils H et al (2015) Chromothripsis from DNA damage in micronuclei. Nature 522:179–184
4. Terradas M, Martín M, Tusell L et al (2009) DNA lesions sequestered in micronuclei induce a local defective-damage response. DNA Repair (Amst) 8:1225–1234
5. Terradas M, Martín M, Hernández L et al (2012) Nuclear envelope defects impede a proper response to micronuclear DNA lesions. Mutat Res 729:35–40
6. Burma S, Chen BP, Murphy M et al (2001) ATM phosphorylates histone H2AX in response to DNA double-strand breaks. J Biol Chem 276:42462–42477
7. Hatch EM, Fischer AH, Deerinck TJ et al (2013) Catastrophic nuclear envelope collapse in cancer cell micronuclei. Cell 154:47–60
8. Huang Y, Hou H, Yi Q et al (2011) The fate of micronucleated cells post X-irradiation detected by live cell imaging. DNA Repair (Amst) 10:629–638

Chapter 14

Study of Telomere Dysfunction in TP53 Mutant LoVo Cell Lines as a Model for Genomic Instability

Oumar Samassekou, Nathalie Bastien, Ju Yan, Sabine Mai, and Régen Drouin

Abstract

Telomere restriction fragment, 3D quantitative FISH on nuclei, and quantitative FISH on metaphases are complementary approaches that explore telomere dysfunction genomically, cellularly, and chromosomally, respectively. We used these approaches to study association between telomere dysfunction and degree of genomic instability related to TP53 mutations in LoVo isogenic cell lines. We found a strong correlation between degree of genomic instability, telomere dysfunction, and specific mutations of *TP53*. The use of complementary approaches to study telomere biology is essential to have a comprehensive understanding of telomere involvement in genomic instability.

Key words Telomere, Genomic instability, Telomere restriction fragment, 3D telomere quantitative FISH, MFISH, Telomere aggregates, Telomere length

1 Introduction

The ends of human chromosomes are termed telomeres. They are formed by a nucleoprotein complex which is composed of a tandem repeat DNA sequence, $(TTAGGG)_n$, measuring 5–15 kb in normal cells; six telomere-specific proteins, named shelterin proteins; telomere-associated proteins; and a special heterochromatin structure [1–4] . Telomeres play a crucial role in genome stability and integrity, and their dysfunction is one of the hallmarks of cancer cells which is mainly characterized by telomere shortening, telomerase activation, and alteration of telomeric nuclear organization [5–7]. Telomere dysfunction and genomic instability are intimately associated. We and others have shown that telomere dysfunction is an early event in tumorigenesis precluding chromosome end-to-end fusion, breakage-fusion-bridge cycle, and genomic instability [5, 8–11].

Telomere shortening can lead to loss of telomere protection and potentially to telomere crisis, which is a state of extensive genome instability that can promote cancer progression [12]. Recent data has

Franck Pellestor (ed.), *Chromothripsis: Methods and Protocols*, Methods in Molecular Biology, vol. 1769,
https://doi.org/10.1007/978-1-4939-7780-2_14,

revealed that telomere crisis can induce numerous cancer-relevant changes, the extent of these rearrangements was unexpected, and their complexity was staggering [12]. These cancer-related multiple chromosomal aberrations include deletions and amplifications, translocations, chromothripsis, kataegis, and tetraploidization [12–15].

Genomic instability is a dynamic process which lead to an accumulation of multiple chromosomal aberrations over several cell cycles and relies mainly on repeated breakage-fusion-bridge cycles causing complex genomic rearrangements which might be amplified by some underlining mutations of genes heavily involved in oncogenesis such as *TP53* [16]. We have shown the degree of genomic instability is associated with particular patterns of telomere dysfunction and specific mutations of *TP53* [17].

We used complementary approaches and a unique cellular model of colon cancer to study an association between telomere dysfunction and degree of genomic instability when common mutations of TP53, occurring during oncogenesis, are present (TP53-A143V, TP53-R175H, and TP53-R273H). We found that various phenotypes define telomere dysfunction and are correlated to high degree of genomic instability and some TP53 mutations. Telomeres can form telomeric aggregates that represent telomere clusters, which at an optical resolution of 200 nm cannot be resolved into single telomere signals. Short telomeres, peripheral nuclear localization of telomeres, high number of telomeric aggregates, clonal elongation, or shortening of individual telomeres were mean features of telomere dysfunction [17] (Figs. 1 and 2).

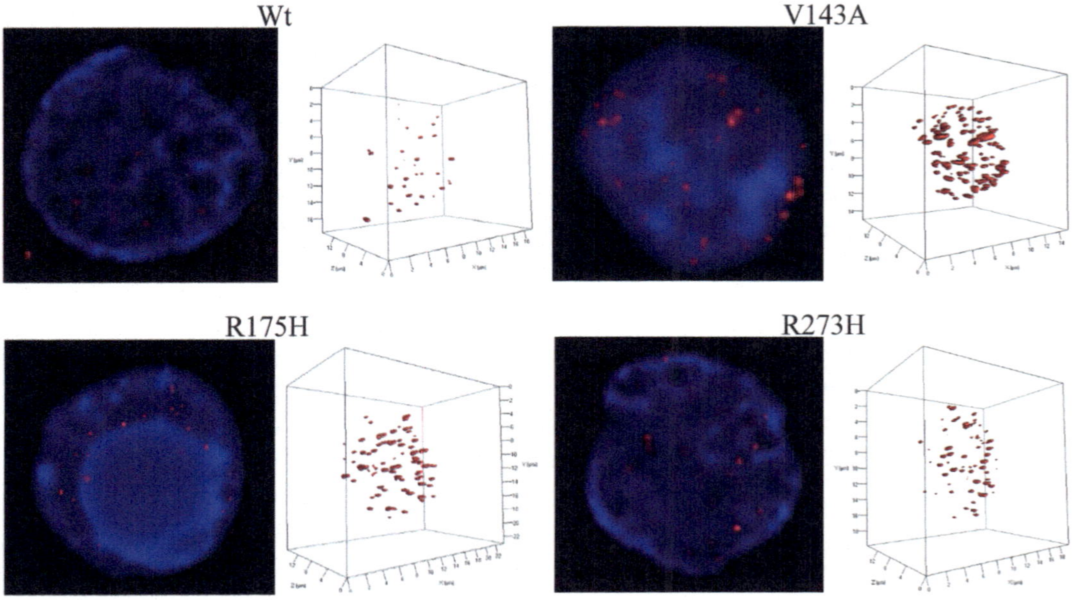

Fig. 1 Representative 2D and 3D images of LoVo cells. Two-dimensional and three-dimensional images of LoVo nuclei (blue) and their telomeres (red) imaged and visualized with AxioVision 4.6. The wild-type TP53 and the mutant TP53-R273H showed less telomere signals and aggregates, whereas the mutant TP53-V143A and TP53-R175H presented more telomere signals and aggregates

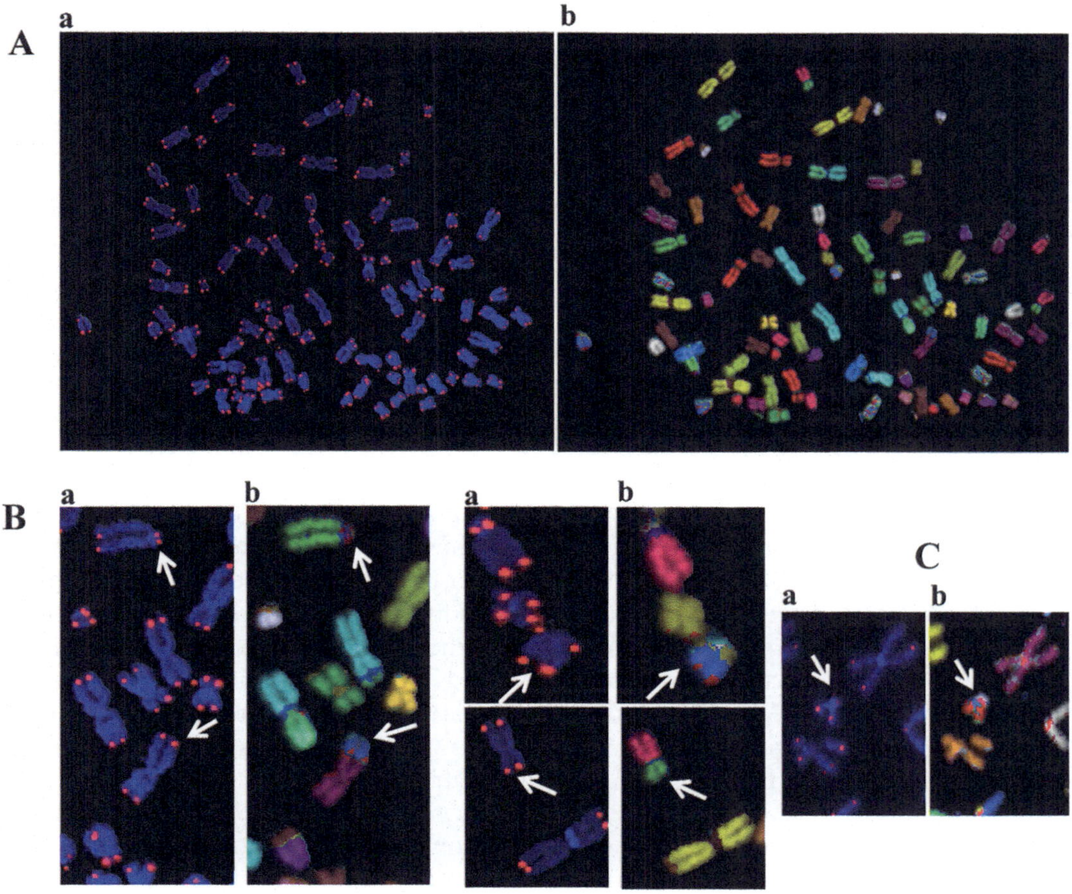

Fig. 2 Telomere-specific *FISH* and corresponding M *FISH*. (a) Telomere-specific *FISH*. (b) Corresponding multi-color *FISH* of (a). (**A**) A metaphase of a near-tetraploid cell of the mutant R175H. (**B**) Partial metaphases of (**A**) showing structurally abnormal chromosomes, pointed out by white arrows. (**C**) Partial metaphases of the mutant R273H, showing telomere signal on the centromeric region of the del(5)(q10), pointed out by white arrows

The complementary approaches we used to assess telomere dysfunction at genomic, cellular, and chromosomal levels and chromosomal aberrations were the telomere restriction fragment (TRF) assay, the 3D quantitative telomere fluorescence in situ hybridization (*FISH*) on nuclei, quantitative telomere *FISH* on metaphases, and spectral karyotype, respectively.

TRF assay evaluates the average length of telomeres and is based on digestion of genomic DNA by restriction enzymes that do not cut telomeric sequences. Subsequently, gel electrophoresis and Southern blot are carried out, and telomere sequences are revealed with a labeled telomeric probe. The resultant DNA size distribution on the gel represents telomere fragments, and the average length is calculated according to size and intensity of the smear [4]. One of the main advantages of the TRF assay is the possibility to measure telomere length in kb, a standard and reference unit, that enables a comparison between studies done in different experimental settings. Moreover, the TRF assay can be complimentary to other approaches such as fluorescence-based methods, so telomere length of a single cell or a specific chromosome arm can be deduced in kb.

At cellular level, we studied telomere dysfunction by assessing telomeric nuclear organization by doing 3D quantitative telomere *FISH* on nuclei. This method is based on in situ hybridization of telomeric-labeled probe on preserved 3D nuclei. Then, 3D images composed of multiple focal plans of nuclei and telomeric signals are captured. Finally, TeloView™ (3D Signatures Holdings Inc.) software measures different parameters which define the 3D nuclear organization of telomeres. These parameters are (1) the number of telomere signals per cell, (2) the intensity of each telomeric signal, (3) the distribution of telomere signals in the nuclear space (*a/c* ratio), (4) the number of telomere aggregates (telomere clusters, telomere fusions, or telomeres in close vicinity that cannot be further resolved into individual telomere signals at an optical resolution of 200 nm), (5) the telomere position relative to nuclear center and periphery, and (6) the nuclear volume [6, 18]. Assessment of 3D telomeric nuclear organization has the advantages to simultaneously evaluate telomere length, telomere dynamic, and genomic instability at single-cell level.

Finally, we used quantitative telomere *FISH* on metaphases to evaluate telomere dysfunction at the chromosomal level. This assay is based on in situ hybridization of fluorescence-labeled telomere probe on metaphases, image capture by a fluorescence microscope, and measure of telomeric signals on each chromosome arm by the ISIS software (Metasystems). This approach has the advantages to measure telomere length on each of the 92 chromosome arms of a normal diploid cell, so length difference between homologous individual telomeres or between telomeres on both ends of the same chromosome could be provided [19–21]. Moreover, it can be combined with spectral karyotyping to study telomere length on arms of aberrant chromosomes.

We used the approaches cited above, as well as isogenic LoVo cell lines (V143A, R175H, and R273H) harboring specific TP53 mutations and LoVo wild type (wt) [22] to study a correlation between degree of genomic instability and telomere dysfunctions when TP53 is mutated. The wild-type LoVo cell line is chromosomally stable in culture and presents only a few chromosomal abnormalities [23, 24], so it represents a suitable model to study genomic instability in context of specific TP53 mutations. We set up cell culture of LoVo cells, and when they reached 20 population doublings, we carried out the different complementary methods described earlier to evaluate telomere dysfunctions.

2 Materials

2.1 Telomere Restriction Fragment Assay

1. Biosafety cabinet.
2. Petri dishes.
3. Cell scrapers.
4. 15 mL conical tubes.
5. 0.6 mL Microtubes.
6. Electrophorese apparatus: wide Mini-Sub Cell GT Systems (Bio-Rad).
7. LoVo cell lines: LoVo TP53-wt, LoVo TP53-V143A, LoVo TP53-R175H, and LoVo TP53-R273H (kindly provided by Pocard et al. [22]).
8. DMEM/F12.
9. Fetal bovine serum.
10. The Gentra Puregene Cell Kit (Qiagen).
11. 10× phosphate-buffered saline (PBS): 1.37 M NaCl, 0.27 M KCl, 0.1 M Na_2HPO_4, 0.018 M KH_2PO_4, pH 7.0.
12. 1× PBS, pH 7.4: diluted from 10× PBS.
13. Nuclease-free water.
14. Distillate water.
15. RsaI 10,000 units/mL (New England Biolabs Inc.).
16. HinfI 10,000 units/mL (New England Biolabs Inc.).
17. NEB buffer 2 (New England Biolabs Inc.).
18. Water bath set at 37 °C.
19. 6× loading buffer: 25% bromophenol blue/0.25% xylene cyanol FF/30% glycerol in H_20.
20. Agarose MP.
21. DNA molecular marker weight III, DIG-labeled 0.12–21.2 kb (Roche) (*see* **Note 1**).
22. 1× TAE buffer: 0.04 M Tris-acetate, 0.001 M EDTA, pH 8.0.
23. UV lamp.
24. Plastic container.
25. Heating block.
26. Nitrocellulose membranes.
27. Deproteinization solution: 0.25 M HCl.
28. Denaturation solution: 0.5 M NaOH, 1.5 M NaCl.
29. Neutralization solution: 0.5 M Tris–HCl, 3 M NaCl, pH 7.5.
30. 20× SSC (saline sodium citrate): 3 M NaCl, 0.3 M sodium citrate, pH 7.0

31. 2× SSC: diluted from 20× SS, pH 7.4.
32. Hybridizer.
33. Whatman 3MM chromatography papers.
34. Whatman 17MM chromatography papers.
35. Absorbent papers.
36. Glassware.
37. Hybridization ovens.
38. DIG Easy Hyb (Roche).
39. Hybridization solution: add 64 mL nanopure water to DIG Easy Hyb and incubate at 37 °C until complete reconstitution.
40. Wash buffer I: 2× SSC, 0.1% SDS.
41. Wash buffer II: 0.2× SSC, 0.1% SDS.
42. Telomere probe DIG-labeled: $(CCCTAA)_3$ (Roche).
43. Hybridization bags.
44. X-ray films.
45. Maleic acid buffer: 0.1 M maleic acid, 0.15 M NaCl, pH 7.5.
46. Wash buffer III: 0.1 M maleic acid, 0.15 M NaCl, 0.3% Tween 20, pH 7.5.
47. 10× blocking buffer: dissolve 10 g of blocking reagent (Roche) in 100 mL of maleic acid buffer under constantly stirring on a heating block (65 °C) or heat in a microwave oven, autoclave. The solution remains opaque and is stored at −20 °C.
48. 1× blocking buffer: dilute 20 mL of 10× blocking buffer in 80 mL of maleic acid buffer.
49. Anti-digoxigenin-alkaline phosphatase conjugate (Roche): 750 mU/μL.
50. Anti-DIG-AP working solution (75 mU/mL): add 5 μL of anti-DIG-AP in 50 mL of blocking buffer. This solution needs to be prepared 5 min before using it.
51. Detection buffer: 0.1 M Tris–HCl, 0.1 M NaCl, pH 9.5.
52. CSPD: 25 mM (Roche).
53. Substrate solution: dilute 30 μL of CSPD in 3 mL of detection buffer.
54. ImageJ software (NIH).

2.2 3D Telomere Quantitative FISH on Nuclei

1. Petri dish.
2. Sterile slides.
3. ThermoBrite®.
4. ThermoBrite® humidity strips.
5. Telomere PNA FISH Kit/Cy3 (Agilent/Dako).

6. Glass slides (25 × 75 × 1 mm) and coverslips (24 × 24 mm) and (24 × 60 mm).
7. Dulbecco's Modified Eagle Medium: Nutrient Mixture F-12 (DMEM/F12).
8. 1× phosphate-buffered saline (PBS): 1.37 M NaCl, 0.27 M KCl, 0.1 M Na_2HPO_4, 0.018 M KH_2PO_4, pH 7.0.
9. 1× PBS, pH 7.4: diluted from 10× PBS.
10. 3.7% formaldehyde/1× PBS: 5 mL of 37% formaldehyde, 5 mL of 10× PBS, 40 mL of nanopure H_2O.
11. Glycerol.
12. 20% glycerol: 20 mL of glycerol, 80 mL of nanopure H_2O.
13. Triton X-100.
14. 1× PBS/0.5% Triton X-100: 5 mL of 10× PBS, 250 μL of Triton X-100, completed to 50 mL with nanopure H_2O, pH 7.2–7.4.
15. 0.1 N HCl: diluted from 1 N HCl.
16. Nitrogen liquid.
17. Deionized formamide.
18. 70% formamide/2× SSC: 35 mL of deionized formamide, 5 mL of 20× SSC, complete to 50 mL with nanopure H_2O, pH 7.2–7.4.
19. 1 M Tris.
20. 70% formamide/10 mM Tris: 70 mL of deionized formamide, 1 mL of 1 M Tris, complete to 100 mL with nanopure H2O, pH 7.2–7.4.
21. 0.1× SSC: diluted from 20× SSC.
22. 2× SSC/0.05% Tween 20: 10 mL of 20× SSC, 50 μL of Tween 20, complete to 100 mL with nanopure H2O, pH 7.2–7.4.
23. DAPI (4′,6-diamidine-2′-phenylindole dihydrochloride) stock solution (10 μg/mL): 10 μg of DAPI, 1 mL of 1× PBS, pH 7.0.
24. DAPI working solution (0.1 μg/mL): 100 μL of DAPI stock solution/10 mL of 1× PBS, pH 7.0.
25. Forceps.
26. Coplin jars.
27. Water bath.
28. Rubber cement.
29. Nanopure water
30. Vectashield.
31. Zeiss epifluorescence microscope equipped with Zeiss AxioVision software or Zen Pro and deconvolution module (Zeiss).
32. TeloView™ (3D Signatures Holdings Inc.).

2.3 Quantitative Telomere FISH on Metaphases

1. Glass slides (25 × 75 × 1 mm) and coverslips (24 × 24 mm and 24× 60 mm). The slides must be cleaned by soaking in a series of ethanol solutions followed by polishing with a clean piece of muslin and wiped dry using lint-free paper before dropping the cells on the slides.
2. 1.5 mL, 0.6 mL, and 0.2 mL microcentrifuge tubes.
3. Plastic and glass coplin jars or other containers suitable for slide washing.
4. 15 mL and 50 mL polypropylene conical tubes.
5. Forceps.
6. Humidified slide chamber.
7. Water bath with accurate temperature control at 37 °C.
8. Centrifuge.
9. Thermotron environmental control unit (CDS-5, Thermotron).
10. Incubator at 37 °C.
11. Fume hood.
12. Shaker.
13. Light microscope.
14. ThermoBrite®.
15. ThermoBrite® humidity strips.
16. Rubber cement.
17. Epifluorescence microscope coupled to Metasystems Imaging System (ISIS, including MFISH and telomere analysis modules).
18. Nanopure water.
19. Telomere PNA FISH Kit/Cy3 (Agilent/Dako).
20. MFISH human cocktail probe (Metasystems).
21. Dulbecco's Modified Eagle Medium: Nutrient Mixture F-12 (DMEM/F12).
22. Hanks' Balanced Salt solution (HBSS).
23. Pepsin solution: 25 mg of pepsin in 1 mL HBSS.
24. Sodium citrate solution: 0.4% sodium citrate.
25. Carnoy: 3 volumes of methanol/1volume acetic acid.
26. NP-40.
27. Tween 20.
28. 10× Tris-buffered saline (TBS): 0.5 M Tris, 1.5 M NaCl, pH = 7.0.
29. 1× TBS: 100 mL of 10× TBS, completed to 1000 mL with nanopure H_2O, pH 7.2–7.4.

30. 20× saline sodium citrate (SSC): 3.0 M NaCl, 0.30 M trisodium citrate, pH 7.0.
31. 1× SSC, pH 7.0–7.5, RT: diluted from 20× SSC.
32. 0.1× SSC, pH 7.0–7.5, 4 °C: diluted from 20× SSC.
33. 2× SSC, pH 7.0–7.5, 70 °C: diluted from 20× SSC.
34. 2× SSC, pH 7.0–7.5, 4 °C: diluted from 20× SSC.
35. 0.4× SSC, pH 7.0–7.5, 72 °C: diluted from 20× SSC.
36. 2× SSCT (2× SSC, 0.05% Tween 20): 10 mL of 20× SSC, 50 μL of Tween 20, complete to 100 mL with nanopure H_2O, pH 7.2–7.4.
37. 37% formaldehyde.
38. 3.7% formaldehyde/1× TBS: 5 mL of 37% formaldehyde, 5 mL of 10× TBS, 40 mL of nanopure H_2O.
39. 0.4× SSC/0.3%NP-40: 2.5 mL of 20× SSC, 300 μL of NP-40, completed to 100 mL with nanopure H_2O, pH 7.2–7.4.
40. 2× SSC/0.3%NP-40: 10 mL of 20× SSC, 100 μL of NP-40, completed to 100 mL with nanopure H_2O, pH 7.2–7.4.
41. 1 N HCl.
42. Pepsin (2500 units/mg). Prepare a stock solution at a concentration of 10 mg/mL in nanopure H_2O.
43. Pepsin/HCl solution: 0.05 mg/mL pepsin, 0.1 N HCl.
44. Series of cold ethanol: 70% ethanol, 90% ethanol, and 100% ethanol (−20 °C).
45. Ethanol series: 100%, 95%, 70%, RT.
46. 0.07 N NaOH.
47. DAPI/antifade (MetaSystems).
48. Trypsin solution.
49. 200 μL tube.
50. Water bath.

3 Methods

3.1 Telomere Restriction Fragment Assay

3.1.1 Cell Culture and DNA Extraction

1. Culture separately the LoVo cell lines (LoVo TP53-wt, LoVo TP53-V143A, LoVo TP53-R175H, and LoVo TP53-R273H) by using DMEM/F12 supplemented with10% FBS.
2. When cell confluence reaches 70%, briefly rinse them in 1× PBS at 37 °C.
3. Detach the cells from the dish by using a cell scraper.
4. Use the Gentra Puregene Cell Kit to extract DNA according to the manufactory instructions (*see* **Note 2**).

3.1.2 DNA Digestion

1. Dilute the extracted DNA with nuclease-free water in a 0.6 mL tube to get 2 μg of purified genomic DNA (per sample) in a final volume of 14 μL.
2. Add 2 μL of NEB buffer 2, 2 μL of Hinf I, and 2 μL of RsaI to the tube containing DNA. The final volume should be 20 μL.
3. Mix and incubate the reaction mixture in water bath at 37 °C for 2 h.
4. Stop the reaction by adding 4 μL of 6× loading buffer and quick-spin the tubes. Keep the tubes at 4 °C for 24–48 h or at −20 °C for up to12 weeks.

3.1.3 Gel Electrophorese

1. Prepare 150 mL of 0.8% of agarose gel (having 10 cm × 10 cm × 1 cm as dimensions) in 1× TAE buffer.
2. Mount the gel on gel apparatus and add enough 1× TAE buffer to cover the gel entirely.
3. Prepare molecular weight mixture by adding up 4 μL of DIG molecular weight marker, 14 μL of nuclease-free water, and 4 μL of loading buffer 6× and mix.
4. Load 10 μL of molecular weight mixture in the first well of the gel.
5. Load 12 μL (equivalent of 1 μg of the digested DNA) of each sample in subsequent wells. To obtain valid quantitative analysis of telomere length, each well should contain the same amount of DNA.
6. Run the gel at 5 volt/cm in 1× TAE buffer until the bromophenol blue tracking dye is separated about 8–9 cm from the starting wells (total run time 4 h). A better DNA separation is obtained when transfer is done longer (6–8 h) by decreasing the voltage.

3.1.4 Non-electrophoretic (Capillary) Transfer

All steps should be performed with gentle agitation.

1. Incubate the gel in 200 mL of deproteinization solution for 10 min at RT. An optimal treatment of this solution is achieved when bromophenol stain changes from blue to yellow.
2. Rinse twice the gel with distilled water at RT.
3. Incubate the gel in:
   (a) 200 mL of denaturation solution at RT for 15 min.
   (b) Change the solution and repeat once.
4. Rinse twice the gel with distilled water at RT.
5. Incubate the gel in:
   (a) 200 mL of neutralization solution at RT for 15 min.
   (b) Change the solution and repeat once.
6. Mount the transfer dispositive as shown in Fig. 3.

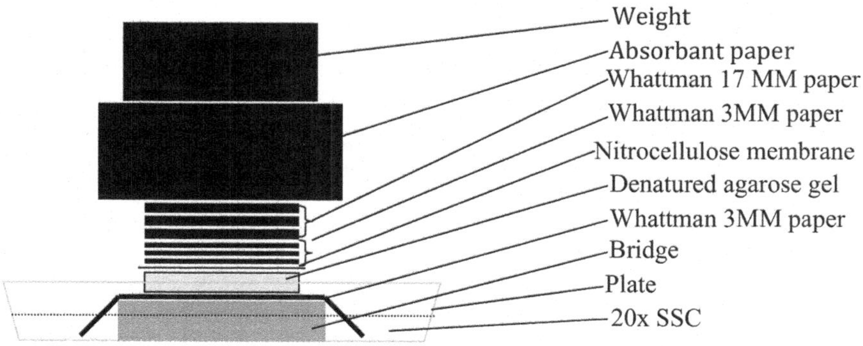

Fig. 3 Dispositive of Southern blot transfer

7. Cut a nitrocellulose membrane to the size of the gel and immerse it in 20× SSC.
8. Lay gently the nitrocellulose membrane on top of the gel.
9. Blot the digested DNA from the gel to the nitrocellulose membrane by capillary transfer using 20× SSC, as the transfer buffer, at RT for overnight (16 h) (*see* **Note 3**).
10. Fix transferred DNA on the blotting membrane by UV cross-linking (120 mJ/cm^2) (*see* **Note 4**).
11. Wash the membrane with 2× SSC at RT for 5 min and let air-dry (*see* **Note 5**).

3.1.5 Hybridization

All steps are done with gentle agitation.

1. Prewarm 25 mL of hybridization solution at 42 °C.
2. Incubate the membrane in 13 mL of hybridization solution at 42 °C for 30–60 min with gentle agitation.
3. Add 2 μL of telomere probe into 10 mL freshly prewarmed hybridization solution.
4. Discard the prehybridization solution completely and immediately add the hybridization solution to the membrane, and incubate at 42 °C for 3 h.

3.1.6 Washing

All wash steps are done with gentle agitation.

1. Discard hybridization solution.
2. Wash the membrane in:
   (a) 200 mL of wash buffer I at RT for 5 min.
   (b) Change the solution and repeat once.
3. Wash the membrane in:
   (a) 200 mL of prewarmed wash buffer II at 50 °C for 15 min.
   (b) Change the solution and repeat once.

3.1.7 Detection

All wash and incubation steps are done with gentle agitation.

1. Rinse the membrane in washing buffer III at RT for 3 min.
2. Incubate the membrane in 100 mL freshly prepared blocking solution at RT for 30 min.
3. Incubate the membrane in 50 mL anti-DIG-AP working solution at RT for 30 min.
4. Wash the membrane in:
   (a) Wash buffer III at RT for 15 min.
   (b) Change the solution and repeat once.
5. Incubate the membrane in 100 mL detection buffer at RT for 5 min.
6. Discard the detection buffer, place the membrane (DNA side up) on an opened hybridization bag, and quickly apply approximately 3 mL of substrate solution on the membrane.
7. Cover the membrane with the second sheet of the hybridization bag to spread the substrate solution homogeneously and without air bubbles over the membrane, and incubate the membrane at RT for 5 min.
8. Squeeze out excess substrate solution and seal the edges of the hybridization bag.
9. Incubate the membrane at 37 ° C for 15 min
10. Expose the membrane to the imaging device or to X-ray film at RT for 20 min (*see* **Note 6**).

3.1.8 Analysis

1. Measure the densitometry of the DNA smears corresponding to the samples by using ImageJ software.
2. Draw a rectangle with 100 rows around each sample and background (empty lane) in ImageJ software.
3. Get the intensity of each of the 100 boxes per sample in ImageJ software and export the values from ImageJ to Excel.
4. In Excel, create columns A and B where data for a sample and the background are plotted, respectively.
5. Determine the optical density for each box without background, ODi, by deducing the background values from those of the sample, and plot those values in column C.
6. Determine the molecular size of each given position (the position of the boxes), MWi.
   (a) Identify the size of each band of the molecular marker (ladder) in bp.
   (b) Enter those values into a column in Excel. From those values, determine the log (base 10) of the band sizes (this is data *Y*).

(c) Measure the distance (in cm) from the top of the gel to each band in the ladder (marker), and enter those values into a column to the left of data *Y*. This is data *X*.

(d) Graph data *X* and *Y* by using a scatterplot, and fit a trendline (linear) which shows the $y = mx + b$ equation (equation of the marker) and the R^2 value on the graph.

(e) Measure the distance between the top of the gel and each box (in cm) for a sample and plot those values in Excel.

(f) Use the equation of the marker to determine the y value for each box where x is the distance between the top of the gel and each box for a sample.

(g) Determine the molecular size of each box, MWi, by getting the antilog of the y value, and plot it in column D.

7. Determine the mean size of telomeres by using the formula $\Sigma \text{ODi} / \Sigma\left(\frac{\text{ODi}}{\text{MWi}}\right)$, where ODi is the optical density at a given position (position of the boxes) on the gel and MWi is the molecular weight at that position (boxes).

3.2 3D Telomere Quantitative FISH on Nuclei

3.2.1 Cell Culture and Harvesting

1. Culture separately the LoVo cell lines (LoVo TP53-wt, LoVo TP53-V143A, LoVo TP53-R175H, and LoVo TP53-R273H) on sterile slides by using DMEM/F12 supplemented with 10% FBS.
2. When cell confluence reaches 70%, briefly rinse the slides with 1× PBS at 37 °C.
3. Fix the cells on slides with 3.7% formaldehyde/1× PBS at RT for 20 min (*see* **Note 7**).
4. Wash the slides in:

   (a) 50 mL of 1× PBS at RT for 5 min.

   (b) Change the solution and repeat.

   (c) Change the solution and repeat again.

3.2.2 Slide Preparation

1. Incubate the slides in 50 mL of 1× PBS/0.5% Triton X-100 at RT for 10 min (*see* **Note 8**).
2. Incubate the slides in 50 mL of 20% glycerol at RT for 1 h (the slides can be left in glycerol solution for overnight).
3. Take the slides one by one with forceps and dip them into liquid nitrogen for 30 s or until the sound "click" is heard.
4. Thaw on a piece of paper towel. As soon as the frozen layer disappears, put the slide back into 20% glycerol, and then repeat freeze-thaw treatment three more times.

5. Wash the slides in:
   (a) 50 mL of 1× PBS at RT for 5 min.
   (b) Change the solution and repeat.
   (c) Change the solution and repeat again.
6. Incubate the slides in 50 mL of 0.1 N HCl at RT for 5 min (*see* **Note 9**).
7. Wash the slides in:
   (a) 50 mL of 1× PBS at RT for 5 min.
   (b) Change the solution and repeat once.

3.2.3 Denaturation and Hybridization

1. Incubate the slides in 70% formamide/2× SSC at RT for 1–2 h (*see* **Note 10**).
2. Take the slides out of the 70% formamide/2× SSC and quickly drain the excess of fluid off.
3. Apply 10 μL of PNA-telomere probe (in the dark) on a predetermined area of each slide, cover the area with a 24 × 24 mm coverslip, and seal the edge of the coverslip by applying rubber cement and let it dry completely.
4. Prepare the hybridization device: open the ThermoBrite® and submerge in water ThermoBrite® humidity strips for 2 min and place them on the ThermoBrite®.
5. Place the slides on the ThermoBrite and start the following program: DNA denaturation at 80 °C for 3 min, followed by hybridization at 30 °C for 2 h.

3.2.4 Post Hybridization Washing and Counterstain

All of the wash steps are done with gentle agitation and

1. Remove rubber cement and the coverslip carefully (*see* **Note 10**).
2. Wash the slides in:
   (a) 50 mL of 70% formamide/10 mM Tris pH 7.4 at RT for 15 min.
   (b) Change the solution and repeat once.
3. Wash in 50 mL of 1× PBS at RT for 1 min.
4. Wash in 50 mL 0.1× SSC at 55 °C for 5 min.
5. Wash the slides in:
   (a) 50 mL of 2× SSC/0.05% Tween 20 at RT for 5 min.
   (b) Change the solution and repeat once.
6. Stain with 60 μL of DAPI (0.1 μg/mL), cover with a 24 × 60 mm coverslip, and incubate at RT for 3 min.
7. Rinse the slides in nanopure water for 1 min.
8. Mount the slides with Vectashield.
9. Keep the slides at −20 °C and proceed to image acquisition the next day.

3.2.5 Image Capture and Processing

1. Use AxioImager Z1 microscope and AxioVision or Zen Pro software for image acquisition (*see* **Note 11**).
2. Set up the imaging system according to the following parameters: DAPI, auto exposure; Cy3, 300 ms (*see* **Note 12**).
3. Capture images of 30–35 nuclei for each specimen at 630× magnification. A 3D image of each nucleus consists of 80 *z*-stacks at a sampling distance of 200 nm along the *z* axis and 102 nm in the *x*, *y* direction. The constrained iterative algorithm needs to be used for deconvolution [25].

3.2.6 Image Analysis

1. Analyze 30 nuclei for each specimen by using the TeloView™ software. Telomere measurements are done with TeloView™ [18] and the six parameters defining the telomere nuclear organization are measured. By choosing a simple threshold for the telomeres, a binary image is determined, and based on that, the center of gravity of intensities is calculated for every object resulting in a set of coordinates (*x*, *y*, *z*).
2. Telomere number is defined as the number of signals detected after thresholding.
3. The integrated intensity of each telomere is calculated, and it is proportional to the telomere length [18].
4. Telomere aggregates are defined as clusters of telomeres that are found in close vicinity and cannot be further resolved as separate entities at the optical resolution limit of 200 nm.
5. The nuclear volume is calculated according to the 3D nuclear DAPI staining and corresponds to the integral of all DAPI-stained regions of the 80 *z*-stacks.
6. The a/c ratio corresponds to the nuclear distribution of all telomeres which is the nuclear space occupied by telomeres is measured by two main axes, *a* and *b*, that are equal in length, and a third axis, *c*, that has a different length. A low *a/c* ratio represents an oblate spheroid shape while a high *a/c* ratio corresponds to disk-like shape [18].
7. The distance of each telomere to center of gravity of the nucleus is calculated for each cell [18].

3.3 Quantitative Telomere FISH on Metaphases

3.3.1 Telomere Hybridization

Metaphase Preparation

1. Culture separately the LoVo cell lines (LoVo TP53-wt, LoVo TP53-V143A, LoVo TP53-R175H, and LoVo TP53-R273H) on slides by using DMEM/F12 supplemented with 10% FBS.
2. When cell confluence reaches 70%, briefly rinse the slides with 10 mL of HBSS.
3. Incubate the cells in 10 mL of trypsin solution at 37 °C for 5 min.
4. Stop the trypsin action by adding culture medium.

5. Detach the cells by gently tapping the petri dish and recover the cells as well as the supernatant in a 15 mL tube.
6. Centrifuge the tubes at 200 × *g* for 10 min.
7. Remove the supernatant, resuspend cell pellet in prewarmed sodium citrate solution, and incubate at 37 °C for 45 min.
8. Fix the cells by adding slowly drop-by-drop 10 mL of cold carnoy (−20 °C).
9. Incubate the cells for 10 min at room temperature, centrifuge at 200 × g for 10 min, and remove the supernatant.
10. Repeat the fixation process three times more.
11. Resuspend the pellet in carnoy and adjust its volume according to the cell pellet.
12. Drop the cell suspension onto precleaned slides in a controlled humidity and temperature (Thermotron) at 25 °C and 60% humidity (*see* **Note 13**).
13. Verify spreading, quality, and number of metaphases under light microscope (*see* **Note 14**).
14. Age the slides at 37 °C for overnight.

Slide Pretreatment

1. Immerse the aged slides in 50 mL of 1× TBS at RT for 2 min.
2. Incubate the slides in 50 mL of 3.7% formaldehyde/1× TBS at RT for 2 min. This incubation is done under fume hood.
3. Wash the slides in:
   (a) 50 mL of 1× TBS at RT on shaker.
   (b) Change the solution and repeat once.
4. Incubate the slides in 50 mL of the prepared pepsin/HCl solution at 37 °C for 10 min. The pepsin is just added to the 0.1 N HCl solution and mix before incubation.
5. Wash the slides in:
   (a) 50 mL of 1× TBS at RT on shaker.
   (b) Change the solution and repeat once.
6. Dehydrate the slides in a series of cold ethanol:
   (a) 70% ethanol at −20 °C for 2 min.
   (b) 90% ethanol at −20 °C for 2 min.
   (c) 100% ethanol at −20 °C for 2 min.
7. Air-dry the slides at RT for 5 min.

Hybridization

1. Apply 10 μL of PNA-telomere probe (in the dark) on a predetermined area of each slide, cover the area with a 24 × 24 mm coverslip, and seal its edge with rubber cement.
2. Prepare the hybridization device: open the ThermoBrite and submerge in water ThermoBrite humidity strips for 2 min, and place them on the ThermoBrite.

3. Denature the DNA of the specimen on a slide at 80 °C for 3 min, followed by hybridization in at 37 °C (*see* **Note 15**).
4. After the hybridization, remove carefully the rubber cement and the coverslips.
5. Wash the slides in 50 mL of 0.4× SSC/0.3%NP-40 at 65 °C for 2 min.
6. Wash the slide in 50 mL of 2× SSC/0.1% NP-40 at RT for 2 min.
7. Dehydrate slides in:
   (a) 70% ethanol for 2 min at RT.
   (b) 90% ethanol for 2 min at RT.
   (c) 100% ethanol for 2 min at RT.
8. Air-dry the slides for 3 min.
9. Apply 20 μL of the DAPI/antifade and overlay with a 24 × 60 mm coverslip.
10. Allow penetration of DAPI/antifade for 10 min at 4 °C.
11. The slides are ready to be immediately imaged, or they can be kept at −20 °C to be imaged the next day.

Image Capture

1. Prescan the slides at 100× magnification using DAPI filter to locate well-spread metaphases (*see* **Note 16**).
2. Capture at least 35 well-spread metaphases using Cy3 (200 ms exposure time) and DAPI (autoexposure) with the appropriate filters at 1000×.
3. Record the *X* and *Y* coordinates for each captured metaphase.

3.3.2 Multicolor FISH

Slide Denaturation

1. Precool a coplin jar with 50 mL of 0.1× SSC and 50 mL of 2× SSC at 4 °C at least for 30 min before use.
2. Prewarm a coplin jar with 50 of 2× SSC at 70 °C at least 30 min before use.
3. Put the same slides used for telomere hybridization (after telomere image capture) with coverslips in 50 mL of 1× PBS for 5 min with gentle agitation.
4. Remove carefully the coverslips if they are still on the slides.
5. Wash the slides in:
   (a) 50 mL of 1× PBS for 5 min with gentle agitation.
   (b) Change the solution and repeat once.
6. Put the slides into prewarmed 2× SSC solution and incubate them at 70 °C (water bath) for 30 min.
7. Remove the coplin jar from water bath; let it cool at RT for 20 min.

8. Transfer the slides into 50 mL of 0.1× SSC at RT for 1 min.
9. Denature the slides in 50 mL of 0.07 N NaOH at RT for 1 min.
10. Put the slides in the precooled solution of 0.1× SSC and keep it at 4 °C for 1 min.
11. Transfer the slides in the precooled solution of 2× SSC and keep it at 4 °C for 1 min.
12. Dehydrate slides in:
    (a) 70% ethanol at RT for 1 min.
    (b) 95% ethanol at RT for 1 min.
    (c) 100% ethanol at RT for 1 min.
13. Let the slides air-dry.

Probe Denaturation and Hybridization

1. Prepare 10 μL of MFISH human cocktail probe in a 200 μL tube.
2. Denature the probe by incubating at 75 °C for 5 min.
3. After denaturation, put the tube with the probe on ice for 1 min.
4. Incubate the tube with the probe at 37 °C for 30 min.
5. Spin briefly the tube to collect the probe.
6. Apply 10 μL of the denatured probe (in the dark) on the same predetermined area used for telomere hybridization, cover the area with a 24 × 24 mm coverslip, and seal it with rubber cement.
7. Incubate the slides in a humidified chamber at 37 °C for 2 days.

Post-hybridization Washing and Counterstain

1. Remove carefully the rubber cement and coverslips.
2. Wash the slides in prewarmed 0.4× SSC at 72 °C for 2 min.
3. Wash the slides in 2× SSCT for 30 s.
4. Wash briefly in double distilled water to avoid crystal formation and let air-dry.
5. Apply 20 μL of the DAPI/antifade and overlay with a 24 × 60 mm coverslip.
6. Allow penetration of DAPI/antifade for 10 min at 4 °C.
7. Store the slides at −20 °C.
8. Proceed with image acquisition and analysis the next day.

Imaging Acquisition and Analysis

1. Acquire images of the same metaphases captured after QFISH according to their image coordinates *XY* by using the same ISIS system.
2. Capture images in autoexposure mode by using the following filters: spectrum aqua, spectrum green, spectrum orange, spectrum red, and near infrared.

3. Karyotype and analyze 30 metaphases according to the International System for Human Cytogenomic Nomenclature.
4. Use the MFISH karyotypes as mirror images to karyotype metaphase images acquired after telomere hybridization.
5. Measure telomere signals by telomere analysis module of ISIS on each chromosome arm of metaphases acquired after telomere hybridization.
6. Export fluorescence intensity data of each telomere for further analysis.
7. Fluorescence intensities can be converted in kb. The median of fluorescence intensity of a sample (all individual telomere intensities) is determined and equated to TRF value of that sample.

4 Notes

1. All of the supplies referred to Roche in this manuscript are currently under Sigma-Aldrich license. To keep uniformity between the original manuscript and the current one, we keep the Roche reference.
2. DNA extraction methods providing long DNA fragments allow maintaining telomere integrity and avoiding extraction bias. Moreover, DNA purity should be high pure, so an accurate estimation of telomere repeat can be done without interference of telomere transcripts. From our experience, the Gentra Puregene Cell Kit is one of the best methods to get long fragments (on average 200 kb) and high quality (high purity) of DNA. Some other methods such as the phenol-chloroform-based methods also can provide high quality and long fragments of DNA, so suitable for TRF. However, column-based DNA extraction methods should be avoided because they break down DNA including telomeres into short fragments, which are not suitable to measure telomeres accurately by TRF method.
3. Both capillary transfer and vacuum transfer using 20× SSC as transfer buffer or alkaline transfer buffer can be done. However, maximum transfer efficiency and sensitivity are obtained with positively charged nitrocellulose membrane using capillary transfer with 20× SSC transfer buffers. Transfer for 6 h has shown to give sufficient sensitivity in most applications. However, for maximum sensitivity and reproducibility of the results, overnight blotting is recommended.
4. DNA cross-liking to the membrane can also be done by baking the membrane at 120 °C for 20 min.

5. The membrane can be stored at 4 °C wrapped in thin plastic film for at least 3 months.
6. Luminescence continues for at least 24 h, and signal intensity increases during the first hours. It is very important to avoid overexposition, so an accurate optical density of telomeric DNA could be measured.
7. Cells can be fixed for 10 min in 1× PBS/3.7% formaldehyde, then dehydrated in ethanol, and kept at −80 °C for months. For any future use of the slides, cells need to be rehydrated in a series of ethanol and fix for additional 10 min in 1× PBS/3.7% formaldehyde.
8. Any drying up of cells during the procedure should be avoided, and all of the steps such as washings, changing incubation media, etc. should be performed as quickly as possible by transferring the slides from one coplin jar to another.
9. Incubation in 0.1 N HCl may be extended up to 10 min for slides with densely grown cells, and the appropriate time should be adjusted according to these requirements. However, it is advised to only vary the incubation time and not the concentration of HCl.
10. The slides may be stored for at least 4 weeks in 70% formamide/2× SSC at 4 °C. Longer storage may result in deterioration of nuclear morphology after denaturation.
11. Zeiss updated AxioVision software and named it Zen Pro. The current updates of Zeiss software (Zen Pro) and hardware (stage, camera, 63× objective) allow to the capture of more cells in a short period of time. The efficiency of the system has been increased by at least tenfold. With this new system, it is possible to analyze 300 cells with this new system in the same time. Moreover, the new system can autocalibrate the fluorescence bulb by increasing the iris of the lamp in accordance with the age of the bulb. Thus, image acquisition can be standardized without worrying about the age of the bulb.
12. The exposure time of Cy3 should be determined empirically by averaging autoexposure on five different fields. Then, a half of this average is used for all of the specimens. A longer exposure time (more than 700 ms) should prompt to change the bulb of the fluorescence microscope if it cannot be autocalibrated.
13. This is a key step of the procedure. Well-spread metaphases allow to easily karyotype and measure individual telomere length accurately. It is difficult to measure telomeres of overlapping chromosomes. An adequate metaphase should have less than three overlapping chromosomes if the cell is not polyploid. Temperature and humidity are the most important factors that need to be adjusted to get successful chromosome spreading. Therefore, a device that controls temperature and

humidity such as Thermotron is essential to get consistent and successful spreading.

14. If the nuclei appear to be surrounded by any trace of cytoplasm when the slides are viewed under phase contrast, repeat the fixation step with carnoy solution. Cytoplasm interferes with the penetration of the probes leading to problems with hybridization and increased background. Always examine the sample under phase contrast microscopy at 100× magnification. Cytoplasm will appear as light gray zones around the nuclei.
15. If no ThermoBrite is available, hot plaque can be used for denaturation, and hybridization can be done at RT using humidified chamber. ThermoBrite presents the advantage to have a well-controlled environment of temperature and humidity, so the procedure can be more reproducible. The hybridization can be extended up to 4 h without any change of its quality.
16. Capturing well-spread metaphases is the key of success for subsequent steps.

References

1. Benetti R, Garcia-Cao M, Blasco MA (2007) Telomere length regulates the epigenetic status of mammalian telomeres and subtelomeres. Nat Genet 39:243–250. https://doi.org/10.1038/ng1952
2. Moyzis RK, Buckingham JM, Cram LS et al (1988) A highly conserved repetitive DNA sequence, (TTAGGG)n, present at the telomeres of human chromosomes. Proc Natl Acad Sci U S A 85:6622–6626
3. Chong L, van Steensel B, Broccoli D et al (1995) A human telomeric protein. Science 270:1663–1667
4. De Lange T, Shiue L, Myers RM et al (1990) Structure and variability of human chromosome ends. Mol Cell Biol 10:518–527
5. Blasco MA, Lee HW, Hande MP et al (1997) Telomere shortening and tumor formation by mouse cells lacking telomerase RNA. Cell 91:25–34
6. Chuang TCY, Moshir S, Garini Y et al (2004) The three-dimensional organization of telomeres in the nucleus of mammalian cells. BMC Biol 2:12. https://doi.org/10.1186/1741-7007-2-12
7. Mai S, Garini Y (2005) Oncogenic remodeling of the three-dimensional organization of the interphase nucleus: c-Myc induces telomeric aggregates whose formation precedes chromosomal rearrangements. Cell Cycle 4:1327–1331
8. Louis SF, Vermolen BJ, Garini Y et al (2005) c-Myc induces chromosomal rearrangements through telomere and chromosome remodeling in the interphase nucleus. Proc Natl Acad Sci U S A 102:9613–9618. https://doi.org/10.1073/pnas.0407512102
9. Hande MP, Samper E, Lansdorp P et al (1999) Telomere length dynamics and chromosomal instability in cells derived from telomerase null mice. J Cell Biol 144:589–601
10. Artandi SE, Chang S, Lee SL et al (2000) Telomere dysfunction promotes non-reciprocal translocations and epithelial cancers in mice. Nature 406:641–645. https://doi.org/10.1038/35020592
11. Mc CB (1939) The behavior in successive nuclear divisions of a chromosome broken at meiosis. Proc Natl Acad Sci U S A 25:405–416
12. Maciejowski J, de Lange T (2017) Telomeres in cancer: tumour suppression and genome instability. Nat Rev Mol Cell Biol 18:175–186. https://doi.org/10.1038/nrm.2016.171
13. Yates LR, Campbell PJ (2012) Evolution of the cancer genome. Nat Rev Genet 13:795–806. https://doi.org/10.1038/nrg3317
14. Davoli T, Denchi EL, de Lange T (2010) Persistent telomere damage induces bypass of mitosis and tetraploidy. Cell 141:81–93. https://doi.org/10.1016/j.cell.2010.01.031
15. Davoli T, de Lange T (2012) Telomere-driven tetraploidization occurs in human cells under-

going crisis and promotes transformation of mouse cells. Cancer Cell 21:765–776. https://doi.org/10.1016/j.ccr.2012.03.044

16. Rausch T, Jones DTW, Zapatka M et al (2012) Genome sequencing of pediatric medulloblastoma links catastrophic DNA rearrangements with TP53 mutations. Cell 148:59–71. https://doi.org/10.1016/j.cell.2011.12.013
17. Samassekou O, Bastien N, Lichtensztejn D et al (2014) Different TP53 mutations are associated with specific chromosomal rearrangements, telomere length changes, and remodeling of the nuclear architecture of telomeres. Genes Chromosomes Cancer 53:934–950. https://doi.org/10.1002/gcc.22205
18. Vermolen BJ, Garini Y, Mai S et al (2005) Characterizing the three-dimensional organization of telomeres. Cytometry A 67:144–150. https://doi.org/10.1002/cyto.a.20159
19. Lansdorp PM, Verwoerd NP, van de Rijke FM et al (1996) Heterogeneity in telomere length of human chromosomes. Hum Mol Genet 5:685–691
20. Londono-Vallejo JA, DerSarkissian H, Cazes L et al (2001) Differences in telomere length between homologous chromosomes in humans. Nucleic Acids Res 29:3164–3171
21. Samassekou O, Ntwari A, Hebert J et al (2009) Individual telomere lengths in chronic myeloid leukemia. Neoplasia 11:1146–1154
22. Pocard M, Chevillard S, Villaudy J et al (1996) Different p53 mutations produce distinct effects on the ability of colon carcinoma cells to become blocked at the G1/S boundary after irradiation. Oncogene 12:875–882
23. Dridi W, Fetni R, Lavoie J et al (2003) The dominant-negative effect of p53 mutants and p21 induction in tetraploid G1 arrest depends on the type of p53 mutation and the nature of the stimulus. Cancer Genet Cytogenet 143:39–49
24. Masramon L, Ribas M, Cifuentes P et al (2000) Cytogenetic characterization of two colon cell lines by using conventional G-banding, comparative genomic hybridization, and whole chromosome painting. Cancer Genet Cytogenet 121:17–21
25. Schaefer LH, Schuster D, Herz H (2001) Generalized approach for accelerated maximum likelihood based image restoration applied to three-dimensional fluorescence microscopy. J Microsc 204:99–107

Chapter 15

Genes, Proteins, and Biological Pathways Preventing Chromothripsis

Martin Poot

Abstract

The highly complex structural genome variations chromothripsis, chromoanasynthesis, and chromoplexy are subsumed under the term chromoanagenesis, which means chromosome rebirth. Precipitated by numerous DNA double-strand breaks, they differ in number of and distances between breakpoints, associated copy number variations, order and orientation of segments, and flanking sequences at joining points. Results from patients with the autosomal dominant cancer susceptibility disorder Li-Fraumeni syndrome implicated somatic *TP53* mutations in chromothripsis. *TP53* participates in the G2/M phase checkpoint, halting cell cycling after premature chromosome compaction during the second half of the S phase, thus preventing chromosome shattering. By experimental *TP53* ablation and micronucleus induction, one or a few isolated chromosomes underwent desynchronized replication and chromothripsis. Secondly, chromothripsis occurred after experimental induction of telomere crisis after which dicentric chromosomes sustained TREX1-mediated resolution of chromosome bridges and kataegis. Third, DNA polymerase Polθ-dependent chromothripsis has been documented. Finally, a family with chromothripsis after L1 element-dependent retrotransposition and Alu/Alu homologous recombination has been reported. Human chromosomal instability syndromes share defects in responses to DNA double-strand breaks, characteristic cell cycle perturbations, elevated rates of micronucleus formation, premature chromosome compaction, and apoptosis. They are also associated with elevated susceptibility to malignant disease, such as medulloblastomas and gliomas in ataxia-telangiectasia, leukemia and lymphoma in Bloom syndrome, and osteosarcoma and soft tissue sarcoma in Werner syndrome. The latter syndrome is characterized by a premature aging-like progressive decline of mesenchymal tissues. In all thus far studied cases, constitutional chromothripsis occurred in the male germline and male patients with defects in the double-strand break response genes *ATM*, *MRE11*, *BLM*, *LIG4*, *WRN*, and *Ku70* show impaired fertility. Conceivably, chromothripsis may, in a stochastic rather than deterministic way, be implicated in germline structural variation, malignant disease, premature aging, genome mosaicism in somatic tissues, and male infertility.

Key words Chromoanagenesis, Chromoanasynthesis, Chromoplexy, Nonhomologous end joining (NHEJ), Microhomology-mediated break-induced repair (MMBIR), TP53, WRN helicase exonuclease

Franck Pellestor (ed.), *Chromothripsis: Methods and Protocols*, Methods in Molecular Biology, vol. 1769, https://doi.org/10.1007/978-1-4939-7780-2_15, © Springer Science+Business Media, LLC, part of Springer Nature 2018

1 Introduction

Viewed through the microscope, the genomes of all eukaryotes are contained within a set of chromosomes with a characteristic number and morphology, collectively known as the "karyotype." A century ago deviations from this karyotype have been observed in samples from some patients and malignant tissue and labeled as "chromosomal aberrations" [1]. The steady improvement in resolution of molecular cytogenetic techniques has widened the scope of "chromosomal aberrations" to what is now known as "structural variations" (SVs). Depending on the number of chromosome breaks, SVs have been classified as "simple," e.g., reciprocal translocations, inversions and copy number variations (CNVs) with one or two breaks, and complex chromosome rearrangements (CCRs) with more than two breaks [2]. CCRs resulting from errors during meiosis or during DNA replication in somatic cells consist of insertion translocations, inversions associated with CNVs, complex translocations, and any combinations thereof and may affect all chromosomes [2–4]. Extremely complex and instable CCRs, which may lead to "chromosome chaos," have been revealed by FISH and spectral karyotyping of tumor cells [5]. Recently a novel class of CCRs with an unprecedented number of breakpoints, but affecting only a limited number of chromosomes, has been described [6–13].

Holland and Cleveland [9] proposed the term "chromoanagenesis," derived from the Greek chromo for chromosomes and anagenesis, to be regenerated, to classify these forms of CCRs independent of their putative provoking mechanisms. Chromoanagenesis encompasses three types: chromothripsis, chromoanasynthesis, and chromoplexy [6–8, 14, 15]. Chromothripsis results from a single event in which one or several chromosomes are broken up in pieces and reassembled in random order and orientation, during which several segments are lost [7]. Chromoanasynthesis, in contrast, produces clusters of rearrangements with altered DNA copy numbers by serial, microhomology-mediated template switching during DNA replication [6]. Chromoplexy generates chains of translocations, involving up to eight chromosomes, with frequent deletions at their breakpoints [14].

The three forms of chromoanagenesis as revealed by multiple pairwise FISH analyses and spectral karyotyping are shown in Fig. 1, assuming that three chromosomes are affected in each case. After chromosome shattering, during chromothripsis, or reshuffling, during chromoanasynthesis, the chromosome fragments are joined together in random order and orientation. During chromoplexy, in contrast, the original chromosomal orientation is retained. After chromothripsis, the de novo junctions show little or no microhomologies [7, 12, 15–17]. After chromoanasynthesis and chromoplexy, the junctions are flanked by regions of microhomol-

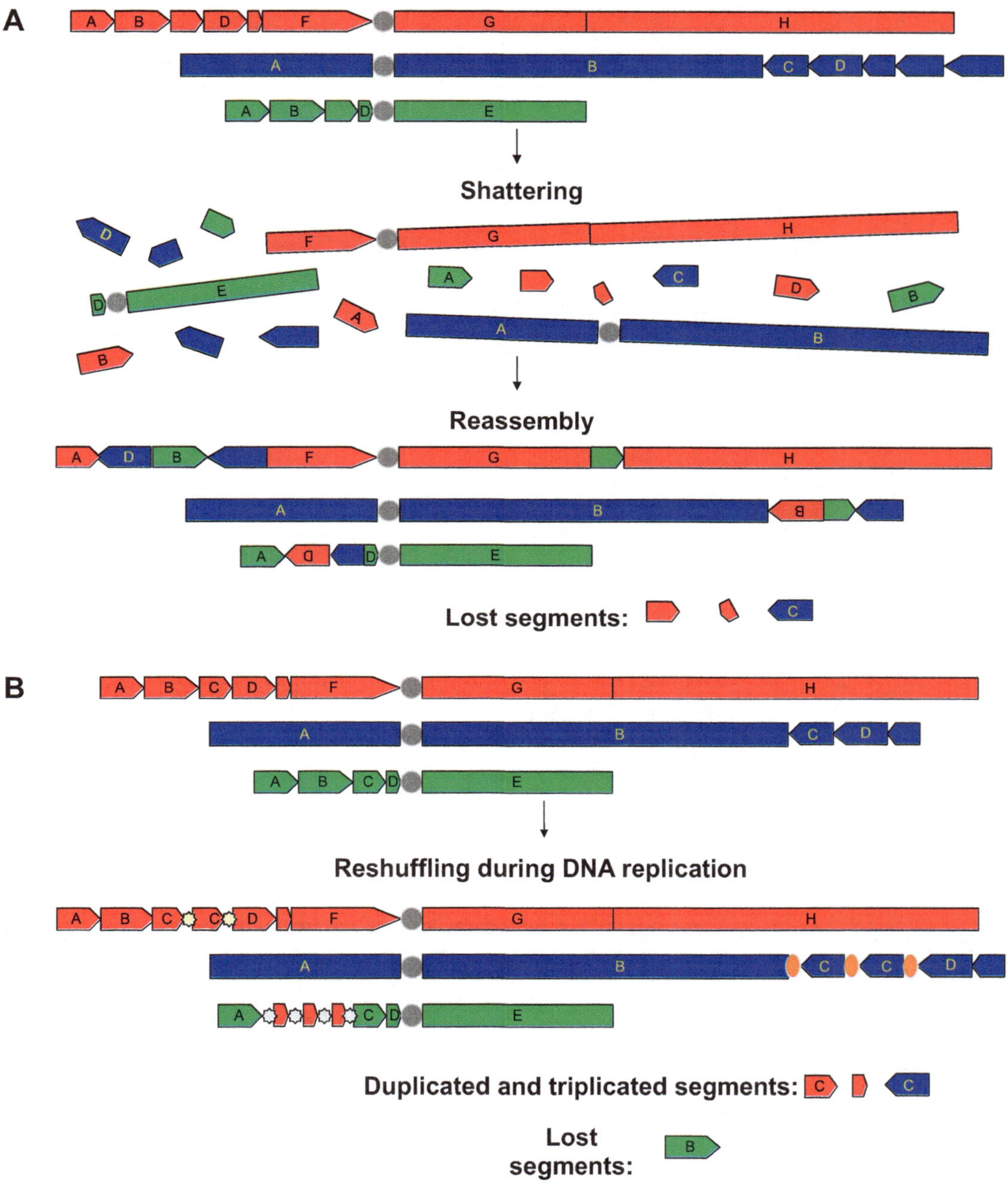

Fig. 1 Schematic depiction of the three types of chromoanagenesis: (**a**) chromothripsis, (**b**) chromoanasynthesis, (**c**) chromoplexy. Each chromosome is drawn with a different color. Arrows indicate affected chromosome segments and their orientation; arrowheads represent segment heads and blunt-end segment tails. Gray ellipses represent centromeres; orange and white symbols represent inserted sequences

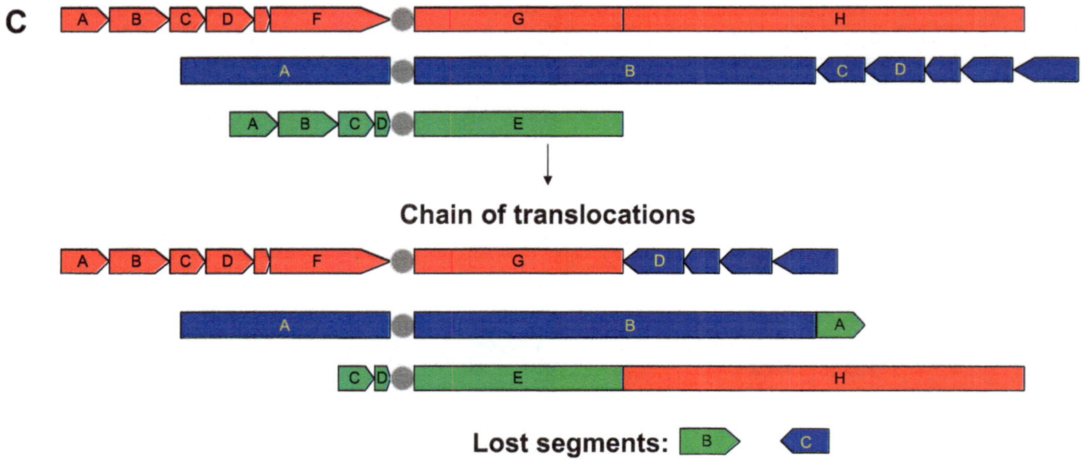

Fig. 1 (continued)

ogy, and often considerable resection at the breakpoints and insertion of non-templated sequences have occurred. This suggests that all three processes are a response to DNA double-strand breaks (DSBs) [18]. While in the case of chromothripsis this response predominantly consists of nonhomologous end joining (NHEJ), chromoanasynthesis and chromoplexy most likely involve DNA replication-based mechanisms such as fork stalling and template switching (FoSTeS) or microhomology-mediated break-induced repair (MMBIR) [4]. Recently, three cases of chromoanasynthesis with duplications and insertion translocations within a single chromosome have been described [19]. Here, the breakpoint junctions consisting of template-dependent microhomologies and non-templated insertions are most likely created by a DNA polymerase Polθ-dependent pathway of alternative NHEJ [19]. In an unrelated study, chromothripsis in 11 siblings of a three-generation family appeared to be mediated L1 element-dependent retrotransposition and Alu/Alu homologous recombination (HR) [20]. The shared and distinguishing features of the thus far known forms of chromothripsis and of chromoanasynthesis, chromoplexy, and atypical chromoanagenesis are summarized in Table 1.

While chromoanasynthesis or chromoplexy is accompanied by both losses and gains, chromothripsis frequently gives rise to losses of acentric chromosomal segments leading to multiple alternations between losses and diploid chromosomal segments, in which heterozygosity is retained (Fig. 2). This excludes loss of one homologue and partial endoduplication of the retained homologue as a possible mechanism of origin of chromothripsis. Thus, based on this difference in alterations of DNA copy number states, chromothripsis can be unambiguously distinguished from chromo-

Table 1
Shared and distinguishing features of chromothripsis, chromoanasynthesis, and chromoplexy

	Chromothripsis			Chromoanasynthesis	Chromoplexy	Atypical chromoanagenesis
Cause or precipitating event	Micronucleus formation	TREX1-mediated fragmentation of chromosome bridges	Retrotransposition	?	Simultaneous DSBs in several chromosomes	?
Number of steps	One-off	One-off	One-off	One-off	In chains	One-off
Number of involved chromosomes	1 up to 5	2 or 3	1 or 2	1	Up to 6 or even more	1
Distribution within chromosome	Throughout the entire chromosome	Within one chromosome arm	Throughout the entire chromosome	Throughout the entire chromosome	Throughout the entire chromosome	Throughout the entire chromosome
Breakpoints	Clustered	Clustered	Clustered	Random	Random	Random
Copy number states	Losses only	Losses + gains	?	Losses + gains	Losses + gains	Gains only
Order of segments	Random	Random	Random	Random	Random	Random
Orientation of segments	Random	Random	Random	Random	?	Random

(continued)

Table 1 (continued)

	Chromothripsis			Chromoanasynthesis	Chromoplexy	Atypical chromoanagenesis
Sequences at joining points	No or very short microhomologies	?	LINEs, SINEs	Microhomologies + insertions	?	Non-templated insertions
Joining mechanism	MMBIR	?	DNA polymerase Polθ driven alt-NHEJ	MMBIR/FoSTeS	Probably canonical NHEJ	DNA polymerase Polθ driven alt-NHEJ
Key references	Zhang et al. [21]	Maciejowski et al. [22]	Nazaryan-Petersen et al. [20]	Liu et al. [6]	Baca et al. [14]	Masset et al., [19]

Question marks indicate not known
For explanation of abbreviations, *see* text

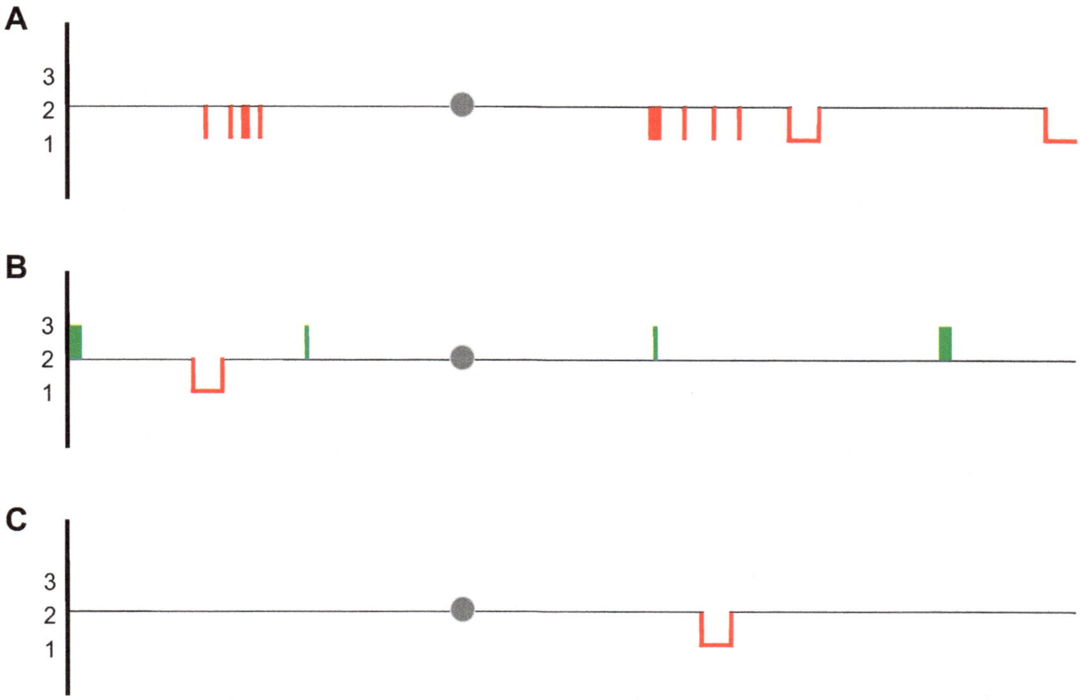

Fig. 2 CNVs associated with chromothripsis (panel **A**), chromoanasynthesis (panel **B**), and chromoplexy (panel **C**). Abscissa chromosome position; ordinate copy number. Red downward symbols represent copy number losses and green upward symbols gains. Gray ellipses represent centromeres

anasynthesis and chromoplexy. These processes are not entirely exclusive, however. Zhang et al. [10] pointed out that a chain of translocations may give rise to chromoplexy, which produces an instable karyotype that subsequently may precipitate further rearrangements and losses of acentic fragments resulting in chromothripsis [10, 15]. A web-based algorithm automatically infers chromothripsis from next-generation sequence data by considering, among other features, closely spaced alternations in DNA copy number states [23]. While this kind of data as such generates important insights, only combined with results from molecular cytogenetic and next-generation sequence analyses chromothripsis can be unambiguously identified [5, 7, 11, 15, 16, 24–27]. While this field of research is still rapidly evolving, several excellent reviews have covered potential mechanisms of origin of CCRs in general and replication-based processes such as chromoanasynthesis and chromoplexy [2, 6, 11–13]. Therefore, this review will focus on the genes, proteins, and biological pathways that guard the genome against chromothripsis.

2 The *TP53* Gene and Specific Mechanisms of Chromothripsis

Chromothripsis was initially described in a patient with chronic lymphocytic leukemia, and its characteristic pattern of alternating losses and retention of diploid chromosomal segments was subsequently detected in primary tumor samples from 20 patients with bone cancer [7]. In some samples, loss of *CDKN2A* and *FBXW7* and a disruption near *WRN* were found, which were a consequence of rather than a prerequisite for chromothripsis. An independent study of a Sonic Hedgehog medulloblastoma brain tumor from a patient with a germline *TP53* mutation (Li-Fraumeni syndrome) and subsequent confirmation in a pediatric medulloblastoma indicate a predisposing role for chromothripsis of the p53 protein, known as the "guardian of the genome" [8, 28]. Since *TP53* participates in controlling the G2/M checkpoint [28], premature chromosome compaction during the second half of the S phase may have provoked shattered chromosomes [29, 30]. Such premature chromosome compaction is frequently observed in chromosomal instability syndromes, which typically show cell cycle perturbations, elevated rates of micronucleus formation, and apoptosis in response to DNA damage [2, 31–33]. Cells from patients with Li-Fraumeni syndrome, an autosomal dominant disorder associated with increased susceptibility to cancer, show shortened telomeres lacking telomere repeats caps [34, 35]. This telomere shortening precipitates chromosome end-to-end fusions and subsequent breakage during a p53-dependent inhibition of the G2/M transition [36]. Thus, the observations of Rausch et al. [8] offered two potential mechanisms of chromothripsis, each of which meriting further experimental investigation.

3 Cell Cycle Desynchronization, Micronucleus Formation, and Chromothripsis

To test the hypothesis that chromosomes, after isolation into micronuclei, may undergo cell cycle perturbation and thus acquire DNA damage, possibly leading to chromosome pulverization, Crasta et al. [30] used nocodazole to block cultures of retinal pigment epithelial cells at the G1/S border of the cell cycle. After release from this blockage and defeating the p53-dependent cell cycle checkpoint by p53 knockdown with siRNA, the cells were allowed to progress during one cell cycle, and subsequently mitotic cells were isolated [30]. After these cells had progressed through another G1 phase and entered the S and G2 phase of the cell cycle, increasing numbers of micronuclei staining positive by TUNEL and for DNA damage-dependent phosphorylation of the histone variant H2AX20 were detected. TUNEL detects DNA DSBs, while the presence of H2AX20 is evidence for DSBs in the micronuclei [37, 38]. Similar results were obtained after knockdown of the kinetochore-associated microtubule depolymerase MCAK26 or the kinetochore protein Nuf227. MCAK26 knockdown does not delay cells in mitosis, which indicates that the DNA damage in micronuclei is independent of mitotic arrest. Since the DNA damage in the micronuclei did not activate caspase-3, induction of apoptosis was excluded. Thus, the DNA damage in the micronuclei resulted from a desynchronization of their cell cycle traverse relative to the corresponding main nucleus. If micronuclei are reintegrated into the main nucleus, the chromosomes from the micronuclei proved to be fragmented and subsequently reassembled by a process bearing hallmarks of chromothripsis [21]. The authors demonstrated that before reintegration into the main nucleus, the under-replicated chromosomal fragments in the micronuclei were most likely joined by MMBIR.

But for the mechanism by which the chromosome fragments were joined, isolation of one or a few chromosomes into a micronucleus and their consequent desynchronized cell cycle traverse is fully congruent with the reported cases of germline chromothripsis [16, 17]. The discrepancy in chromosome fragment joining mechanisms can be explained, if we take into account that germline chromothripsis has taken place during meiosis, whereas the experimental form of chromothripsis in continuously cycling somatic cells in which the p53-dependent cell cycle checkpoint has been defeated. In cycling cells DSBs are preferentially repaired by DNA replication-dependent mechanisms such as FoSTeS and MMBIR [39]. In contrast, during meiosis two rounds of chromatid division and a single round of DNA replication occur. After the second meiotic division, no homologous chromosomes are available, and therefore DSBs can only be repaired by NHEJ. Since in cases of germline chromothripsis no or very short homologies flanking the joining points and no insertions have been observed, this form of chromothripsis had to have taken place after meiosis has been completed. Since NHEJ is downregulated during spermatogenesis, DSBs, resulting from SPO11-endonuclease activity or other causes [3], will only poorly be repaired before fertilization. This limited DSB repair capacity and the elevated sensitivity of spermatids to oxidative stress make the male germline particularly sensitive to formation of CCRs and chromothripsis [2, 40, 41]. Only if, after fertilization, DSBs in the sperm nucleus are repaired by proteins from the oocyte, a proliferation-competent zygote can be formed, albeit that the genome of this zygote may bear the hallmarks of chromothripsis. This hypothetical mechanism may explain why chromothripsis has only been observed in the male germline [16, 17].

4 Telomere Crisis Resulting in Chromothripsis Combined with Kataegis

In a hTERT-expressing derivative of a RPE-1 retinal pigment epithelial cell line in which both the Rb and the TP53 pathways were disabled, telomere crisis was induced by expressing a dominant negative allele of TRF2, which deprotects telomeres and induces telomere fusions [22]. As expected, 53BP1-containing foci indicative of telomere dysfunction emerged; ATM kinase signaling was activated and cell proliferation impaired. No micronuclei, but metaphases with telomere fusions, were readily observed. The dicentric chromosomes persisted through anaphase to become long chromatin bridges that induced rupture of the nuclear envelopes. The chromatin bridges were coated by Replication Protein A (RPA). Subsequently, the cytoplasmic 3′->-5′ exonuclease TREX1 converted the chromatin bridges into single-stranded DNA, thus facilitating their resolution. The authors isolated cell clones and confirmed in clones with aneuploidy and/or marker chromosomes that telomere fusions had indeed taken place. By whole-genome sequencing clusters of genomic rearrangements, affecting one or more chromosomes and exhibiting hallmarks of chromothripsis were found [15]. These chromothripsis events consisted in particular of terminal deletions, indicating telomere fusion. Also more proximal chromothripsis-associated changes, being limited to a single chromosome arm, were found. Never an entire chromosome was affected. These characteristics are consistent with TREX1-mediated chromothripsis after a telomere crisis during which dicentric chromosomes underwent breakage-fusion-bridge cycles [22]. A telomere crisis may also precipitate reciprocal translocations and even chromoplexy [42]. The latter process generates several possibly instable derivative chromosomes, which may subsequently trigger chromothripsis [10].

The RPA-coated chromatin bridges contained extended stretches of ssDNA, which makes them a target substrate for APOBEC enzymes that convert clusters of cytosines to thymidines or uracils in stretches of ssDNA formed in the vicinity of DSBs and stalled DNA replication forks [43]. This process is known as kataegis. Maciejowski et al. [22] found 29 clusters of point mutations in 7 out of 10 cell clones, which were absent from the parental cell lines. These clusters exhibited the cardinal features of kataegis as observed earlier in human cancers. First, the clusters were often found within a kilobase or two from genomic breakpoints and predominantly associated with chromothripsis. Second, the clusters showed a pronounced preference for C > T and C > G mutations within a TpC context, which is the classic signature of APOBEC3A/B activity [44]. Third, the mutation clusters were processive, which indicates that the damage occurred on a single strand of DNA.

The close association of this form of chromothripsis with kataegis and the finding that it involves telomere crisis and chromatin bridges, but no micronucleus formation, clearly sets it apart from the micronucleus-dependent mechanism of chromothripsis [21]. For chromothripsis after telomere crisis to occur, both the Rb and the TP53 pathways have to be disabled, whereas the micronucleus-dependent mechanism of chromothripsis requires inactivation of TP53 only [21, 22, 42]. Except for patients with Li-Fraumeni syndrome, this is unlikely to occur in the germline but may very well take place in somatic cells during malignant disease [8, 42, 45, 46].

5 Retrotransposition-Mediated Chromothripsis

In postmitotic somatic cells, retrotransposition of DNA segments flanked by long interspersed elements (LINEs) may mediate CNV by nonallelic homologous recombination (NAHR) [47]. The human genome contains some 17,000 LINEs spaced less than 10 Mb apart and probably involved in germline CNVs [48]. Recently a retrotransposition-mediated, germline chromothripsis event involving six breaks in chromosome arm 3q and 1 break in 5q was reported [20, 49]. At two breakpoint-joining sites, no homologies, deletions, insertions, and inversions were found, which is consistent with NHEJ involving the Ku70/80 heterodimer and the DNA-ligase-IV/XRCC4 complex [50]. At a second pair of breakpoint-joining sites, involving a paracentric inversion, two very short microhomologies were detected, which indicates that these breakpoints were ligated by MMBIR. The remaining two breakpoint-joining sites harbor a 177 base pair homology and an inserted retrotransposition. This has not previously been reported in cases of chromothripsis. Conceivably, Alu elements residing at four breakpoints may have triggered formation of two large hairpin loops, which then facilitated physical interactions between distally located DNA segments on chromosome arms 3q and 5q. The presence of inverted Alu elements (AluSq and AluJb) at one pair of breakpoints is in agreement with Alu/Alu recombination-mediated DSB repair. Thus, L1 endonuclease activity may have produced several DNA breaks and have mediated retrotransposition of SINE-VNTR-Alu elements close to a deletion. L1 endonuclease and SINE-VNTR-Alu elements are expressed in human germinal vesicles, and de novo retrotransposition events have been found in human primary oocytes [51]. While this is the first report of chromothripsis involving L1 endonuclease-mediated retrotransposition, some 20 simple pathogenic rearrangements sharing four characteristics with the present chromothripsis event have been reported [20]. First, all insertions occurred at L1 endonuclease cleavage sites. Second, the 5′-ends of almost all insertions

(15 out of 17) are extremely truncated with some having retained only poly-A tails. Third, large deletions always occur at the 5′-end of the insertion. Fourth, none of the cases showed duplications at their target sites. Thus, retrotransposition-mediated chromothripsis and CNV formation represent a replication and micronucleus-independent mechanism of chromosome rearrangements, which is not limited to the germline and may also produce somatic deletion mosaicism in, by example, the brain [52, 53].

6 DSB Repair Pathways Guarding the Genome Against Chromothripsis

DSBs are detected by DNA-dependent protein kinases, which thereupon form complexes with heterodimers of Ku70 and Ku80 [54]. During the G1 and early S phase of the cell cycle, a complex of MRE11, RAD50, and NBS1 binds to these DSBs and recruits ATM, which then phosphorylates nucleosomal targets, such as H2AX, resulting in γH2AX20 [55]. Subsequently, γH2AX20 recruits another ATM target, MDC1, which upon phosphorylation recruits the E3 ubiquitin ligase RNF8, and after recruitment of a second E3 ubiquitin ligase, RNF168, histone H2A is ubiquitinated. Histone H2A ubiquination, together with H4K20 methylation, recruits 53BP1. Thereupon, 53BP1 is phosphorylated at its N terminus by ATM creating docking sites for RIF1 and PTIP. 53BP1 blocks resection at DSBs, thereby channeling DSB repair toward Ku70/80-, WRN-, XRCC4-, and DNA-ligase-IV-dependent canonical NHEJ. Long-range resection is also controlled by HELB binding to RPA and inhibition of Exo1-BLM-DNA2-mediated end resection. In concert these exonucleases and helicases perform limited DNA end resection, which paves the way for the PARP1-, CtIP-, MRE11-, and ligase I/III-dependent DNA polymerase θ-mediated alternative NHEJ [55]. This latter pathway is likely to be involved in retrotransposition-mediated chromothripsis and in the atypical form of chromoanagenesis described recently [19, 20]. This particular form of chromoanagenesis characteristically involves a single chromosome and is accompanied by duplications and non-templated insertions, which distinguishes it from chromothripsis.

7 Molecular Switches Governing DSB Repair Pathway Choice

The cellular DSB response mechanisms are not only of a bewildering complexity; they also harbor several molecular switchers between alternative, mutually compensating pathways. If the DSB contains short single-strand overhangs, their repair is shifted toward the PARP1-, CTIP-, MRE11-, and ligase I/III-dependent DNA polymerase θ-mediated alternative NHEJ pathway. This inserts 5–25 base pair long stretches of microhomologous DNA before DSB ligation [55]. If the DSB ends are blunt, the Ku70/80 complex binds to the DSB and inhibits homologous recombination and the alternative NHEJ pathway [54, 56]. In that case, the WRN-, XRCC4-, and DNA-ligase-IV-dependent canonical NHEJ pathway repairs the DSB by adding 0–5 base pair microinsertions before ligation.

The WRN protein activates the ATR CHK1-induced S-phase checkpoint, recruits and stabilizes Rad51, and suppresses the activity of the MRE11-encoded exonuclease [57, 58]. Thus, in the absence of WRN, the MRE11-encoded 3′-5 ′ exonuclease degrades the DNA at DSBs [59]. This may account for the losses found at ligation sites of DSBs in WRN-deficient cells [60]. CDK1-dependent phosphorylation later in S phase is also required for WRN-DNA2-mediated DNA resection as part of HR. Thus, in the second half of the S phase, WRN phosphorylation serves as a molecular switch between the HR and NHEJ pathways for DSB repair [61]. The BLM protein, on the other hand, forms a complex with topoisomerase IIIα, RMI1, and RMI2 and cooperates with DNA2 helicase-exonuclease and RPA to promote DNA end resection and to prevent alternative NHEJ involving CtIP, Mre11, PARP1, ligase I/III, and DNA polymerase θ [62–65]. In addition, both BLM and WRN interact physically with DNA2 and coordinate their enzymatic activities to mediate 5′-3′ DNA end resection [65].

8 Chromothripsis in Somatic Cells

After the pioneering discovery in chronic lymphocytic leukemia and bone cancer samples, chromothripsis has been well documented in many cases of human malignant disease [18, 45, 46]. Analyzing data from 18,394 genome-wide CNV screens, Cai et al. [45] found 918 cases with chromothripsis-like patterns. Three cancer types strongly surpassed the average of 4.99% of chromothripsis-like patterns: lipoid sarcomas (53.5%), fibroid tumors (23.7%), and "other sarcomas" (22.9%). In a recent review, the preponderance of chromothripsis among soft tissue tumors was largely confirmed [46]. A few other types, such as Sonic Hedgehog medulloblastoma with mutant *TP53* (100%), acute lymphoblastic leukemia with intrachromosomal amplification of chromosome 21 (88.9%), acute myeloid leukemia with mutant *TP53* (47.1%), invasive bladder carcinoma (60%), lung adenocarcinoma (33.3%), and glioblastoma (38.9%) showed even higher prevalences of chromothripsis [46]. These results suggest that chromothripsis is particularly prevalent in soft tissue tumors and those with mutant *TP53* [8, 46]. Since *TP53* is pivotal for cellular DSB response pathways, this suggests that loss of *TP53* function may be a prerequisite for chromothripsis. Patients with a complete loss of *WRN*, another gene controlling cellular DSB response pathways (*see* above), show an excess of soft tissue sarcoma, osteosarcoma, myeloid disorders, and benign meningioma, which resembles prevalences of chromothripsis among tumor types [45, 46, 66]. Lymphoblastoid cells from WRN patients show hypersensitivity to DNA cross-linking drugs by undergoing S-phase apoptosis, whereas fibroblasts responded to these challenges with impaired cell proliferation [67–69]. Reduced proliferation of damaged mesenchymal cells in WRN patients may lead to earlier proliferative senescence, and their failure to undergo apoptosis in response to DNA damage may promote genomic instability. Thus, persistence of mesenchymal cells carrying DNA damage may account for the increased risk of sarcoma and loss of proliferative homeostasis in WRN patients [68, 70]. Extrapolating these findings, it is conceivable that some types of human malignant disease will undergo apoptosis, and therefore possible chromothripsis, as a result of DNA damage, may go undetected. Others, such as mesenchymal cells and those with *TP53* defects, escape apoptosis and will persist with genomes bearing the hallmarks of chromothripsis.

Cells from patients with ataxia-telangiectasia, due to functional loss of both alleles of the *ATM* gene, exhibit less efficient joining of DSBs, which was accompanied by insertions or complex mutations at the joining sites [71]. Acute lymphoblastic leukemias from patients with ataxia-telangiectasia showed an increased prevalence of chromothripsis [18]. These findings again link defective DSB responses and loss of ATM function in particular to chromothripsis.

If chromothripsis were to be caused by a random process, e.g., oxidative DNA damage [3], the likelihood of being affected would be proportional to the size of a chromosome. Contrary to this expectation, not the largest chromosome, number 1, but chromosome 17 followed by 8, 12, 11, and 6 turn out to be the most frequently involved in chromothripsis of tumor cells [45]. The cell cycle checkpoint, DNA damage response, and apoptosis controlling *TP53* gene are located within region 17p13.1. In addition, *TP53* losses were found in 438 out of 918 chromothripsis-like cases which corroborates its pivotal role in regulating cellular response to DNA damage in connection with chromothripsis, cell cycle arrest, or apoptosis [45]. In contrast to chromothripsis in malignant disease, germline chromothripsis affects chromosome 5 the most often (eight times in 62 affected chromosomes), followed by chromosome 7 (six times), and 3 as well as 9 (five times each) [12]. This may be due to different mechanisms of origin or selective pressures in somatic cells vs. the germline.

In malignant disease 76% of cases with a chromothripsis-like pattern affected a single chromosome, while in 24% of cases, two or more chromosomes were affected [45]. Among germline chromothripsis in only 3 out of 22 cases, a single chromosome was affected [12]. Cai et al. [45] inferred that this makes micronucleus formation accompanied by cell cycle desynchronization less likely for tumor chromothripsis, but rather supports a process like aborted apoptosis. This difference in number of affected chromosomes suggests that chromothripsis in the germline may be mediated by different processes than those operating in somatic cells.

9 Human and Mouse Mutants as Suspects for Germline Chromothripsis

Surprisingly, for only a few genes involved in DSB response pathways, i.e., *ATM*, *MRE11*, *BLM*, *LIG4*, and *WRN*, human and mouse mutants and their phenotypes have been described [2, 40, 72, 73]. Most likely loss of key regulatory genes, such as DNA-protein kinases, may not be compatible with life [54]. Patients with defects in *ATM*, *MRE11*, *BLM*, *LIG4*, and *WRN*, present with syndromes such as ataxia-telangiectasia (OMIM 208900; gene ATM), ataxia-telangiectasia-like disorder 1 (OMIM 604391; gene MRE11), Bloom syndrome (OMIM 210900; gene BLM), LIG4 syndrome (OMIM 606593; gene, LIG4), and Werner syndrome (OMIM 277700; gene WRN) and exhibit spontaneous chromosomal instability (ATLD1, BLM, WRN) or elevated sensitivity to ionizing radiation (ATM, LIG4). Absence of the WRN-encoded 3′-5′ helicase-3′-5′ exonuclease causes variegated translocation mosaicism in fibroblast cultures, spontaneous deletions in vivo and in vitro, prolongation of the S phase of the cell cycle, hypersensitivity to DNA cross-linking and DSBs provoking agents, elevated frequencies of micronuclei, and a reduction in recombinational DSB repair [31–33, 67–69, 74–78]. Cells from BLM patients and from Ku70 knockout mice showed a spontaneously increased rate of sister chromatid exchange [79–81]. In addition, chicken DT40 cells in which both BLM alleles have been deleted are hypersensitive to the DNA topoisomerase II inhibitor etoposide, DSB causing bleomycin, and DNA cross-linking 4-nitroquinoline-1-oxide, whereas they show normal sensitivity to X-ray irradiation and hydroxyurea [82]. While SV-40-transformed fibroblasts from ATM patients show less efficient DSB joining and a higher mutation frequency, cells from WRN patients show normal DSB joining frequencies but higher mutation rates, which were mainly due to deletions [60, 71]. These findings underscore the different functions of ATM and WRN in the cellular DSB response and probably chromothripsis.

Assuming that chromothripsis may be an outcome of defective DSB responses, it seems surprising that as yet no chromothripsis has been reported for patients with ATM, ATLD1, BLM, and WRN. Since patients with these disorders are very rare, they may have not yet been investigated. Thus far, all cases of constitutional chromothripsis have taken place in the male germline [12, 17]. The limited capacity to respond to DSBs and the hypersensitivity of spermatids to oxidative stress may make the male germline particularly sensitive to CCRs and chromothripsis [2, 40, 41]. Moreover, male patients with ATM, ATLD1, BLM, and WRN show impaired fertility, which may also be the result of defective cellular DSB response. Thus, hypogonadism and impaired spermatogenesis in ATM, hypospadias and cryptorchidism in ATLD1, azoospermia and cryptorchidism in BLM, complete azoospermia in WRN, and impaired spermatogenesis in epididymidies of Ku70 −/− mice may preclude chromothripsis in male patients with these disorders [2, 40, 70].

10 In Lieu of Conclusions: Ramifications of Chromothripsis and Future Prospects

To differentiate between chromothripsis, chromoanasynthesis, and chromoplexy, numbers of and distances between breakpoints, associated copy number states, order and orientation of segments at joining points, and flanking sequences have been used (Fig. 1) [10, 12, 13, 15]. The micronucleus-mediated form of chromothripsis adheres to all criteria of Korbel and Campbell [15], except for the breakpoint-joining mechanism, being MMBIR instead of NHEJ. Chromothripsis resulting from TREX1-mediated resolution of chromosome bridges [22] is accompanied by kataegis and both losses and gains of chromosome fragments. The latter feature it shares with chromoanasynthesis and chromoplexy. Thus, chromothripsis after TREX1-mediated resolution of chromosome bridges can be distinguished from the other forms of chromothripsis based on the nature of the associated copy number states. In contrast to chromoanasynthesis and chromoplexy, the TREX1-mediated form of chromothripsis affects only one arm of a chromosome, while the breaks and junctions in all other forms of chromothripsis, chromoanasynthesis, and chromoplexy and the atypical form of chromoanagenesis reported by Masset et al. [19] are distributed throughout the entire chromosome. The absence of microhomologies at the chromosome segment joining points, which has originally been taken as a key feature of chromothripsis [16], has become considerably less clear-cut in subsequent studies [12, 17, 20]. Conceivably, both canonical and alternative NHEJ may be involved in the breakpoint-joining process [17, 83]. Nevertheless, the patterns emerging after numbers of and distances between breakpoints, associated copy number states, order and orientation of segments at joining points, and flanking sequences have been determined and allow to distinguish chromothripsis, chromoanasynthesis, and chromoplexy and the atypical form of chromoanagenesis reported by Masset et al. [19]. These patterns also allow to dissect the distinct mechanisms that precipitate chro-

Table 2
Cause or precipitating event, defeated cell cycle checkpoints, and involved genes for the three documented and the one inferred mechanisms of chromothripsis

Cause or precipitating event	Micronucleus dependent	Dicentric chromosomes	Retrotransposition	Chromoplexy?
Affected cells	Cycling	Cycling	Postmitotic	Meiotic
Defeated checkpoints	TP53	TP53 + Rb	Not known	Not known
DSB generating mechanism	Chromosome underreplication	Chromosome bridge fragmentation by TREX1	Chromosome looping?	SPO11?
Joining mechanism	MMBIR	?	Alternative NHEJ	Canonical NHEJ
Involved genes	PIF1 helicase, DNA polymerases Poζ, and REV1L	?	PARP1, CTIP, MRE11, DNA polymerase θ, ligase I/III	WRN, XRCC4, DNA-ligase-IV
Sequences at joining points	No or very short microhomologies	?	LINEs, SINEs	No or very short microhomologies
Key references	Zhang et al. [21]	Maciejowski et al. [22]	Nazaryan-Petersen et al. [20]	Zhang et al. [10]

Question marks indicate not known. For explanation of abbreviations, *see* text

mothripsis and to uncover the genes, proteins, and pathways guarding against this form of chromoanagenesis.

Because of technical limitations, some CCRs may not be recognized as the result of chromothripsis, while some cases of CCRs may have been attributed to chromothripsis without thorough investigation of their structure or mechanism of origin. Thus, the variegated translocation mosaicism found in cultured cells from patients with Werner syndrome may, in hindsight, very well be a manifestation of chromothripsis resulting from defective DSB response due to loss of WRN function [10, 74]. On the other hand, in some cases chromoplexy may induce a state of chromosomal instability prompting chromothripsis [10]. Also some forms of aborted apoptosis may have been inadvertently labeled as chromothripsis. These questions can now be addressed with recently developed novel molecular techniques [27].

The recent gains of insight notwithstanding, not all mechanisms of chromothripsis may be known. As of now three types of chromothripsis can be distinguished depending on whether the affected cells are actively cycling, postmitotic or meiotic (Table 2). In cycling somatic cells, chromothripsis is a possible outcome of impaired DSB response in combination with loss of TP53-dependent cell cycle checkpoint control and may proceed via micronucleus formation or telomere erosion followed by anaphase bridges consisting of fused chromosomes. In postmitotic cells, L1-mediated retrotransposition may produce chromothripsis. The mechanisms underlying germline chromothripsis are less clear. Chromothripsis may be a secondary outcome of a previous CCR, such as chromoplexy. Alternatively, chromothripsis may have resulted from an imbalance in DSB formation and repair during alignment of homologous chromosomes in meiosis [2].

The outcome of chromothripsis also depends on whether the affected cells are actively cycling or undergoing meiosis. In malignant disorders chromothripsis may be a mechanism leading to, or aggravating, genomic instability. Germline chromothripsis may thus underlie some forms of CCRs. Chromothripsis mostly takes place in the male germline, but male patients with impaired DSB responses, due to defects of ATM, MRE11, BLM, LIG4, WRN, or Ku70, show impaired fertility. It is tempting to hypothesize that complete loss of these genes may preclude chromothripsis. Yet, a stochastic, rather than deterministic, involvement of these genes in chromothripsis should be considered. In postmitotic tissues and organs, such as the brain, L1-associated genomic regions are hotspots for CNVs and chromothripsis [53, 84]. Thus, chromothripsis may contribute to somatic mosaicism, e.g., in the brain, or to proliferative decline of somatic tissues during aging. Defects in the DSB response gene *WRN* typically lead to proliferative decline in mesenchymal tissues in vivo and cultured fibroblasts together with variegated translocation mosaicism of the karyotype [70, 74]. It is conceivable that these phenomena are interrelated and may result from a common cause. It remains to be seen, however, to what extent chromothripsis may be involved in Werner syndrome and in normal aging. Malignant disease, somatic mosaicism, and aging are characterized by stochastic genomic instability. Chromothripsis is a relatively rare phenomenon, which often affects only part of the cells in a tissue, which agrees with the notion that chromothripsis may be a stochastic rather than a deterministic process, and that the genes protecting against this cataclysmic form of CCR may fail to do so in a stochastic manner. The novel single-cell genomic technologies now becoming available [53] may enable us to further elucidate this point.

References

1. Boveri T (1914) Zur Frage der Entstehung maligner Tumoren. Fischer, Jena
2. Poot M, Haaf T (2015) Mechanisms of origin, phenotypic effects and diagnostic implications of complex chromosome rearrangements. Mol Syndromol 6:110–134
3. Pellestor F, Anahory T, Lefort G et al (2011) Complex chromosomal rearrangements: origin and meiotic behavior. Hum Reprod Update 17:476–494
4. Liu P, Carvalho CM, Hastings PJ et al (2012) Mechanisms for recurrent and complex human genomic rearrangements. Curr Opin Genet Dev 22:211–220
5. Liu G, Stevens JB, Horne SD et al (2014) Genome chaos: survival strategy during crisis. Cell Cycle 13:528–537
6. Liu P, Erez A, Nagamani SC et al (2011) Chromosome catastrophes involve replication mechanisms generating complex genomic rearrangements. Cell 146:889–903
7. Stephens PJ, Greenman CD, Fu B et al (2011) Massive genomic rearrangement acquired in a single catastrophic event during cancer development. Cell 144:27–40
8. Rausch T, Jones DT, Zapatka M et al (2012) Genome sequencing of pediatric medulloblastoma links catastrophic DNA rearrangements with TP53 mutations. Cell 148:59–71
9. Holland AJ, Cleveland DW (2012) Chromoanagenesis and cancer: mechanisms and consequences of localized, complex chromosomal rearrangements. Nat Med 18:1630–1638
10. Zhang CZ, Leibowitz ML, Pellman D (2013) Chromothripsis and beyond: rapid genome evolution from complex chromosomal rearrangements. Genes Dev 27:2513–2530
11. Pellestor F, Gatinois V, Puechberty J et al (2014) Chromothripsis: potential origin in gametogenesis and preimplantation cell divisions. A review. Fertil Steril 102:1785–1796
12. Nazaryan-Petersen L, Tommerup N (2016) Chromothripsis and human genetic disease. eLS:1–10. https://doi.org/10.1002/9780470015902.a0024627
13. Fukami M, Shima H, Suzuki E et al (2017) Catastrophic cellular events leading to complex chromosomal rearrangements in the germline. Clin Genet 91:653–660
14. Baca SC, Prandi D, Lawrence MS et al (2013) Punctuated evolution of prostate cancer genomes. Cell 153:666–677
15. Korbel JO, Campbell PJ (2013) Criteria for inference of chromothripsis in cancer genomes. Cell 152:1226–1236
16. Kloosterman WP, Guryev V, van Roosmalen M et al (2011) Chromothripsis as a mechanism driving complex de novo structural rearrangements in the germline. Hum Mol Genet 20:1916–1924
17. Kloosterman WP, Tavakoli-Yaraki M, van Roosmalen MJ (2012) Constitutional chromothripsis rearrangements involve clustered double-stranded DNA breaks and nonhomologous repair mechanisms. Cell Rep 1:648–655
18. Ratnaparkhe M, Hlevnjak M, Kolb T et al (2017) Genomic profiling of acute lymphoblastic leukemia in ataxia telangiectasia patients reveals tight link between ATM mutations and chromothripsis. Leukemia 31(10):2048–2056. https://doi.org/10.1038/leu.2017.55
19. Masset H, Hestand MS, Van Esch H et al (2016) A distinct class of chromoanagenesis events characterized by focal copy number gains. Hum Mutat 37:661–668
20. Nazaryan-Petersen L, Bertelsen B, Bak M et al (2016) Germline chromothripsis driven by L1-mediated retrotransposition and Alu/Alu homologous recombination. Hum Mutat 37:385–395
21. Zhang CZ, Spektor A, Cornils H et al (2015) Chromothripsis from DNA damage in micronuclei. Nature 522:179–184
22. Maciejowski J, Li Y, Bosco N et al (2015) Chromothripsis and kataegis induced by telomere crisis. Cell 163:1641–1654
23. Yang J, Liu J, Ouyang L et al (2016) CTLPScanner: a web server for chromothripsis-like pattern detection. Nucleic Acids Res 44:W252–W258
24. Nazaryan L, Stefanou EG, Hansen C et al (2014) The strength of combined cytogenetic and mate-pair sequencing techniques illustrated by a germline chromothripsis rearrangement involving FOXP2. Eur J Hum Genet 22:338–343
25. Weckselblatt B, Rudd MK (2015) Human structural variation: mechanisms of chromosome rearrangements. Trends Genet 31:587–599
26. Weckselblatt B, Hermetz KE, Rudd MK (2015) Unbalanced translocations arise from diverse mutational mechanisms including chromothripsis. Genome Res 25:937–947
27. Marcozzi A, Pellestor F, Kloosterman WP (2017) The genomic characteristics and origin of chromothripsis. In: Pellestor F (ed) Chromothripsis. Springer, New York
28. Kastan MB, Bartek J (2004) Cell-cycle checkpoints and cancer. Nature 432:316–323

29. Meyerson M, Pellman D (2011) Cancer genomes evolve by pulverizing single chromosomes. Cell 144:9–10
30. Crasta K, Ganem NJ, Dagher R et al (2012) DNA breaks and chromosome pulverization from errors in mitosis. Nature 482:53–58
31. Poot M, Hoehn H, Rünger TM et al (1992) Impaired S-phase transit of Werner syndrome cells expressed in lymphoblastoid cell lines. Exp Cell Res 202:267–273
32. Poot M, Gollahon KA, Rabinovitch PS (1999) Werner syndrome lymphoblastoid cells are sensitive to camptothecin-induced apoptosis in S phase. Hum Genet 104:10–14
33. Honma M, Tadokoro S, Sakamoto H et al (2002) Chromosomal instability in B-lymphoblasotoid cell lines from Werner and Bloom syndrome patients. Mutat Res 520:15–24
34. Trkova M, Prochazkova K, Krutilkova V et al (2007) Telomere length in peripheral blood cells of germline TP53 mutation carriers is shorter than that of normal individuals of corresponding age. Cancer 110:694–702
35. Tusell L, Pampalona J, Soler D et al (2010) Different outcomes of telomere-dependent anaphase bridges. Biochem Soc Trans 38:1698–1703
36. Thanasoula M, Escandell JM, Martinez P et al (2010) p53 prevents entry into mitosis with uncapped telomeres. Curr Biol 20:521–526
37. Haaf T, Raderschall E, Reddy G et al (1999) Sequestration of mammalian Rad51-recombination protein into micronuclei. J Cell Biol 144:11–20
38. Darzynkiewicz Z, Li X, Bedner E (2001) Use of flow and laser-scanning cytometry in analysis of cell death. Methods Cell Biol 66:69–109
39. Sakofsky CJ, Ayyar S, Deem AK et al (2015) Translesion polymerases drive microhomology-mediated break-induced replication leading to complex chromosomal rearrangements. Mol Cell 60:860–872
40. Ahmed EA, Sfeir A, Takai H et al (2013) Ku70 and non-homologous end joining protect testicular cells from DNA damage. J Cell Sci 126:3095–3104
41. Den Boer PJ, Poot M, Verkerk A et al (1990) Glutathione-dependent defence mechanisms in isolated round spermatids from the rat. Int J Androl 13:26–38
42. Maciejowski J, de Lange T (2017) Telomeres in cancer: tumour suppression and genome instability. Nat Rev Mol Cell Biol 18(3):175–186
43. Roberts SA, Sterling J, Thompson C et al (2012) Clustered mutations in yeast and in human cancers can arise from damaged long single-strand DNA regions. Mol Cell 46:424–435
44. Roberts SA, Gordenin DA (2014) Hypermutation in human cancer genomes: footprints and mechanisms. Nat Rev Cancer 14:786–800
45. Cai H, Kumar N, Bagheri HC et al (2014) Chromothripsis-like patterns are recurring but heterogeneously distributed features in a survey of 22,347 cancer genome screens. BMC Genomics 15:82. https://doi.org/10.1186/1471-2164-15-82
46. Rode A, Maass KK, Willmund KV et al (2016) Chromothripsis in cancer cells: an update. Int J Cancer 138:2322–2333
47. Beck CR, Collier P, Macfarlane C et al (2010) LINE-1 retrotransposition activity in human genomes. Cell 141:1159–1170
48. Startek M, Szafranski P, Gambin T et al (2015) Genome-wide analyses of LINE-LINE-mediated nonallelic homologous recombination. Nucleic Acids Res 43:2188–2198
49. Bertelsen B, Nazaryan-Petersen L, Sun W et al (2016) A germline chromothripsis event stably segregating in 11 individuals through three generations. Genet Med 18:494–500
50. Lieber MR (2010) The mechanism of double-strand DNA break repair by the nonhomologous DNA end-joining pathway. Annu Rev Biochem 79:181–211
51. Georgiou I, Noutsopoulos D, Dimitriadou E et al (2009) Retrotransposon RNA expression and evidence for retrotransposition events in human oocytes. Hum Mol Genet 18:1221–1228
52. McConnell MJ, Lindberg MR, Brennand KJ et al (2013) Mosaic copy number variation in human neurons. Science 342:632–637
53. Erwin JA, Paquola AC, Singer T et al (2016) L1-associated genomic regions are deleted in somatic cells of the healthy human brain. Nat Neurosci 19:1583–1591
54. Rulten SL, Grundy GJ (2017) Non-homologous end joining: common interaction sites and exchange of multiple factors in the DNA repair process. Bioessays 39. https://doi.org/10.1002/bies.201600209
55. Hustedt N, Durocher D (2016) The control of DNA repair by the cell cycle. Nat Cell Biol 19:1–9
56. Pierce AJ, Jasin M (2001) NHEJ deficiency and disease. Mol Cell 8:1160–1161
57. Patro BS, Frøhlich R, Bohr VA et al (2011) WRN helicase regulates the ATR-CHK1-induced S-phase checkpoint pathway in response to topoisomerase-I-DNA covalent complexes. J Cell Sci 124:3967–3979

58. Su F, Mukherjee S, Yang Y et al (2014) Nonenzymatic role for WRN in preserving nascent DNA strands after replication stress. Cell Rep 9:1387–1401
59. Shamanna RA, Lu H, de Freitas JK et al (2016) WRN regulates pathway choice between classical and alternative non-homologous end joining. Nat Commun 7:13785. https://doi.org/10.1038/ncomms13785
60. Rünger TM, Bauer C, Dekant B et al (1994) Hypermutable ligation of plasmid DNA ends in cells from patients with Werner syndrome. J Invest Dermatol 102:45–48
61. Palermo V, Rinalducci S, Sanchez M et al (2016) CDK1 phosphorylates WRN at collapsed replication forks. Nat Commun 7:12880. https://doi.org/10.1038/ncomms12880
62. Wu L, Hickson ID (2001) Molecular biology. DNA ends ReQ-uire attention. Science 292:229–230
63. Cejka P, Cannavo E, Polaczek P et al (2010) DNA end resection by Dna2-Sgs1-RPA and its stimulation by Top3-Rmi1 and Mre11-Rad50-Xrs2. Nature 467:112–116
64. Grabarz A, Guirouilh-Barbat J, Barascu A et al (2013) A role for BLM in double-strand break repair pathway choice: prevention of CtIP/Mre11-mediated alternative nonhomologous end-joining. Cell Rep 5:21–28
65. Sturzenegger A, Burdova K, Kanagaraj R et al (2014) DNA2 cooperates with the WRN and BLM RecQ helicases to mediate long-range DNA end resection in human cells. J Biol Chem 289:27314–27326
66. Goto M, Miller RW, Ishikawa Y et al (1996) Excess of rare cancers in Werner syndrome (adult progeria). Cancer Epidemiol Biomark Prev 5:239–246
67. Poot M, Yom JS, Whang SH et al (2001) Werner syndrome cells are sensitive to DNA cross-linking drugs. FASEB J 15:1224–1226
68. Poot M, Gollahon KA, Emond MJ et al (2002) Werner syndrome diploid fibroblasts are sensitive to 4-nitroquinoline-N-oxide and 8-methoxypsoralen: implications for the disease phenotype. FASEB J 16:757–758
69. Poot M, Jin X, Hill JP et al (2004) Distinct functions for WRN and TP53 in a shared pathway of cellular response to 1-beta-D-arabinofuranosylcytosine and bleomycin. Exp Cell Res 296:327–336
70. Epstein CJ, Martin GM, Schultz AL et al (1966) Werner's syndrome a review of its symptomatology, natural history, pathologic features, genetics and relationship to the natural aging process. Medicine (Baltimore) 45:177–221
71. Rünger TM, Poot M, Kraemer KH (1992) Abnormal processing of transfected plasmid DNA in cells from patients with ataxia telangiectasia. Mutat Res 293:47–54
72. O'Driscoll M, Jeggo PA (2006) The role of double-strand break repair -insights from human genetics. Nat Rev Genet 7:45–54
73. Scott SP, Pandita TK (2006) The cellular control of DNA double-strand breaks. J Cell Biochem 99:1463–1475
74. Salk D, Au K, Hoehn H, Martin GM (1981) Cytogenetics of Werner's syndrome cultured skin fibroblasts: variegated translocation mosaicism. Cytogenet Cell Genet 30:92–107
75. Fukuchi K, Martin GM, Monnat RJ Jr (1989) Mutator phenotype of Werner syndrome is characterized by extensive deletions. Proc Natl Acad Sci U S A 86:5893–5897
76. Ogburn CE, Oshima J, Poot M et al (1997) An apoptosis-inducing genotoxin differentiates heterozygotic carriers for Werner helicase mutations from wild-type and homozygous mutants. Hum Genet 101:121–125
77. Kamath-Loeb AS, Loeb LA, Johansson E et al (2001) Interactions between the Werner syndrome helicase and DNA polymerase delta specifically facilitate copying of tetraplex and hairpin structures of the d(CGG)n trinucleotide repeat sequence. J Biol Chem 276:16439–16446
78. Dhillon KK, Sidorova J, Saintigny Y et al (2007) Functional role of the Werner syndrome RecQ helicase in human fibroblasts. Aging Cell 6:53–61
79. Rosin MP, German J (1985) Evidence for chromosome instability in vivo in Bloom syndrome: increased numbers of micronuclei in exfoliated cells. Hum Genet 71:187–191
80. Li GC, Ouyang H, Li X (1998) Ku70: a candidate tumor suppressor gene for murine T cell lymphoma. Mol Cell 2:1–8
81. Suzuki T, Yasui M, Honma M (2016) Mutator phenotype and DNA double-strand break repair in BLM helicase-deficient human cells. Mol Cell Biol 36:2877–2889
82. Imamura O, Fujita K, Shimamoto A et al (2001) Bloom helicase is involved in DNA surveillance in early S phase in vertebrate cells. Oncogene 20:1143–1151
83. Simsek D, Jasin M (2010) Alternative end-joining is suppressed by the canonical NHEJ component Xrcc4-ligase IV during chromosomal translocation formation. Nat Struct Mol Biol 17:410–416
84. Baillie JK, Barnett MW, Upton KR et al (2011) Somatic retrotransposition alters the genetic landscape of the human brain. Nature 479:534–537

Check for updates

Chapter 16

Expression of Genes Associated with Telomere Homeostasis in TP53 Mutant LoVo Cell Lines as a Model for Genomic Instability

Oumar Samassekou, Nathalie Bastien, Ju Yan, Sabine Mai, and Régen Drouin

Abstract

We describe a method that assesses the impact of specific mutations of *TP53* and genomic instability on gene expression of the most important genes involved in telomere length and structure homeostasis. The approaches consist of using a reverse transcriptase method and a quantitative PCR that were applied to isogenic cell lines from a colon cancer.

Key words Telomere, Genomic instability, Telomerase, Shelterin, Telomere-associated gene, Reverse transcriptase, Quantitative PCR

1 Introduction

Telomeres are the ends of chromosomes are formed by repeated DNA sequences, (TTAGGG)n; six telomere-specific proteins, named shelterin proteins; telomere-associated proteins; and a special heterochromatin structure [1–4]. Telomeres end with an overhang at their 3′-end which serves as an anchor for *TERC*, the telomerase RNA subunit [5, 6]. From that anchor, the telomerase catalytic component, TERT, elongates the overhang by adding telomeric sequences [7, 8]. Furthermore, the G-overhang serves as a binding site for a telomere-binding protein, protection of telomere-1 (POT1), which participates in the formation of a higher telomere structure order, the telomere loop [9, 10]. G-overhang bends on itself with the rest of telomeric DNA to form a telomere loop (T-loop), which protects telomeres from being recognized as double-stranded breaks by sequestering the telomeric 3′-overhang. Therefore, this higher-order telomere structure is a first layer of telomere protection and confers the architectural structure to telomeres to resolve

Franck Pellestor (ed.), *Chromothripsis: Methods and Protocols*, Methods in Molecular Biology, vol. 1769,
https://doi.org/10.1007/978-1-4939-7780-2_16,

the end protection problem [11, 12]. The T-loop is maintained and protected by shelterin and telomere-associated proteins which act in concert to form the telomere structure. The human shelterin proteins are TERF1 and TERF2 (telomere repeat-binding factor-1 and factor-2), POT1 (protection of telomere-1), TINF2 (TERF1interacting nuclear factor 2), TERF2IP (TERF2 interacting protein), and TPP1 (tripeptidyl-peptidase 1). They protect telomeres by maintaining telomere structure, repress the DNA repair machinery at telomeres, and regulate telomere length. Moreover, these six proteins interact with other proteins, called telomere-associated proteins, to maintain telomere length and structure homeostasis [4, 13–16].

Disruption of length and structure homeostasis can lead to telomere shortening and/or loss of telomere protection and potentially to telomere crisis, which is a state of extensive genome instability that can promote cancer progression [17] or cell death and senescence in primary cells [18]. Recent data has revealed that telomere crisis can induce numerous cancer-relevant changes, and the extent of these rearrangements was unexpected, and their complexity was staggering [17]. These cancer-related multiple chromosomal aberrations include deletions and amplifications, translocations, chromothripsis, kataegis, and tetraploidization [17, 19–21]. The link between telomere crisis and chromothripsis is controversial and not accepted by everybody; indeed, Righolt and Mai strongly argued against this single cell cycle idea [22].

Genomic instability is a dynamic process which lead to an accumulation of multiple chromosomal aberrations over several cell cycles and relies mainly on repeated breakage-fusion-bridge cycles causing complex genomic rearrangements which might be amplified by some underlining mutations of genes heavily involved in oncogenesis, such as *TP53* [23]. We have shown that the degree of genomic instability is associated with particular patterns of telomere dysfunction and specific mutations of *TP53* [24].

We investigated the impact of *TP53* mutations and genomic instability on the gene expression of telomerase enzyme components (*TERT*, *TERC*, and *DKCI*), shelterin genes, and some telomere-associated genes (*ATM*, *ATR*, *MRE11*, *Ku80*, and *CDKN1A*).

We found that the gene of the catalytic component of telomerase, *TERT*, was highly expressed in all LoVo cell lines, and the wild type presented significantly higher expression than the mutants. Except for *TTP1*, which was upregulated, the other shelterin genes were downregulated or had a normal expression in all of the cell lines (Fig. 1a). *TERF1*, *POT1*, and *TERF2IP* were repressed in all cell lines, except for the mutant TP53-R273H, and this repression was more pronounced in the wild type. *TERF2* was repressed in the mutants TP53-V143A and TP53-R273H, whereas it had normal induction in the wild-type TP53 and the mutant TP53-R175H. In summary, the wild-type TP53 and the mutant TP53-R175H presented similar gene expression profiles, while

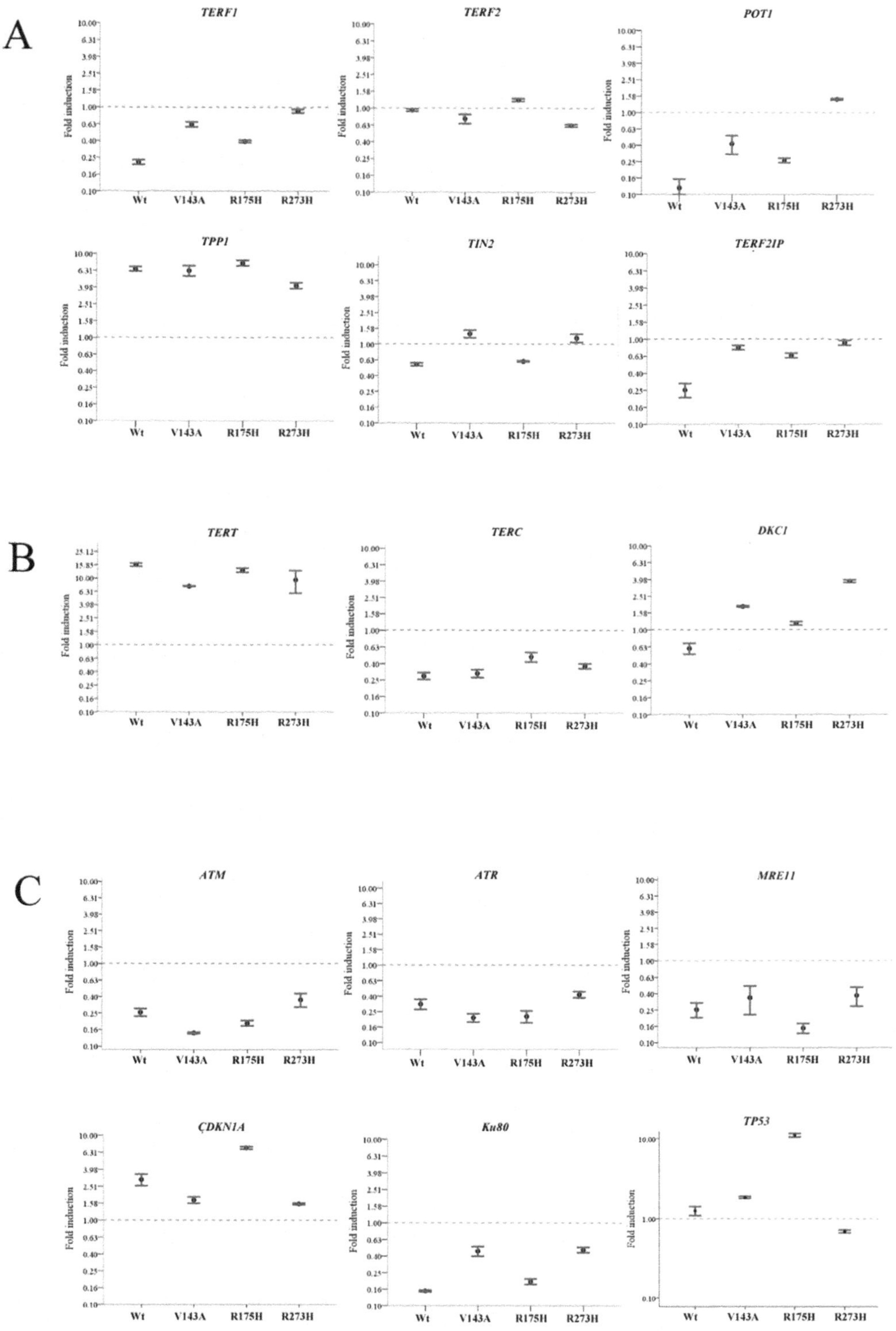

Fig. 1 Gene expression profile. Mean of relative mRNA induction of three experiments. The mRNA levels of the target genes in LoVo cell lines were normalized to those of primary fibroblasts. Error bars are standard deviation. (**a**) Shelterin genes. (**b**) Telomerase component genes. (**c**) *TP53* upstream and downstream target genes and telomere-associated genes

those of the mutants TP53-V143A and TP53-R273H, the most genomically unstable, were alike (Fig. 1b). Finally, we studied the expression of *TP53* and some upstream and downstream gene targets of TP53, which are also telomere-associated gene. All the studied sensor genes of TP53 were downregulated (Fig. 1c). Importantly, we found that the *ATM/ATR* genes were less expressed in the unstable genomic lines (TP53-R175H and TP53-V143A) (Fig. 1c). In conclusion, the expression of telomerase-, shelterin-, and telomere-associated genes is associated to genomic instability and TP53 mutations [24].

2 Materials

1. Biosafety cabinet.
2. Petri dishes.
3. 15 mL conical tubes.
4. 0.2 mL, 0.6 mL, and 1.5 mL microtubes.
5. LoVo cell lines: LoVo TP53-wt, LoVo TP53-V143A, LoVo TP53-R175H, and LoVo TP53-R273H (kindly provided by Pocard et al. [25]).
6. SW480 (American Type Culture Collection).
7. Fibroblasts (American Type Culture Collection).
8. DMEM/F12.
9. DMEM.
10. McCoy's 5A.
11. Antibiotics (penicillin/streptomycin).
12. Fetal bovine serum.
13. Nuclease-free water.
14. TRIzol™ Reagent (Invitrogen).
15. Centrifuge and rotor capable of reaching 12,000 × *g* and 4 °C.
16. Chloroform.
17. Water bath or heat block at 55–60 °C.
18. Isopropanol.
19. Ethanol, 75%.
20. RNase-free water, 0.1 mM EDTA.
21. DEPC-treated water.
22. TaqMan® Reverse Transcription Reagents (Applied Biosystems).
23. SYBR Green Master Mix (Applied Biosystems).
24. Hot block.
25. NanoDrop™.

26. Primer 3 (http://primer3.ut.ee).
27. Primers (Table 1) (Integrated DNA Technologies).
28. Rotor-Gene 3000 (Qiagen).

Table 1
Primers of target genes and the reference gene (in bold) and their sequences and concentration used in quantitative PCR to quantify mRNA

Primers	Sequences	Concentration
ATM-forward	5′- CAG CAG CTG TTA CCT GTT TG -3′	300 nM
ATM-reverse	5′- TAG ATA GGC CAG CAT TGG AT -3′	500 nM
ATR-forward	5′- TGT CTG TAC TCT TCA CGG CAT GTT -3′	500 nM
ATR-reverse	5′- AAG AGG TCC ACA TGT CCG TCT T -3′	500 nM
DKC1-forward	5′- CTG CTA TGG GGC CAA GAT TA -3′	500 nM
DKC1-reverse	5′- CCA TGG TCG CAG GTA GAG AT -3′	500 nM
GAPDH-forward	**5′- AAG GTC GGA GTC AAC GGA T -3′**	**1.78 ng/μL**
GAPDH-reverse	**5′- GGA AGA TGG TGA TGG GAT TTC -3′**	**0.894 ng/μL**
Ku80-forward	5′- CCC CCAA TTC AGC AGC ATA TT -3′	500 nM
Ku80-reverse	5′- CCT TCA GCC AGA CTG GAG AC -3′	300 nM
MRE11-forward	5′- GCC TTC CCG AAA TGT CAC TA -3′	300 nM
MRE11-reverse	5′- TTC AAA ATC AAC CCC TTT CG -3′	300 nM
CDKN1A-forward	5′- GGC AGA CCA GCA TGA CAG ATT -3′	0.894 ng/uL
CDKN1A-reverse	5′- GGC GGC CAG GGT ATG TA -3′	1.78 ng/uL
POT1-forward	5′- TGG GTA TTG TAC CCC TCC AA -3′	500 nM
POT1-reverse	5′- GAT GAA GCA TTC CAA CCA CGG -3′	500 nM
TERC-forward	5′- GTG GTG GCC ATT TTT TGT CTA AC -3′	500 nM
TERC-reverse	5′- TGC TCT AGA ATG AAC GGT CCA A -3′	500 nM
TERF1-forward	5′- GCT GTT TGT ATG GAA AAT GGC -3′	500 nM
TERF1-reverse	5′- CCG CTG CCT TCA TTA GAA AG -3′	500 nM
TERF2-forward	5′- GAC CTT CCA GCA GAA GAT GCT -3′	300 nM
TERF2-reverse	5′- GTT GGA GGA TTC CGT AGC TG -3′	300 nM
TERF2IP-forward	5′- CGG GGA ACC ACA GAA TAA GA -3′	500 nM
TERF2IP -reverse	5′- CTC AGG TGT GGG TGG ATC AT -3′	500 nM
TERT-forward	5′- TGT TTC TGG ATT TGC AGG TG -3′	500 nM
TERT-reverse	5′- GTT CTT GGC TTT CAG GAT GG -3′	500 nM

(continued)

Table 1 (continued)

Primers	Sequences	Concentration
TIN2-forward	5′- GGA GTT TCT GCG ATC TCT GC -3′	150 nM
TIN2-reverse	5′- GAT CCC GCA CTA TAG GTC CA -3′	300 nM
TP53-forward	5′-TCC ACT ACA ACT ACA TGT GTA AC -3′	300 nM
TP53-reverse	5′- GTG AAA TAT TCT CCA TCC AGT G -3′	500 nM
TPP1-forward	5′- CCC GGA GAG TTC TAT CTC CA -3′	300 nM
TPP1-reverse	5′- GGA CAG TGA TAG GCC TGC AT -3′	500 nM

3 Methods

3.1 RNA Extraction

1. Culture separately the cell lines in 10 cm diameter petri dishes.
2. Maintain the LoVo cell lines (LoVo TP53-wt, LoVo TP53-V143A, LoVo TP53-R175H, and LoVo TP53-R273H) in DMEM/F12 supplemented with 10% FBS and 1% antibiotics (penicillin/streptomycin).
3. Culture the SW480 cell line in McCoy's 5A supplemented with 10% FBS and 1% antibiotics (penicillin/streptomycin).
4. Maintain the fibroblasts in DMEM supplemented with 10% FBS and 1% antibiotics (penicillin/streptomycin).
5. Remove the growth media when cells reach 70% confluence.
6. Add 1 mL of TRIzol™ in each flask.
7. Pipet the lysate up and down several times to homogenize (*see* **Notes 1 and 2**).
8. Transfer the lysate in a 1.5 mL tube and incubate it for 5 min to permit complete dissociation of the nucleoproteins complex.
9. Add 0.2 mL of chloroform and then securely cap the tube.
10. Centrifuge (12,000 × *g*) the tube for 15 min at 4 °C. The mixture separates into a lower red phenol-chloroform, an interphase, and a colorless upper aqueous phase.
11. Transfer the aqueous phase containing the RNA to a new tube by angling the tube at 45° and pipette the solution out.
12. Add 0.5 mL of isopropanol to the aqueous phase and incubate for 10 min.
13. Centrifuge (12,000 × *g*) for 10 min at 4 °C.
14. Discard the suspension of isopropanol, and total RNA precipitate forms a white gel-like pellet at the bottom of the tube.

15. Resuspend the pellet in 1 mL of 75% ethanol (*see* **Note 3**).
16. Vortex the tube briefly, and then centrifuge (7500 × *g*) at 4 °C for 5 min.
17. Discard the supernatant with a micropipette and air dry the RNA pellet at room temperature for 5–10 min (*see* **Note 4**).
18. Resuspend the pellet in 50 μL of RNase-free water, 0.1 mM EDTA by pipetting up and down.
19. Incubate in a water bath or heat block set at 55–60 °C for 10–15 min.
20. Measure the concentration of RNA. An absorbance ratio 280/260 = 2 indicate a good quality of RNA.

3.2 Reverse Transcription

1. Run each sample in triplicate.
2. Dilute the RNA at concentration 200 ng/μL in DEPC-treated water.
3. Thaw all TaqMan® Reverse Transcription Reagents and keep them on ice.
4. Mix and briefly centrifuge the reagents.
5. Combine the following components in each reaction tube (*see* **Note 5**):
   (a) 2.0 μL of 10X RT Buffer
   (b) 1.4 μL of 25 mM $MgCl_2$
   (c) 4.0 μL of 10 mM dNTP mix (2.5 mM each)
   (d) 1 μL of 100 mM DTT
   (e) 1 μL of RNase Inhibitor (20 U/μL)
   (f) 1 μL of MultiScribe™ RT (50 U/μL)
   (g) 1 μL of 50 μM random hexamers
   (h) 3.0 μL of template RNA
   (i) 5.6 μL DEPC-treated water
6. Cap, mix, and briefly centrifuge the components.
7. Execute the following cycling parameters:
   (a) 25 °C for 10 min
   (b) 37 °C for 30 min
   (c) 95 °C for 5 min
   (d) 4 °C for 30 min
8. Briefly centrifuge and pool the triplicate from the same sample in a tube.
9. Dilute the cDNA in RNAse-free water at the concentration of 15 ng/μL.

3.3 Quantitative PCR

1. Run each sample in triplicate.
2. Prepare each set of forward and reversed primers (*see* **Note 6**).
3. For each sample, prepare the following mix containing:
   (a) 25 μL l of 2X SYBR
   (b) 5 μL of cDNA (corresponding to 75 ng of cDNA)
   (c) 1 μL of forward primer (*see* Table 1 for the corresponding concentration)
   (d) 1 μL of reversed primer (*see* Table 1 for the corresponding concentration)
   (e) 18 μL of RNAse-free water
4. Prepare controls for each gene without cDNA template which is replaced by water. The same mix as mentioned is prepared without cDNA.
5. Place the tubes on Rotor-Gene 3000
6. Execute the following cycling parameters:
   (a) 95 °C for 10 min
   (b) 95 °C for 15 s and 60 °C for 1 min (40 cycles)
   (c) 4 °C for 60 min

3.4 Analysis

1. Determine the Ct (for each replicate in each sample) after establishing the threshold.
2. Calculate the average Ct for each sample.
3. Assess the relative difference in gene expression using the $2^{-\Delta\Delta Ct}$ method.
4. Relative fold change in gene expression = $2^{-\Delta\Delta Ct}$.
5. Where: ΔΔCt = ΔCt (LoVo cell lines) – ΔCt (fibroblast for the other except those of telomerase) and ΔCt = (Ct target gene – Ct reference gene- *GAPDH*). For telomerase gene, the following formula was used: ΔΔCt = ΔCt (LoVo cell lines) – ΔCt (SW480) (*see* **Note** 7).

4 Notes

1. Do not wash cells before addition TRIzol™ Reagent to avoid mRNA degradation.
2. Samples can be stored at 4 °C overnight or at—20 °C for up to a year.
3. The RNA can be stored in 75% ethanol for at least 1 year at −20 °C or at least 1 week at 4 °C.
4. Do not let the RNA pellet dry, to ensure total solubilization of the RNA. Partially dissolved RNA samples have an A230/280 ratio = 1.6 which indicates a contaminated RNA.

5. Prepare a master mix of components common to all reactions to minimize pipetting error, and then dispense appropriate volumes into each 0.2 mL PCR reaction tube prior to adding template RNA and primers.
6. For each stock primer, the stock concentration is adjusted, so 1 μL of that stock corresponds to the concentration indicated in the Table 1.
7. The expressions of telomerase genes, specifically TERT, were so high in LoVo lines that it was informative to compare their expression to that of a well-characterized cancer cell line such as SW480.

References

1. Benetti R, Garcia-Cao M, Blasco MA (2007) Telomere length regulates the epigenetic status of mammalian telomeres and subtelomeres. Nat Genet 39:243–250. https://doi.org/10.1038/ng1952
2. De Lange T, Shiue L, Myers RM et al (1990) Structure and variability of human chromosome ends. Mol Cell Biol 10:518–527
3. Moyzis RK, Buckingham JM, Cram LS et al (1988) A highly conserved repetitive DNA sequence, (TTAGGG)n, present at the telomeres of human chromosomes. Proc Natl Acad Sci U S A 85:6622–6626
4. Chong L, van Steensel B, Broccoli D et al (1995) A human telomeric protein. Science 270:1663–1667
5. Lejnine S, Makarov VL, Langmore JP (1995) Conserved nucleoprotein structure at the ends of vertebrate and invertebrate chromosomes. Proc Natl Acad Sci U S A 92:2393–2397
6. Counter CM, Avilion AA, LeFeuvre CE et al (1992) Telomere shortening associated with chromosome instability is arrested in immortal cells which express telomerase activity. EMBO J 11:1921–1929
7. Nakamura TM, Morin GB, Chapman KB et al (1997) Telomerase catalytic subunit homologs from fission yeast and human. Science 277:955
8. Zhao YM, Li JY, Lan JP et al (2008) Cell cycle dependent telomere regulation by telomerase in human bone marrow mesenchymal stem cells. Biochem Biophys Res Commun 369:1114–1119. https://doi.org/10.1016/j.bbrc.2008.03.011
9. Stansel RM, de Lange T, Griffith JD (2001) T-loop assembly in vitro involves binding of TRF2 near the 3′ telomeric overhang. EMBO J 20:5532
10. Munoz-Jordan JL, Cross GA, de Lange T, Griffith JD (2001) T-loops at trypanosome telomeres. EMBO J 20:579
11. De Lange T (2004) T-loops and the origin of telomeres. Nat Rev Mol Cell Biol 5:323
12. Sfeir A, de Lange T (2012) Removal of shelterin reveals the telomere end-protection problem. Science 336:593–597. https://doi.org/10.1126/science.1218498;
13. Celli GB, Denchi EL, de Lange T (2006) Ku70 stimulates fusion of dysfunctional telomeres yet protects chromosome ends from homologous recombination. Nat Cell Biol 8:885–890. https://doi.org/10.1038/ncb1444
14. Denchi EL, de Lange T (2007) Protection of telomeres through independent control of ATM and ATR by TRF2 and POT1. Nature 448:1068–1071. https://doi.org/10.1038/nature06065
15. Sfeir A, Kosiyatrakul ST, Hockemeyer D et al (2009) Mammalian telomeres resemble fragile sites and require TRF1 for efficient replication. Cell 138:90–103. https://doi.org/10.1016/j.cell.2009.06.021
16. Takai KK, Kibe T, Donigian JR et al (2011) Telomere protection by TPP1/POT1 requires tethering to TIN2. Mol Cell 44:647–659. https://doi.org/10.1016/j.molcel.2011.08.043;
17. Maciejowski J, de Lange T (2017) Telomeres in cancer: tumour suppression and genome instability. Nat Rev Mol Cell Biol 18:175–186. https://doi.org/10.1038/nrm.2016.171
18. Romanov SR, Kozakiewicz BK, Holst CR et al (2001) Normal human mammary epithelial cells spontaneously escape senescence and acquire genomic changes. Nature 409:633–637. https://doi.org/10.1038/35054579

19. Davoli T, de Lange T (2012) Telomere-driven tetraploidization occurs in human cells undergoing crisis and promotes transformation of mouse cells. Cancer Cell 21:765–776. https://doi.org/10.1016/j.ccr.2012.03.044
20. Davoli T, Denchi EL, de Lange T (2010) Persistent telomere damage induces bypass of mitosis and tetraploidy. Cell 141:81–93. https://doi.org/10.1016/j.cell.2010.01.031
21. Yates LR, Campbell PJ (2012) Evolution of the cancer genome. Nat Rev Genet 13:795–806. https://doi.org/10.1038/nrg3317
22. Righolt C, Mai S (2012) Shattered and stitched chromosomes-chromothripsis and chromoanasynthesis-manifestations of a new chromosome crisis? Genes Chromosomes Cancer 51:975–981. https://doi.org/10.1002/gcc.21981
23. Rausch T, Jones DTW, Zapatka M et al (2012) Genome sequencing of pediatric medulloblastoma links catastrophic DNA rearrangements with TP53 mutations. Cell 148:59–71. https://doi.org/10.1016/j.cell.2011.12.013
24. Samassekou O, Bastien N, Lichtensztejn D et al (2014) Different TP53 mutations are associated with specific chromosomal rearrangements, telomere length changes, and remodeling of the nuclear architecture of telomeres. Genes Chromosomes Cancer 53:934–950. https://doi.org/10.1002/gcc.22205
25. Pocard M, Chevillard S, Villaudy J et al (1996) Different p53 mutations produce distinct effects on the ability of colon carcinoma cells to become blocked at the G1/S boundary after irradiation. Oncogene 12:875–882

Part IV

The Bioinformatic Tools for Chromothripsis Analysis

Check for updates

Chapter 17

Chromothripsis Detection and Characterization Using the CTLPScanner Web Server

Jian Yang, Bo Liu, and Haoyang Cai

Abstract

Accurate detection of chromothripsis event is important to study the mechanisms underlying this phenomenon. CTLPScanner (http://cgma.scu.edu.cn/CTLPScanner/) is a web-based tool for identification and annotation of chromothripsis-like pattern (CTLP) in genomic array data. In this chapter, we illustrate the utility of CTLPScanner for screening chromosome pulverization regions and give interpretation of the results. The web interface offers a set of parameters and thresholds for customized screening. We also provide practical recommendations for effective chromothripsis detection. In addition to the user data processing module, CTLPScanner contains more than 50,000 preprocessed oncogenomic arrays, which allow users to explore the presence of chromothripsis signatures from public data resources.

Key words Chromothripsis detection, CTLPScanner, Web server, Metadata analysis, SNP array, Array CGH, Chromosome pulverization

1 Introduction

The key characteristics of chromothripsis are clustering of chromosomal breakpoints, a small number of oscillating copy number states and localized rearrangements involving one or a few chromosomes [1, 2]. The complex rearrangements and distinct copy number alteration patterns make it challenging to detect chromothripsis reliably [3]. Since the initial report of chromothripsis, a number of following studies also identified this phenomenon in many tumor types [4–9]. However, a lot of studies identified chromothripsis by visual inspection. Korbel and Campbell [10] have introduced six rigorous criteria for chromothripsis, which represent the most detailed definition so far. Up to now, only a few tools have been developed to detect chromothripsis event. For example, an algorithm named ShatterProof [11] can be used to detect chromothripsis from next-generation sequencing (NGS) data. In this chapter, we introduce our web server, CTLPScanner, a user-friendly website for inferring chromothripsis-like pattern in array-based copy number profiling data [12].

Franck Pellestor (ed.), *Chromothripsis: Methods and Protocols*, Methods in Molecular Biology, vol. 1769,
https://doi.org/10.1007/978-1-4939-7780-2_17,

According to recent studies, both sequencing and microarray have been successfully used to identify chromothripsis phenomenon. Next-generation sequencing technologies are widely used for identifying mutations, whereas genomic array platforms are commonly employed for detecting copy number alterations (CNAs). At the moment, the high cost of NGS is still a major limitation for large-scale sequencing of whole genomes for scientific research. And arrays offer the most cost-effective solutions for producing genome-wide copy number alteration profiles. The further advantages of microarray-based methods compared to NGS include faster turnaround time and lower computational complexity. CTLPScanner takes DNA copy number segmentation data as input. These data can be generated by array comparative genomic hybridization (aCGH) or single-nucleotide polymorphism (SNP) array platforms.

CTLPScanner implements an algorithm that employs scan statistics to identify clustering of copy number status changes in the genome. Scan statistics are commonly used to detect clusters over space and time [13]. In our implementation, a fixed-sized window is moved along the genome, and for each window, the likelihood ratio based on the Poisson model is computed from observed and expected copy number status change times [14]. This scanning procedure is repeated for different window sizes to detect the zone that is most likely to be a cluster. The window with the highest likelihood ratio is selected as a candidate of chromothripsis for each chromosome. In this way, CTLPScanner is able to detect both the location and the size of the genomic region affected by chromothripsis.

In addition to the user data processing module, CTLPScanner allows users to explore potential chromothripsis events from public data sources with self-defined constraints. CTLPScanner integrates more than 50,000 preprocessed tumor samples, which are mainly collected from The Cancer Genome Atlas (TCGA) [15] and NCBI Gene Expression Omnibus (GEO) [16]. All data are converted into normalized and segmented DNA copy number data. Since the incidence of chromothripsis is relatively low in most cancer types (probably occurs in only 2–3% of all cancers) [1], public datasets are valuable resources to perform metadata analysis. The user-friendly interface enables chromothripsis screening, visualization, and annotation in a single step, without carrying out any local data processing.

2 Materials

2.1 Overview of Web Server

The CTLPScanner website (http://cgma.scu.edu.cn/CTLPScanner/) can be accessed using an Internet browser. The website is tested to ensure functionality across different browsers, including Safari, Microsoft Internet Explorer, Firefox, Opera, and Google Chrome. The web interface is developed with HTML,

PHP, and JavaScript and is hosted on an Apache web server. The back-end data analysis and visualization are implemented using R and Perl programming language. CTLPScanner contains two main modules: (1) chromothripsis detection and visualization with user input data and annotation of the detected chromosome pulverization region and the associated genes and (2) public array data exploration.

2.2 Input Data

CTLPScanner identifies chromothripsis event based on DNA copy number segmentation data, which can be generated by aCGH or SNP array platforms. The segmented copy number data can be obtained by many bioinformatics tools, such as circular binary segmentation (CBS) [17], Gain and Loss Analysis of DNA (GLAD) [18], and Copynumber [19]. For the input file, the minimum required data contents include sample name, chromosome number, start and stop positions of each segment, and log normalized signal intensity (Table 1). The data should be tab delimited with an optional first line identifying the columns. The "Download" page provides sample data for users to confirm data formatting requirements. Moreover, there are several links on the "Scan" page that enable automatic loading and try out sample data.

2.3 Public Data Content

The public data exploration module allows users to interactively explore over 50,000 preprocessed genomic array data. All raw array data are collected from TCGA and GEO and processed using an in-house pipeline. In total, CTLPScanner integrates more than 50,000 tumor samples representing 159 cancer types. A summary of cancer types and array numbers is given in Table 2. The normalized GEO array data are obtained from arrayMap database [20],

Table 1
The minimum required data contents for an input file

Column	Description
Sample name	The sample name will be displayed in figures on the results page
Chromosome number	The chromosome of the segment. This column must be numeric. Chromosomes X and Y should be represented as 23 and 24, respectively
Start position	The start genomic position of the segment
Stop position	The stop genomic position of the segment
Signal intensity	The normalized tumor/normal fluorescence intensity ratios (usually log2 transformed) of the segment

Table 2
Public array data currently available in CTLPScanner

Data type	GEO	TCGA
Arrays	29,334	22,419
Series	457	33
Cancer types	159	33
Array platforms	223	1

which is a curated database providing copy number profiling data in human cancer. The genomic positions of array probes are mapped to the human reference genome assembly NCBI36/hg18. The copy number profiles are segmented using the CBS algorithm.

3 Methods

The "Scan" page allows users to either upload a file or paste formatted segmentation data for chromothripsis detection. The "Search" page is designed to explore preprocessed array data. We will use the dataset GSE27560 from GEO as an example, which can be downloaded from the web server. This dataset contains ten multiple myeloma samples that are assayed with Affymetrix Genome-Wide Human SNP 6.0 Array.

3.1 Chromothripsis Detection with User Input Data

1. Navigate to http://cgma.scu.edu.cn/CTLPScanner. The browser will display "Scan" interface as the default page (Fig. 1).
2. Click "Choose File" button, and browse your text file that contains copy number segmentation data. The supported file types include .txt, .csv, and .tab (*see* **Note 1**). Alternatively, the server allows users to copy and paste copy number segmentation data directly into the input text field (*see* **Note 2**) (Fig. 1a).
3. To perform batch data analysis, a compressed zip archive file that contains multiple sample files can be uploaded (*see* **Note 3**). The server will automatically decompress the zip file, and all samples will be processed. The file names will be used as sample identifiers in the results page.
4. Select the human genome assembly corresponding to the input data (Fig. 1b). The server supports both the current and previous versions of the reference human genome assembly (build from NCBI34/hg16 to GRCh38/hg38).

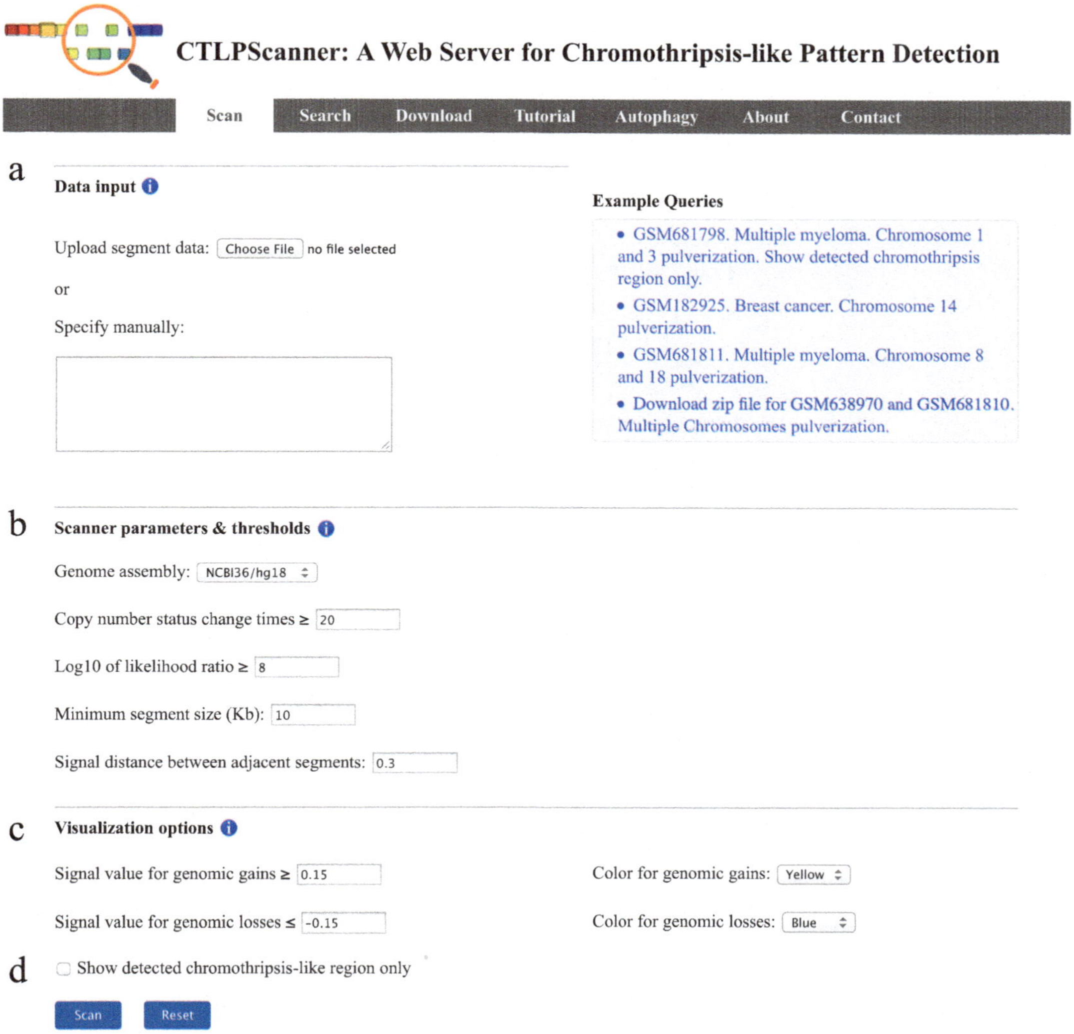

Fig. 1 The Scan page of CTLPScanner. (**a**) The data input interface and query examples. (**b**) The thresholds and parameters for chromothripsis detection. The default values are shown. (**c**) The parameters are used to call copy number aberrations and customize data visualization. (**d**) The checkbox is set to only display detected chromothripsis case

5. Set a threshold for copy number status change times. One of the most striking hallmarks of chromothripsis is the presence of copy number oscillations between two and three states along the chromosome (*see* **Note 4**).
6. Set a threshold for the log 10 of likelihood ratio score. It represents the strength of the evidence for the detected chromothripsis event (*see* **Note 5**).
7. Set a threshold for the minimum segment size in kilobase pair (kb). This threshold is used to eliminate the background noise of copy number data, which occurs during the hypersegmentation process (*see* **Note 6**).

8. Set a threshold for signal distance between adjacent segments. It is used to remove artificial copy number status change (*see* **Note 7**).
9. Set signal value for calling genomic gain and loss (Fig. 1c). They are cutoffs from which a segment is considered a genomic gain or loss (*see* **Note 8**).
10. Choose colors for plotting genomic gain and loss (Fig. 1c). They allow for a customized result visualization.
11. Select the checkbox to show detected chromothripsis case only (Fig. 1d). This step is optional. It is especially useful when performing large-scale data screening.
12. Once all the thresholds and parameters are set (*see* **Note 9**), click the "Scan" button to begin the screening. The server will generate a result page once the processing is done (*see* **Note 10**). Please refer to "Subheading 3.3" for an explanation of the results.

3.2 Public Data Exploration

1. From http://cgma.scu.edu.cn/CTLPScanner, click on the "Search" link in the top bar (Fig. 2).
2. The interface allows users to search either by a single array or by a data series. For search by single array, input the array accession ID into the "Sample ID" text field (*see* **Note 11**) (Fig. 2a).
3. For search by series, select tumor type in the "Cancer Type" selection box (Fig. 2b). These cancer types are classified by tumor loci. Then, select series accession IDs in the "Batch Accession" selection box (*see* **Note 12**). The content of the batch accession list is dynamically determined by the selection of cancer type.
4. Set thresholds and parameters for chromothripsis detection and visualization (Fig. 2c, d). The definitions and default values for these thresholds are the same as the ones introduced in "Subheading 3.1". Please refer to Subheading 3.1 for details.
5. Once all the thresholds and parameters are set, click the "Search" button to begin the screening. The server will generate a result page once the processing is done. Please refer to "Subheading 3.3" for an explanation of the results.

3.3 About Result Data

Once a task is completed, the web server generates a results page. It includes several kinds of information: (1) user-specified thresholds and parameters, (2) detailed information of chromothripsis region, (3) data visualization, (4) annotation of affected cancer genes, and (5) a scatter plot to assess the significance of decision parameters.

1. "Parameters and Thresholds" panel shows the thresholds that were specified by the user to detect chromothripsis (Fig. 3).

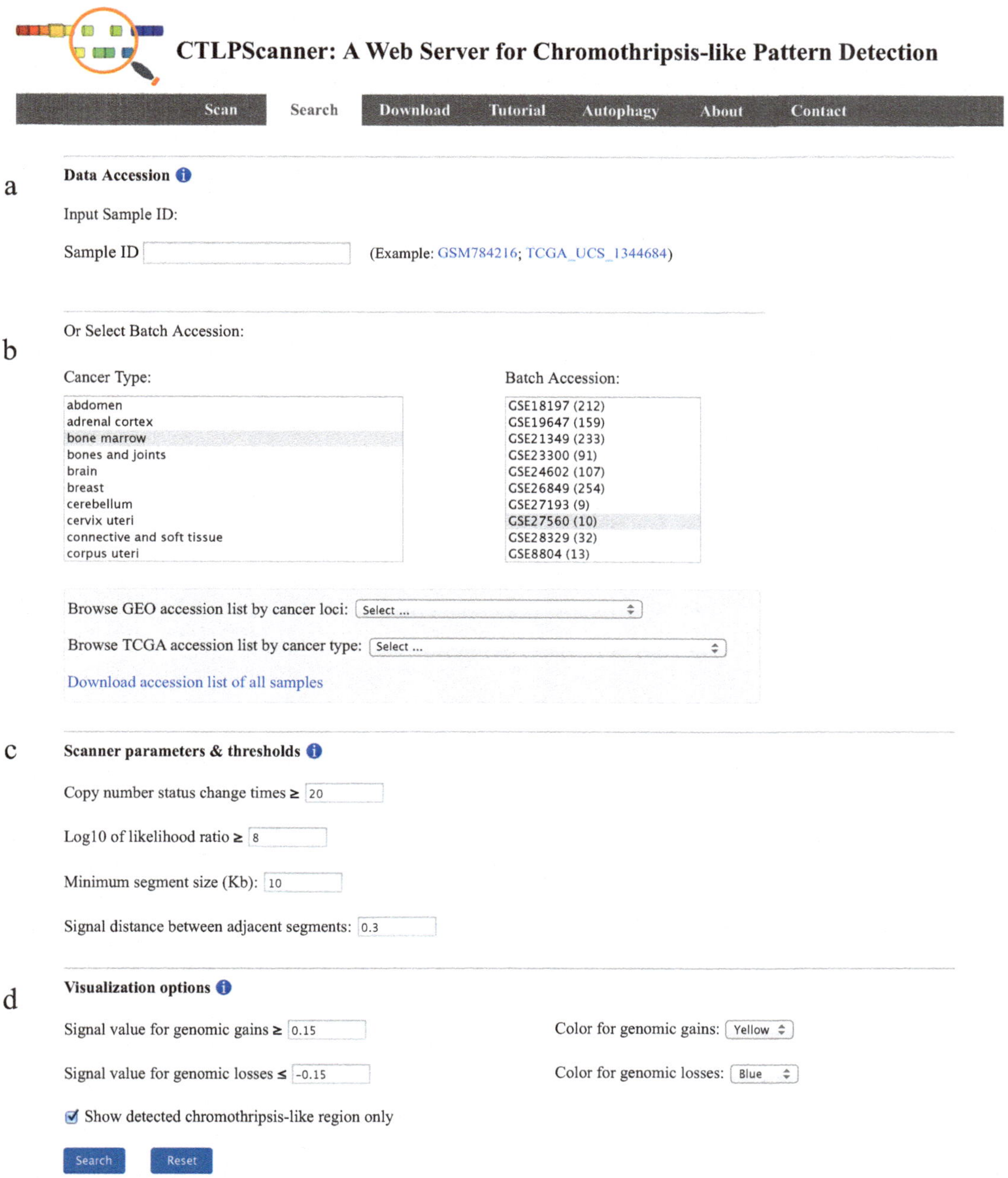

Fig. 2 The Search page of CTLPScanner. (**a**) The sample and series ID input interface. (**b**) Select series data by cancer type. It allows batch data analysis. (**c**) The thresholds and parameters for chromothripsis detection. (**d**) The parameters are used to call copy number aberrations and customize data visualization

The "ReScan" button can perform the screening procedure again with changed thresholds (*see* **Note 13**).

2. If a chromosome shattering event is detected, the "Chromothripsis Region" panel will display detailed information of chromothripsis region (*see* **Note 14**), including length, genomic location, copy number change times, and log10 of

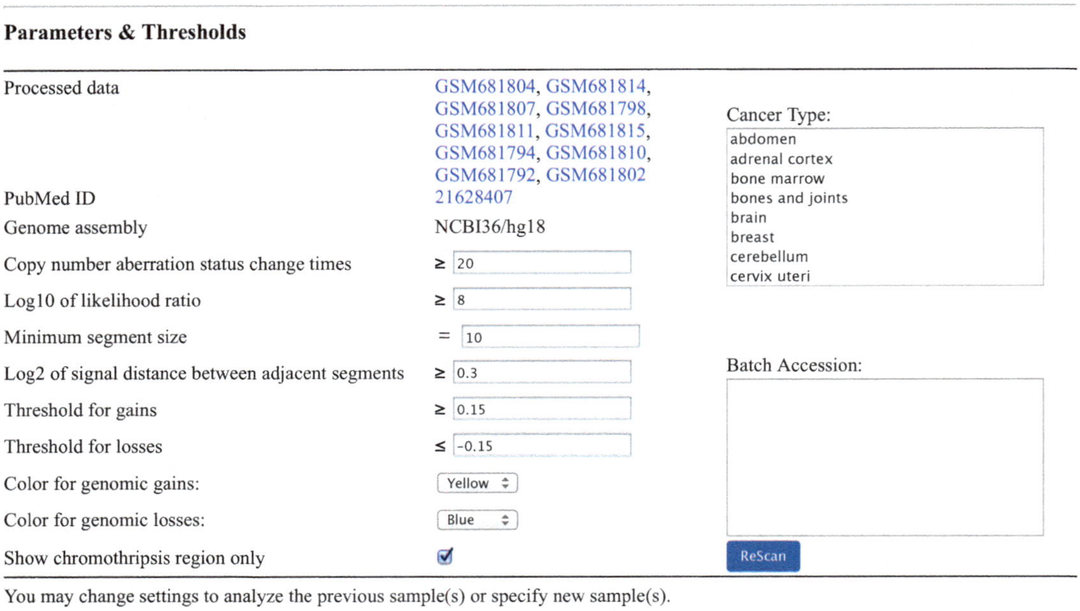

Fig. 3 The thresholds that are specified by the user to detect chromothripsis. The interface allows data reanalysis with changed thresholds

likelihood ratio (Fig. 4). These data can be downloaded as a flat file for further analysis via the "Download result file" hyperlink.

3. "Chromothripsis and Copy Number Aberrations Plots" panel displays genomic CNA profile of each sample. An overall copy number profile plot shows genomic gains and losses by colored lines (Fig. 5). The horizontal axis represents genomic position along 22 chromosomes, while vertical axis is signal intensity ratios on log2 scale. In chromosome level plot, if a chromosome shattering event is detected, the chromothripsis region will be highlighted with a light orange rectangle (Fig. 6a). Moreover, the web server provides a histogram of copy number segment switches to facilitate the identification of chromothripsis event arising in conjunction with telomere crises (*see* **Note 15**) (Fig. 6b). For some SNP arrays in GEO, a B allele frequency plot is provided and can be used to identify loss of heterozygosity (LOH) regions (Fig. 6c). The Catalogue of Somatic Mutations in Cancer (COSMIC) database [21] annotated cancer genes located in chromothripsis regions are also presented in the plot (Fig. 6a). Detailed information can be obtained by clicking on gene symbols or CNA segments (*see* **Note 16**). Users can also zoom in specific genomic regions by inputting the start and stop locations.

Chromothripsis Region

Array ID	CTLP Region (Mb)	Chromosome	Start	Stop	CNA status change times	Likelihood ratio (log10)
GSM681804	30	3	55000001	85000000	21	19
GSM681804	40	7	15000001	55000000	21	19
GSM681814	106.37	1	140881134	247249719	53	50
GSM681807	49.69	3	135000001	184691432	61	58
GSM681798	62.44	1	160000001	222435964	26	15
GSM681798	40	3	135000001	175000000	30	24
GSM681811	62.44	8	80000001	142435964	60	57
GSM681811	30	18	35000001	65000000	26	26
GSM681815	46.94	16	41882931	88827254	77	98
GSM681794	46.94	16	41882931	88827254	46	31
GSM681810	46.94	10	60000001	106944323	21	13
GSM681792	199.5	2	10000001	209501827	38	8
GSM681802	46.94	16	41882931	88827254	43	42
GSM681802	46.94	17	1	46944323	23	15

Download result file

Fig. 4 Summary of detected chromothripsis regions

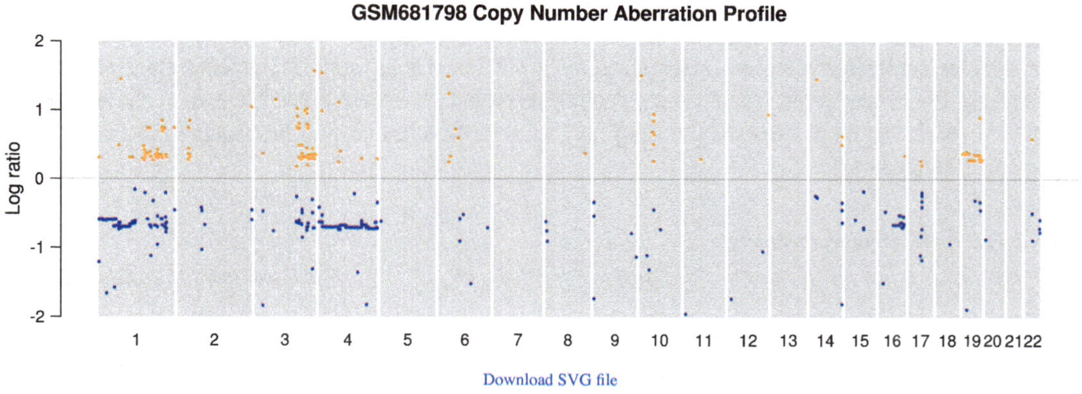

Fig. 5 An example of overall genomic CNA profile. Yellow and blue lines represent genomic gains and losses, respectively

4. "Cancer Genes in Chromothripsis Region" panel shows all the cancer genes located in chromothripsis regions and the related information (Fig. 7). The "Segment Size" column provides the information about whether the mutation is driven by a focal copy number alteration or a large chromosomal rearrangement (*see* **Note 17**). These data can be downloaded as a flat file for further in-depth analysis.

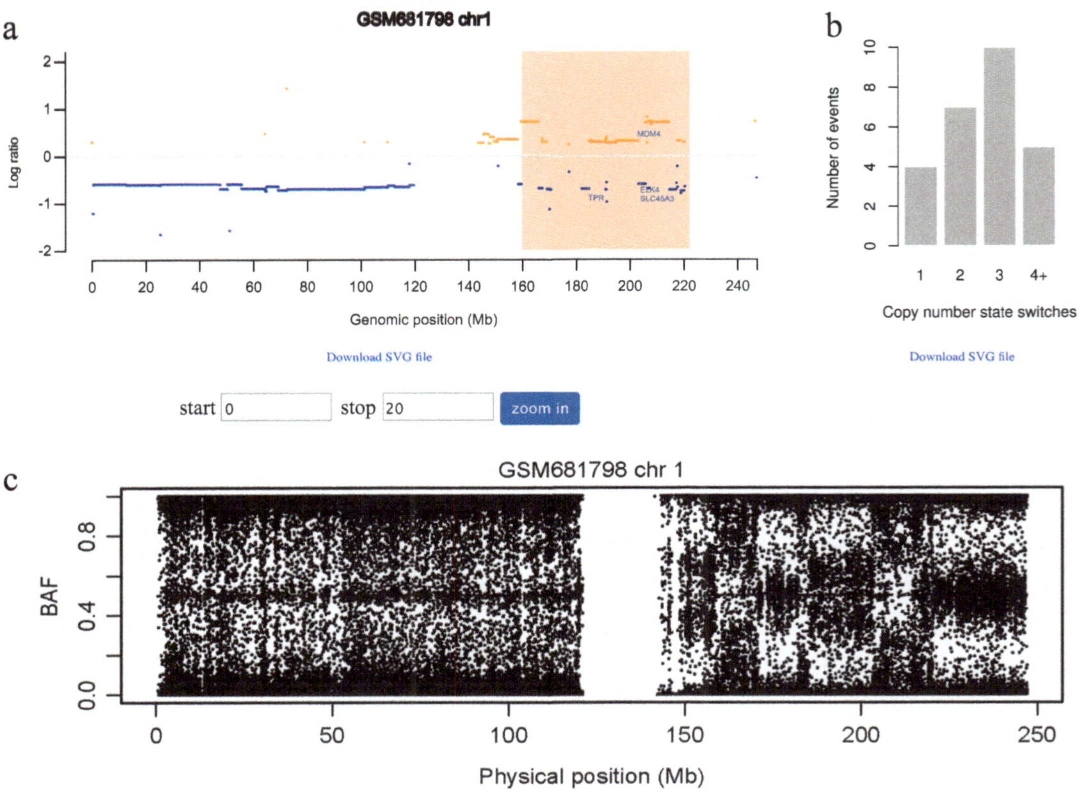

Fig. 6 The results of CTLPScanner. (**a**) An example of detected chromothripsis event. The light orange rectangle represents the chromosomal pulverization region. Yellow and blue lines represent genomic gains and losses, respectively. The cancer genes located in chromothripsis region are shown. A specific area of interest can be zoomed in. (**b**) Histogram of the number of copy number state switches. (**c**) B allele frequency plot of chromosome 1 of a tumor sample

Cancer Genes in Chromothripsis Region

Array ID	Gene Symbol	Gene Name	CNA	Entrez Gene ID	Segment Size (Kb)
GSM681804	MITF	microphthalmia-associated transcription factor	Gain	4286	570
GSM681804	FOXP1	forkhead box P1	Gain	27086	1368
GSM681804	FOXP1	forkhead box P1	Gain	27086	166
GSM681814	PDE4DIP	phosphodiesterase 4D interacting protein	Gain	9659	1125
GSM681814	ARNT	aryl hydrocarbon receptor nuclear translocator	Gain	405	6996
GSM681814	BCL9	B-cell CLL/lymphoma 9	Gain	607	6996
GSM681814	MUC1	mucin 1, cell surface associated	Gain	4582	2322
GSM681814	NTRK1	neurotrophic tyrosine kinase, receptor, type 1	Gain	4914	2322

Fig. 7 An example of COSMIC database annotated cancer genes that are located in identified chromothripsis regions

Scatter Plot

- Copy number status change times compared to the likelihood ratio
- Each point represents the window with the highest LR for each chromosome
- The dashed lines indicate the user-defined thresholds. CN, copy number; LR, likelihood ratio
- The candidates falling in the orange area are identified CTLP cases

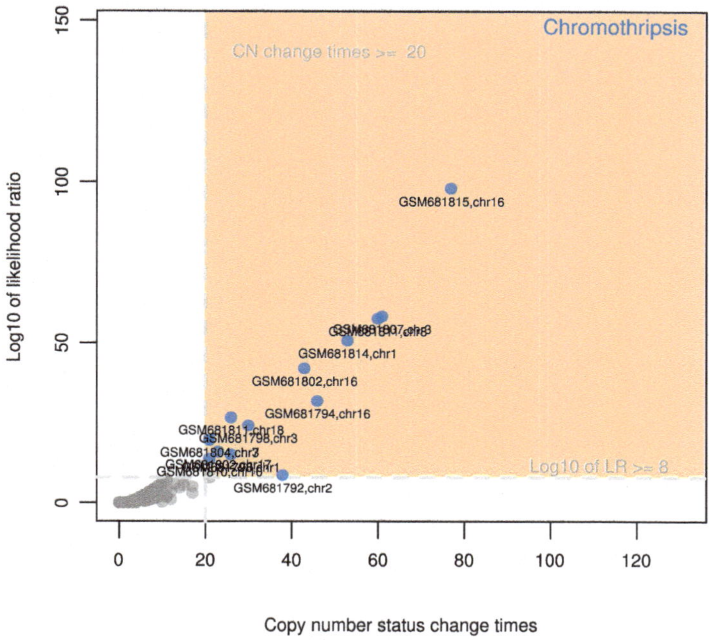

Fig. 8 The scatter plot is used to assess the significance of two decision parameters. Each point represents a chromosome, and the point falling in the orange area indicates chromothripsis event. The ***x***- and ***y***-axis represent copy number status change times and likelihood ratio, respectively. The user-specified thresholds are indicated with dashed lines

5. "Scatter Plot" panel can be used to compare the two decision parameters, i.e., copy number status change times and likelihood ratio (Fig. 8). Each point in the plot represents a chromosome, and the candidates falling in the orange area are inferred chromothripsis cases. The closer the point is to the top-right corner, the more accurately the server can distinguish chromothripsis from non-chromothripsis cases.

4 Notes

1. Make sure the file name does NOT contain special characters, such as dot, colon, dash, and space. These characters may confuse the server program and terminate the data processing pipeline.

2. If you copy and paste copy number segmentation data directly into the input text field, the first column (sample name) will be used as sample identifiers in the results page.
3. As an example, a zip file with ten tumor samples in GSE27560 can be downloaded from the "Download" page. All the samples present chromothripsis phenomenon.
4. The default value of copy number status change times is set to more than 20. According to our experience, this value will effectively identify most reported chromothripsis cases. And this threshold value is adopted by many studies [1, 14]. However, it can be adjusted for operational definitions.
5. This likelihood ratio is calculated based on scan statistics. The higher this value, the more likely the detected region represents chromothripsis event.
6. The hypersegmentation of copy number data will cause a large number of very small copy number alterations and discontinuous segments. The threshold value can be adjusted according to different array platforms with various probe densities. For example, Affymetrix Genome-Wide Human SNP 6.0 Array contains 1.8 million polymorphic and non-polymorphic markers with the mean inter-marker distance of 1.7 kb. It provides a practical resolution of 10–20 kb. For data generated by this array platform, segments smaller than 10 kb are suggested to be removed.
7. Due to the existence of background noise, the copy number profiles occasionally exhibit subtle striation patterns, which are unlikely to be a real biological phenomenon. The definition of this threshold is the sum of the absolute values to call genomic gains and losses. If the distance of two adjacent segments differed by less than this threshold value, the copy number status change will not be considered.
8. Copy number aberrations are called based on the normalized probe signal log2 ratios. The default values (0.15 for gain and −0.15 for loss) are empirical thresholds that are derived from most types of aCGH and SNP arrays. The optimum thresholds may vary across platforms. Both values can be tuned by users to process data derived from different array platforms.
9. All threshold values must be numeric; otherwise, the web server may return an error message.
10. It may take a while to upload your file to the server. While the procedure is running, the textual processing progress bar (below the "Scan" button) indicates the completed samples and the total sample number. In most cases, it takes dozens of seconds to process a single sample. The website employs ActiveX controls to display information in the processing progress bar. If the progress information does not show up, the user needs to enable the "ActiveX controls and plug-ins."

11. The entire list of array and series accession IDs can be downloaded from the web interface. Array IDs can be browsed by subgroups, which are classified by data sources, cancer types and cancer loci.
12. The number in parentheses indicates the number of samples in the series.
13. The “ReScan” button allows users to reuse previously defined parameters for sample screening or analyze the same sample with newly defined parameters.
14. For multiple samples analysis, all detected chromothripsis events will be displayed in this table.
15. According to published studies, the patterns and distribution of copy number state changes indicate the emergence and temporal order of the major genomic rearrangement events [7, 22]. There are many samples with different patterns according to the results obtained from CTLPScanner. For example, some samples show a stepwise increase in copy number toward the chromosome end, followed by a sharp drop. It indicates that chromothripsis may occur subsequent to initiating breakage fusion bridge cycles (BFBs) [7, 22–24]. For some other examples, the chromothripsis patterns show copy number oscillates among three or more states. This type of pattern indicates that chromothripsis may be the initiating event.
16. If there are many genes present in a small region, this region can be zoomed in to display only a portion of the genes.
17. Unlike arm-level copy number aberrations, focal CNAs are the result of many selection events during the evolution of cancer genomes. Candidate driver genes in focal CNAs may give tumor cells a selective growth advantage. It is useful for the identification of novel cancer driver genes in cancer genome studies.

Acknowledgments

This work was funded by the National Natural Science Foundation of China [grant number 31571314 and U1603120].

References

1. Stephens PJ, Greenman CD, Fu B et al (2011) Massive genomic rearrangement acquired in a single catastrophic event during cancer development. Cell 144:27–40
2. Forment JV, Kaidi A, Jackson SP (2012) Chromothripsis and cancer: causes and consequences of chromosome shattering. Nat Rev Cancer 12:663–670
3. Liu P, Erez A, Nagamani SC et al (2011) Chromosome catastrophes involve replication mechanisms generating complex genomic rearrangements. Cell 146:889–903
4. Rausch T, Jones DT, Zapatka M et al (2012) Genome sequencing of pediatric medulloblastoma links catastrophic DNA rearrangements with TP53 mutations. Cell 148:59–71

5. Kloosterman WP, Hoogstraat M, Paling O et al (2011) Chromothripsis is a common mechanism driving genomic rearrangements in primary and metastatic colorectal cancer. Genome Biol 12:R103
6. Molenaar JJ, Koster J, Zwijnenburg DA et al (2012) Sequencing of neuroblastoma identifies chromothripsis and defects in neuritogenesis genes. Nature 483:589–593
7. Li Y, Schwab C, Ryan SL et al (2014) Constitutional and somatic rearrangement of chromosome 21 in acute lymphoblastic leukaemia. Nature 508:98–102
8. Magrangeas F, Avet-Loiseau H, Munshi NC et al (2011) Chromothripsis identifies a rare and aggressive entity among newly diagnosed multiple myeloma patients. Blood 118:675–678
9. Hirsch D, Kemmerling R, Davis S et al (2013) Chromothripsis and focal copy number alterations determine poor outcome in malignant melanoma. Cancer Res 73:1454–1460
10. Korbel JO, Campbell PJ (2013) Criteria for inference of chromothripsis in cancer genomes. Cell 152:1226–1236
11. Govind SK, Zia A, Hennings-Yeomans PH et al (2014) ShatterProof: operational detection and quantification of chromothripsis. BMC Bioinformatics 15:78
12. Yang J, Liu J, Ouyang L et al (2016) CTLPScanner: a web server for chromothripsis-like pattern detection. Nucleic Acids Res 44(W1):W252–W258
13. Naus JI (1965) The distribution of the size of the maximum cluster of points on a line. J Am Stat Assoc 60:532–538
14. Cai H, Kumar N, Bagheri HC et al (2014) Chromothripsis-like patterns are recurring but heterogeneously distributed features in a survey of 22,347 cancer genome screens. BMC Genomics 15:82
15. Weinstein JN, Collisson EA, Mills GB et al (2013) The cancer genome atlas pan-cancer analysis project. Nat Genet 45:1113–1120
16. Barrett T, Wilhite SE, Ledoux P et al (2013) NCBI GEO: archive for functional genomics data sets–update. Nucleic Acids Res 41:D991–D995
17. Olshen AB, Venkatraman ES, Lucito R et al (2004) Circular binary segmentation for the analysis of array-based DNA copy number data. Biostatistics 5:557–572
18. Hupé P, Stransky N, Thiery JP et al (2004) Analysis of array CGH data: from signal ratio to gain and loss of DNA regions. Bioinformatics 20:3413–3422
19. Nilsen G, Liestol K, Van Loo P et al (2012) Copynumber: efficient algorithms for single- and multi-track copy number segmentation. BMC Genomics 13:591
20. Cai H, Gupta S, Rath P et al (2015) arrayMap 2014: an updated cancer genome resource. Nucleic Acids Res 43:D825–D830
21. Forbes SA, Beare D, Gunasekaran P et al (2015) COSMIC: exploring the world's knowledge of somatic mutations in human cancer. Nucleic Acids Res 43:D805–D811
22. Mardin BR, Drainas AP, Waszak S et al (2015) A cell-based model system links chromothripsis with hyperploidy. Mol Syst Biol 11:828
23. Ernst A, Jones DT, Maass KK et al (2016) Telomere dysfunction and chromothripsis. Int J Cancer 138:2905–2914
24. Harrison CJ (2015) Blood Spotlight on iAMP21 acute lymphoblastic leukemia (ALL), a high-risk pediatric disease. Blood 125:1383–1386

Chapter 18

ChromothripsisDB: A Curated Database for the Documentation, Visualization, and Mining of Chromothripsis Data

Haoyang Cai

Abstract

ChromothripsisDB (http://cgma.scu.edu.cn/ChromothripsisDB) is a manually curated database containing a unified description of published chromothripsis cases and relevant genomic aberrations. Available data includes copy number alterations, chromosome structural variations, and gene annotations. The criteria used for detecting chromothripsis in each study are also provided. At present, the molecular mechanisms involved in chromothripsis phenomenon are not fully understood. Thus, further studies with large number of identified chromothripsis samples are needed. The current release of ChromothripsisDB contains more than 400 patient samples, representing over 100 research articles. It represents an extraordinary resource for mining the existing knowledge of chromothripsis.

Key words Database, Copy number alteration, Chromosome structural variation, Genomic aberration, Cancer genome, Data mining

1 Introduction

The ChromothripsisDB database is a manually curated database, which aims at cataloging all published chromothripsis articles and samples [1]. The phenomenon of chromothripsis was first reported by Stephens and colleagues in 2011 [2]. Since then, it has received sustained scholarly attention because it may reveal a new way that cancer genomes can evolve [3, 4]. Until now, hundreds of studies have reported chromothripsis events in various human diseases [5–14]. However, the relevant data acquisition is a challenge, since the experimental data are scattered throughout the scientific literature and are heterogeneous in their contents and formats.

We extracted and normalized genomic aberration data into standardized formats and developed a web interface for data visualization and mining. All data are manually curated from the main text, tables, figures, or supplementary materials and can be easily traced back to the original publication. Moreover, the signature of

Franck Pellestor (ed.), *Chromothripsis: Methods and Protocols*, Methods in Molecular Biology, vol. 1769, https://doi.org/10.1007/978-1-4939-7780-2_18, © Springer Science+Business Media, LLC, part of Springer Nature 2018

chromothripsis was found in only 2–3% of all cancers [2, 15, 16]. Thus, large public datasets are valuable resources for chromothripsis study. A meta-analysis based on large-scale chromothripsis samples will uncover the molecular mechanisms leading to this event and will substantially enhance our understanding of tumor initiation and progression [17, 18]. ChromothripsisDB helps researchers to browse publications effectively and facilitates future research to identify key features of chromothripsis event.

2 Materials

2.1 Data Content

To retrieve all publications that report chromothripsis cases, we conducted a PubMed search using the keyword "chromothripsis." All articles were reviewed for eligibility and classified into three categories: Research, Methodology, and Review and Opinion. The research articles contain experimental data from clinical samples. Three criteria were applied to define eligible samples: (1) the sample must be explicitly declared as chromothripsis by the authors; (2) the clinical diagnosis and pathology of the sample must be clearly stated; (3) each sample is assigned a specific case ID or can be traced back to the original publication. For each case, copy number alterations (CNAs), chromosome structural variations (SVs), relevant genes, genome assembly version, and clinical information were extracted from the main text, tables, figures, and supplementary materials, if available.

The current version of ChromothripsisDB contains more than 400 chromothripsis samples manually curated from over 100 research articles on two species: human and mouse. These data represent 52 disease types, mainly cancers and congenital abnormalities. The chromothripsis cases are identified by next-generation sequencing, array comparative genomic hybridization (aCGH), or single nucleotide polymorphism (SNP) arrays. The data content of ChromothripsisDB is shown in Table 1.

Table 1
Summary of chromothripsis data in ChromothripsisDB

	Human			
	Cancers	Other diseases	Other species	Total
Chromothripsis sample	388	53	13	454
Disease type	45	7	1	53
Research article	88	25	3	116
Review article				46
Methodology article				22

2.2 The Definition of Chromothripsis

To date, the most detailed definition of chromothripsis comes from Korbel and Campbell [19]. They introduced six rigorous criteria for chromothripsis. However, usually not all the criteria for chromothripsis detection can be met due to the limitations of detection technology. For example, the chromosome structural variations cannot be identified by array-based technologies. Thus, for many studies, an operational definition was used to identify chromothripsis, and only some of the criteria were met [5, 15, 17, 20]. We reviewed and extracted the definition of chromothripsis used in each study. Since the definition is given in various descriptions of different publications, we have tried to provide as much information as possible. According to the original description of chromothripsis phenomenon, we selected the three most significant features shared by different operational chromothripsis definitions and curated and normalized the data:

1. The minimum number of close-by breakpoints. The number of breakpoints in the pulverized chromosome is counted. This information is usually extracted according to the breakpoint junctions or chromosome rearrangements.
2. The number of copy number states. The value of this feature is given as a range or a single number.
3. The random joining of chromosome fragments. The value of this feature is set to "Yes" or "NA." It is difficult to decide whether the end-joining process is random or not. However, the orientation of breakpoint junctions (e.g., head-to-head, tail-to-tail, and tail-to-head) is a signature corresponding to chromothripsis. If the orientations are provided by the authors, the chromosomal fragments were considered to be randomly assembled. If this information cannot be obtained, for example, the rearrangement data is generated by aCGH or SNP array platform, the value of this feature was set to "NA."

The definition of chromothripsis was extracted from the main text, tables, supplementary materials, and also figures. The users can check this information in data of research articles when browsing the database by studies.

3 Methods

3.1 Search by Disease

The interface of "Search by Disease" is designed for users to explore disease-specific chromothripsis patterns. It allows users to search by species, detection technology, and disease type. Here we query the database for several types of brain tumors as an example.

1. Navigate to http://cgma.scu.edu.cn/ChromothripsisDB. Click the "Search" tab in the top menu bar.
2. Select "Homo sapiens" in the "Species" field (*see* **Note 1**) (Fig. 1).

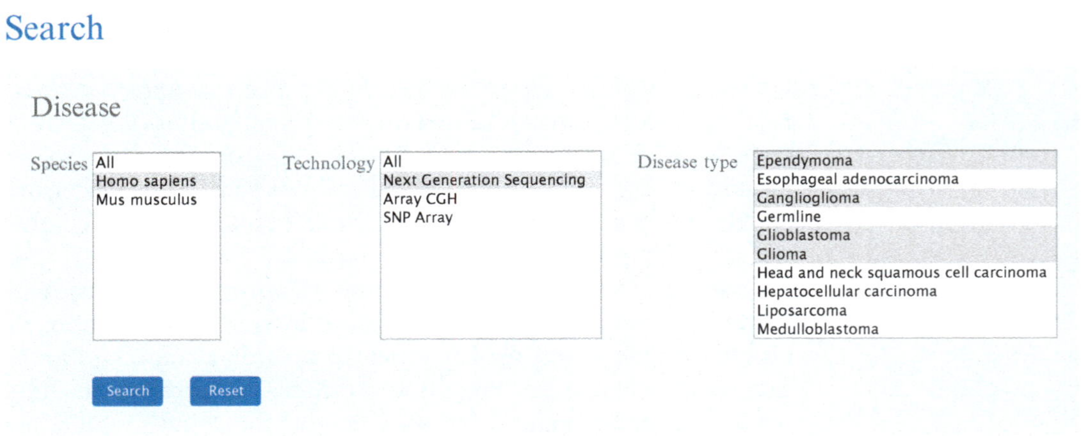

Fig. 1 The Search by Disease interface

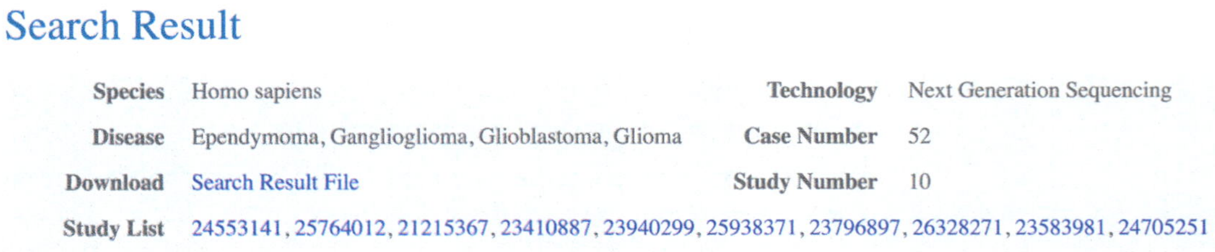

Fig. 2 The summary table of search result

3. Choose "Next Generation Sequencing" in the "Technology" field.
4. Simultaneously select "Ependymoma," "Ganglioglioma," "Glioblastoma," and "Glioma" in the "Disease type" field.
5. Click the "Search" button to execute the query.
6. Once the query is completed, the web server returns a summary table and a visualization of the results. In this example, 52 chromothripsis cases collected from 10 publications are shown (*see* **Note 2**). The summary table includes an overview of the query inputs and outputs, as well as links to related studies (Fig. 2). The links for each study can be used to transfer to specific study for further exploration.
7. The result data can be downloaded as a tab-delimited file (*see* **Note 3**).
8. The bar plot displays the distribution of pulverized chromosomes across the genome (*see* **Note 4**) (Fig. 3a).
9. The pie chart shows the percentage distribution of clinical samples among the disease types (Fig. 3b).
10. The chromothripsis case visualization interface allows users to explore each sample individually. In the data panel, the detailed information of the chromothripsis case is shown (Fig. 4a). If

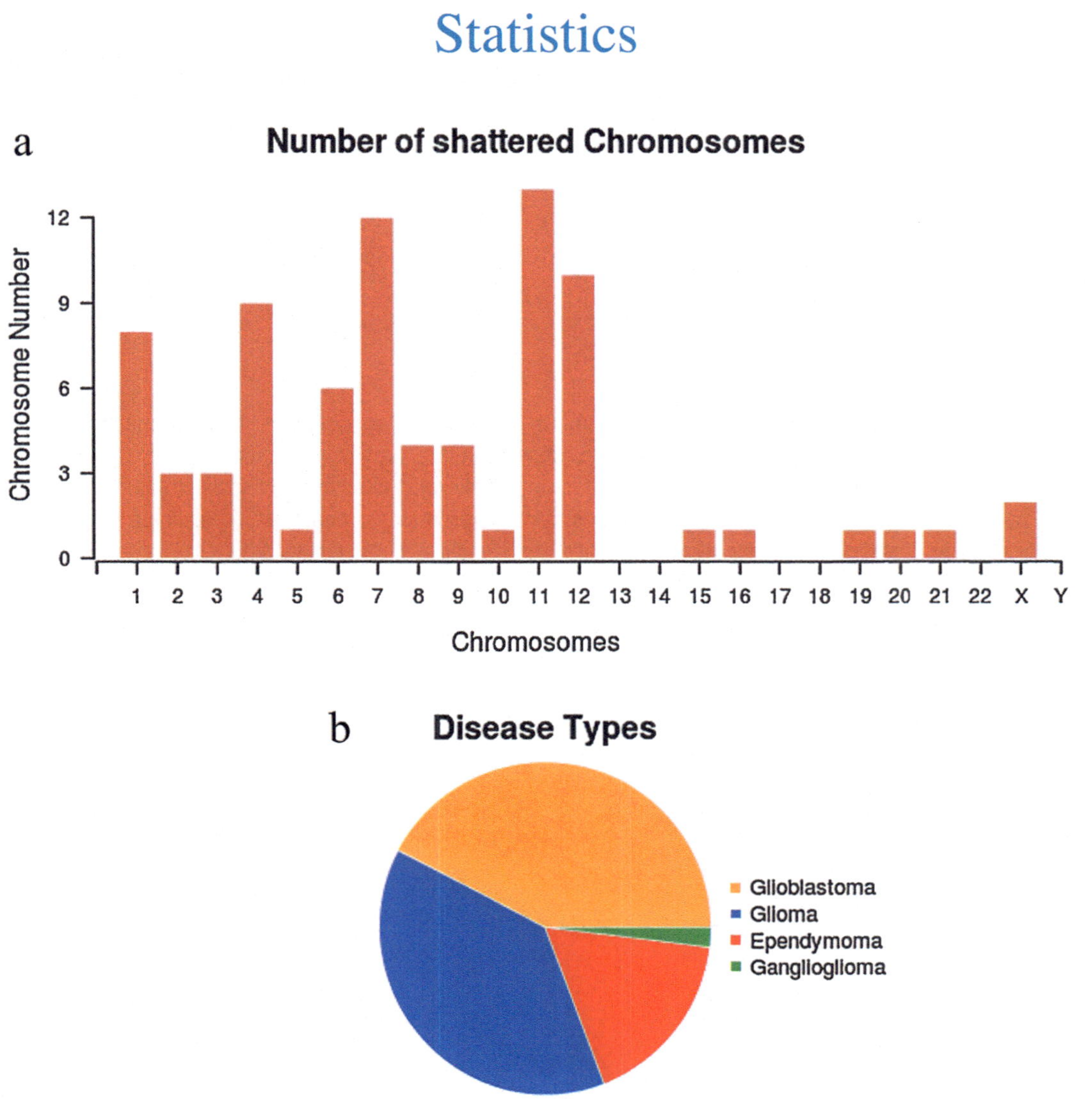

Fig. 3 Statistics of the search result. (**a**) The bar plot shows the distribution of pulverized chromosomes across the genome. (**b**) The pie chart represents the distribution of samples among the queried disease types

copy number aberration data is available, the chromosome idiogram shows the distribution of CNAs across the genome (Fig. 4b). If structural variation data is available, the Circos plot intuitively displays the chromosome breakpoints and copy number profiles (*see* **Note 5**) (Fig. 4c).

11. In the chromothripsis case visualization interface, the "Gene List" link in the "Affected Gene" field displays gene result page, if this data is available (Fig. 4a). In this page, the "Fusion Gene" table lists the locations of two genes that are fused together (*see* **Note 6**) (Fig. 5a). The "Affected Gene" table contains the genes that are located in chromothripsis region (Fig. 5b).

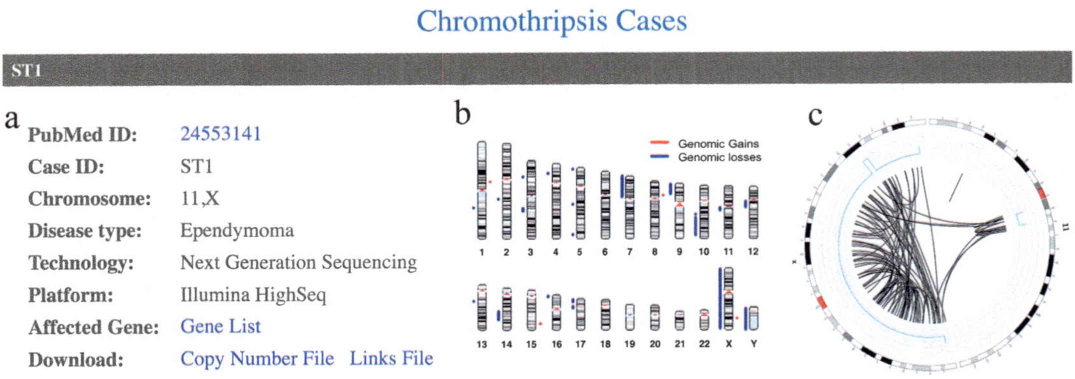

Fig. 4 The visualization of chromothripsis case. (**a**) The data panel displays the related information of chromothripsis case. (**b**) The idiogram of chromosomes shows the summary of copy number alterations. The red and blue lines represent genomic gains and losses, respectively. (**c**) The Circos plot shows structural variations and copy number alterations. Both the inter- and intra-chromosomal rearrangements are shown

a Fusion Gene

PubMed ID: 24553141 **Case ID:** ST1 **Disease:** Ependymoma **Genome Assembly:** GRCh37/hg19

Gene Symbol 1	Chromosome1	Description 1	Gene Symbol 2	Chromosome2	Description 2
C1orf95	1	chromosome 1 open reading frame 95	RELA	11	v-rel avian reticuloendotheliosis viral oncogene homolog A
HDHD1	X	haloacid dehalogenase-like hydrolase domain containing 1	MACROD1	11	MACRO domain containing 1
PRG2	11	proteoglycan 2, bone marrow (natural killer cell activator, eosinophil granule major basic protein)	C11orf95	11	chromosome 11 open reading frame 95

b Affected Gene

PubMed ID: 24553141 **Case ID:** ST1 **Disease:** Ependymoma **Genome Assembly:** GRCh37/hg19

Gene Symbol	Entrez Gene	Chromosome	Start	End	Description
RS1	6247	X	18639688	18672103	retinoschisin 1
SLC1A5	6510	19	47278140	47291851	solute carrier family 1 (neutral amino acid transporter), member 5
P4HA1	5033	10	74766975	74856732	prolyl 4-hydroxylase, alpha polypeptide I
ALG13	79868	X	111680742	111764164	ALG13, UDP-N-acetylglucosaminyltransferase subunit
ZNF827	152485	4	146678779	146859787	zinc finger protein 827

Fig. 5 The detailed information of affected genes. (**a**) The list of fusion genes. (**b**) The list of genes that are involved in chromothripsis event

12. In the chromothripsis case visualization interface, the "Download" field provides links to download related data of this case, if these data are available (*see* **Note 7**) (Fig. 4a).

3.2 Search by Gene

The "Search by Gene" interface enables the query of affected genes in various disease types. It facilitates to narrow the gene list to driver genes that are involved in the chromosome pulverization event. As an example, here we identify the affected genes between breast cancer and prostate cancer.

1. Navigate to http://cgma.scu.edu.cn/ChromothripsisDB. Click the "Search" tab in the top menu bar.
2. In the "Search by Gene" interface, select "Breast cancer" and "Prostate cancer" in the first and second "disease" field, respectively (*see* **Note 8**) (Fig. 6). Choose "AND" as the Boolean operator.
3. Click the "Search" button to execute the query.

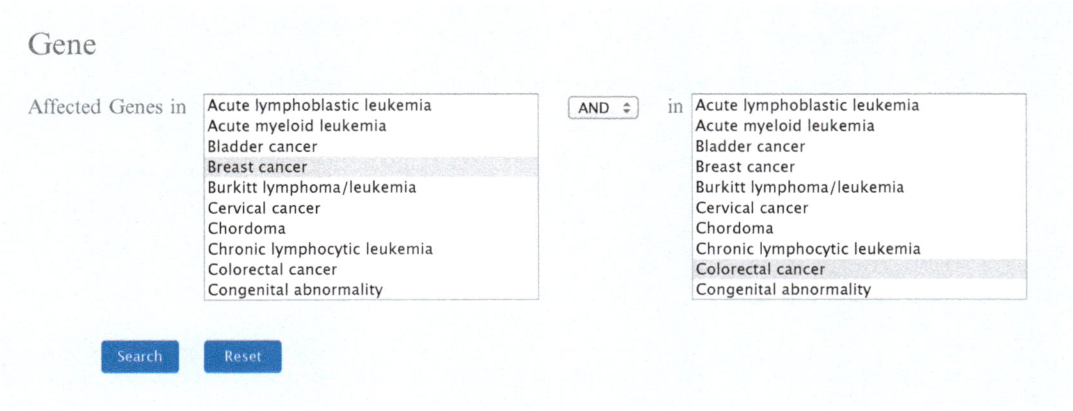

Fig. 6 The Search by Gene interface

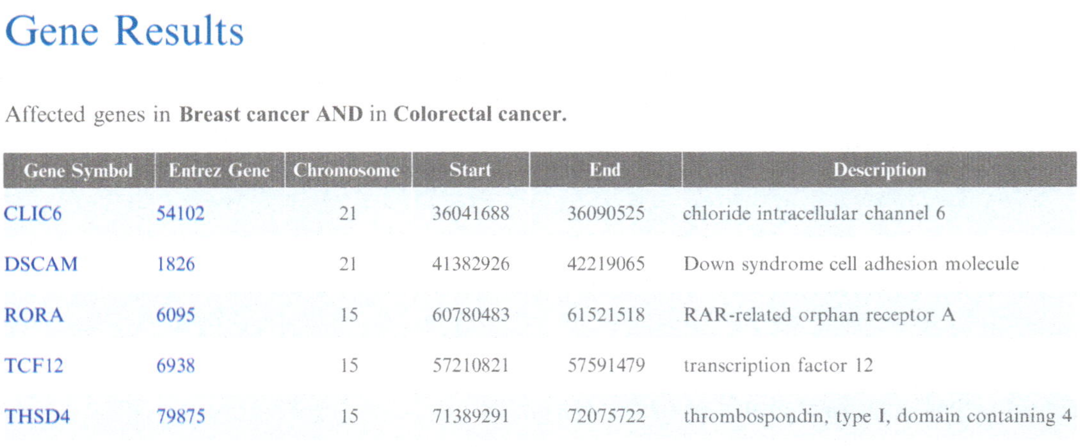

Gene Results

Affected genes in **Breast cancer AND** in **Colorectal cancer.**

Gene Symbol	Entrez Gene	Chromosome	Start	End	Description
CLIC6	54102	21	36041688	36090525	chloride intracellular channel 6
DSCAM	1826	21	41382926	42219065	Down syndrome cell adhesion molecule
RORA	6095	15	60780483	61521518	RAR-related orphan receptor A
TCF12	6938	15	57210821	57591479	transcription factor 12
THSD4	79875	15	71389291	72075722	thrombospondin, type I, domain containing 4

Fig. 7 The result page of Search by Gene function

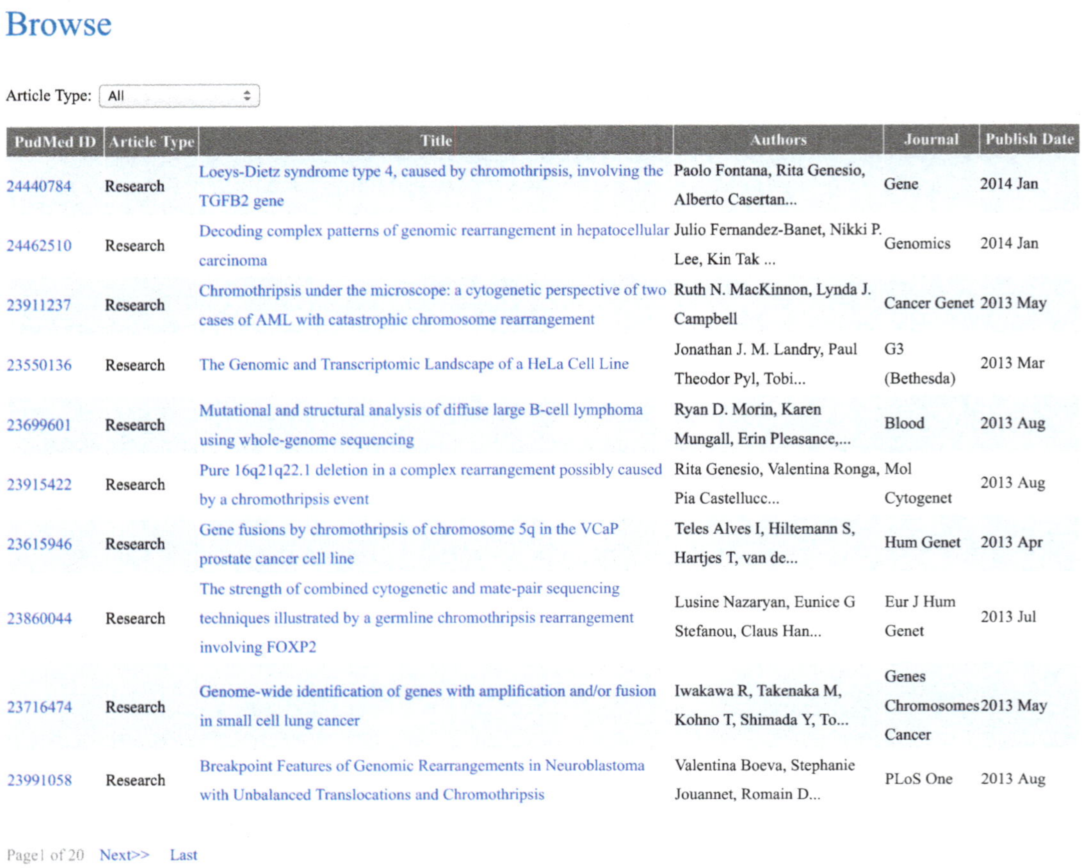

Browse

Article Type: All

PudMed ID	Article Type	Title	Authors	Journal	Publish Date
24440784	Research	Loeys-Dietz syndrome type 4, caused by chromothripsis, involving the TGFB2 gene	Paolo Fontana, Rita Genesio, Alberto Casertan...	Gene	2014 Jan
24462510	Research	Decoding complex patterns of genomic rearrangement in hepatocellular carcinoma	Julio Fernandez-Banet, Nikki P. Lee, Kin Tak ...	Genomics	2014 Jan
23911237	Research	Chromothripsis under the microscope: a cytogenetic perspective of two cases of AML with catastrophic chromosome rearrangement	Ruth N. MacKinnon, Lynda J. Campbell	Cancer Genet	2013 May
23550136	Research	The Genomic and Transcriptomic Landscape of a HeLa Cell Line	Jonathan J. M. Landry, Paul Theodor Pyl, Tobi...	G3 (Bethesda)	2013 Mar
23699601	Research	Mutational and structural analysis of diffuse large B-cell lymphoma using whole-genome sequencing	Ryan D. Morin, Karen Mungall, Erin Pleasance,...	Blood	2013 Aug
23915422	Research	Pure 16q21q22.1 deletion in a complex rearrangement possibly caused by a chromothripsis event	Rita Genesio, Valentina Ronga, Pia Castellucc...	Mol Cytogenet	2013 Aug
23615946	Research	Gene fusions by chromothripsis of chromosome 5q in the VCaP prostate cancer cell line	Teles Alves I, Hiltemann S, Hartjes T, van de...	Hum Genet	2013 Apr
23860044	Research	The strength of combined cytogenetic and mate-pair sequencing techniques illustrated by a germline chromothripsis rearrangement involving FOXP2	Lusine Nazaryan, Eunice G Stefanou, Claus Han...	Eur J Hum Genet	2013 Jul
23716474	Research	Genome-wide identification of genes with amplification and/or fusion in small cell lung cancer	Iwakawa R, Takenaka M, Kohno T, Shimada Y, To...	Genes Chromosomes Cancer	2013 May
23991058	Research	Breakpoint Features of Genomic Rearrangements in Neuroblastoma with Unbalanced Translocations and Chromothripsis	Valentina Boeva, Stephanie Jouannet, Romain D...	PLoS One	2013 Aug

Page1 of 20 Next>> Last

Fig. 8 The data browsing interface

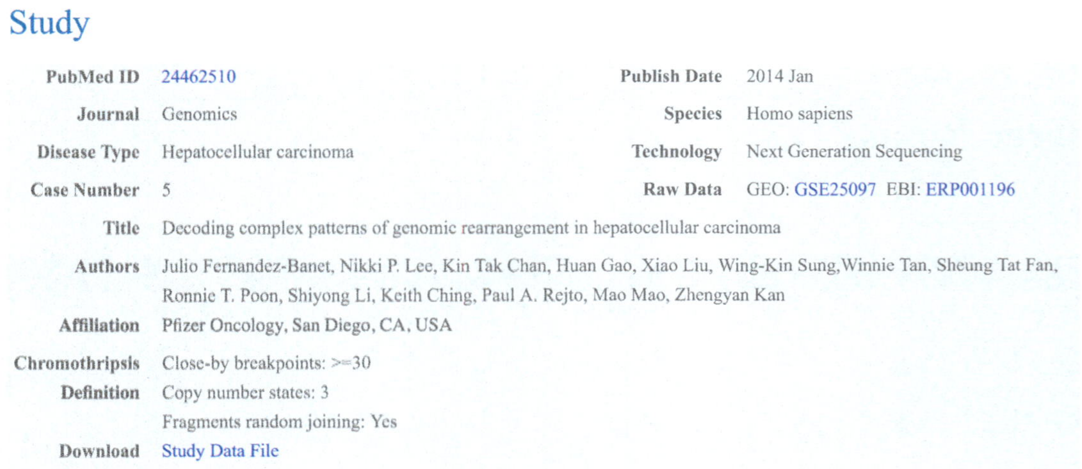

Study

PubMed ID	24462510	Publish Date	2014 Jan
Journal	Genomics	Species	Homo sapiens
Disease Type	Hepatocellular carcinoma	Technology	Next Generation Sequencing
Case Number	5	Raw Data	GEO: GSE25097 EBI: ERP001196
Title	Decoding complex patterns of genomic rearrangement in hepatocellular carcinoma		
Authors	Julio Fernandez-Banet, Nikki P. Lee, Kin Tak Chan, Huan Gao, Xiao Liu, Wing-Kin Sung,Winnie Tan, Sheung Tat Fan, Ronnie T. Poon, Shiyong Li, Keith Ching, Paul A. Rejto, Mao Mao, Zhengyan Kan		
Affiliation	Pfizer Oncology, San Diego, CA, USA		
Chromothripsis Definition	Close-by breakpoints: >=30 Copy number states: 3 Fragments random joining: Yes		
Download	Study Data File		

Fig. 9 The summary table of curated information of a study

4. Once the query is completed, the web server generates a results page. It shows the qualified genes and related annotations (*see* **Note 9**) (Fig. 7). It also contains links to Catalogue of Somatic Mutations in Cancer (COSMIC) database [21] and NCBI Entrez Gene page for details.

3.3 Data Browsing

The data browsing interface allows users to browse the database by studies.

1. Navigate to http://cgma.scu.edu.cn/ChromothripsisDB. Click the "Browse" tab in the top menu bar.
2. All publications that report chromothripsis cases are shown (*see* **Note 10**) (Fig. 8). Clicking on the article title, the web server generates a result page, which includes a table with detailed information of the study and a visualization of chromothripsis cases (Fig. 9).
3. The definition of chromothripsis that is used in the study is shown (*see* **Note 11**) (Fig. 9).
4. For research article, the curated data can be downloaded as a tab-delimited file (*see* **Note 12**).
5. For each chromothripsis case in the study, the result page provides a graphical summary of genomic aberrations in an idiogram of chromosomes, and SVs and CNAs are illustrated by a Circos plot, if these data are available (*see* **Note 13**).

4 Notes

1. All fields support multiple selections, and the field content changes dynamically according to the user's selection.
2. For each case, the published sample ID is used as the case identifier in ChromothripsisDB. This ID allows users to trace the case back to the original publication. If the sample ID is not available, a unique identifier consisting of the PubMed ID and a case number is assigned.
3. The relevant information of each chromothripsis case is extracted from the original publication, including pulverized chromosome IDs, detection technology and platform, disease type, and so on. Note that, for some cases, certain data may not be provided by the authors, such as copy number aberrations, breakpoints, fusion genes, and genome assembly.
4. Some disease types may have specific chromosome pulverization patterns, which can be distinguished from other diseases. These patterns will promote the discovery of disease driver genes.

5. Only the pulverized chromosomes are shown in Circos plot. According to curated data, one or several chromosomes may be pulverized in each sample.
6. All the fusion genes or rearranged genes were extracted from the original publications. The interface contains links to NCBI Gene page and Entrez Gene page, which allow users to investigate each gene in detail. These data are consistent with the authors' original descriptions. This page will facilitate users to identify specific mutated genes that are predisposed to chromothripsis.
7. The copy number and link files contain CNA and structural variation data, respectively. These files can be used to generate the Circos plot.
8. The Boolean search option allows users to search for affected genes in a certain disease and/or not involved in any other disease. Three Boolean search operators are available in the drop-down options box: AND, OR, and NOT. This function allows users to search for common affected genes between two disease types or for specific genes in particular diseases.
9. All the affected genes were extracted from the original publications, and are consistent with the authors' original descriptions.
10. The "Article Type" drop-down list can be used to filter publications by types.
11. If the value of a certain feature is set to "NA," it means this information cannot be obtained from the publication.
12. The raw experimental data can be accessed through hyperlinks, if available.
13. Please refer to "Subheading 3.1" for the description of chromothripsis case visualization.

Acknowledgments

This work was funded by the National Natural Science Foundation of China [grant numbers 31571314 and U1603120].

References

1. Yang J, Deng G, Cai H (2016) ChromothripsisDB: a curated database of chromothripsis. Bioinformatics 32:1433–1435
2. Stephens PJ, Greenman CD, Fu B et al (2011) Massive genomic rearrangement acquired in a single catastrophic event during cancer development. Cell 144:27–40
3. Forment JV, Kaidi A, Jackson SP (2012) Chromothripsis and cancer: causes and consequences of chromosome shattering. Nat Rev Cancer 12:663–670
4. Meyerson M, Pellman D (2011) Cancer genomes evolve by pulverizing single chromosomes. Cell 144:9–10

5. Rausch T, Jones DT, Zapatka M et al (2012) Genome sequencing of pediatric medulloblastoma links catastrophic DNA rearrangements with TP53 mutations. Cell 148:59–71
6. Kloosterman WP, Hoogstraat M, Paling O et al (2011) Chromothripsis is a common mechanism driving genomic rearrangements in primary and metastatic colorectal cancer. Genome Biol 12:R103
7. Molenaar JJ, Koster J, Zwijnenburg DA et al (2012) Sequencing of neuroblastoma identifies chromothripsis and defects in neuritogenesis genes. Nature 483:589–593
8. Nones K, Waddell N, Wayte N et al (2014) Genomic catastrophes frequently arise in esophageal adenocarcinoma and drive tumorigenesis. Nat Commun 5:5224
9. Kloosterman WP, Guryev V, van Roosmalen M et al (2011) Chromothripsis as a mechanism driving complex de novo structural rearrangements in the germline. Hum Mol Genet 20:1916–1924
10. Northcott PA, Shih DJ, Peacock J et al (2012) Subgroup-specific structural variation across 1,000 medulloblastoma genomes. Nature 488:49–56
11. Wu C, Wyatt AW, McPherson A et al (2012) Poly-gene fusion transcripts and chromothripsis in prostate cancer. Genes Chromosomes Cancer 51:1144–1153
12. Hirsch D, Kemmerling R, Davis S et al (2013) Chromothripsis and focal copy number alterations determine poor outcome in malignant melanoma. Cancer Res 73:1454–1460
13. Bassaganyas L, Beà S, Escaramís G et al (2013) Sporadic and reversible chromothripsis in chronic lymphocytic leukemia revealed by longitudinal genomic analysis. Leukemia 27:2376–2379
14. Cai H, Kumar N, Bagheri HC et al (2014) Chromothripsis-like patterns are recurring but heterogeneously distributed features in a survey of 22,347 cancer genome screens. BMC Genomics 15:82
15. Zack TI, Schumacher SE, Carter SL et al (2013) Pan-cancer patterns of somatic copy number alteration. Nat Genet 45:1134–1140
16. Magrangeas F, Avet-Loiseau H, Munshi NC et al (2011) Chromothripsis identifies a rare and aggressive entity among newly diagnosed multiple myeloma patients. Blood 118:675–678
17. Kim TM, Xi R, Luquette LJ et al (2013) Functional genomic analysis of chromosomal aberrations in a compendium of 8000 cancer genomes. Genome Res 23:217–227
18. Li Y, Schwab C, Ryan SL et al (2014) Constitutional and somatic rearrangement of chromosome 21 in acute lymphoblastic leukaemia. Nature 508:98–102
19. Korbel JO, Campbell PJ (2013) Criteria for inference of chromothripsis in cancer genomes. Cell 152:1226–1236
20. Malhotra A, Lindberg M, Faust GG et al (2013) Breakpoint profiling of 64 cancer genomes reveals numerous complex rearrangements spawned by homology-independent mechanisms. Genome Res 23:762–776
21. Forbes SA, Beare D, Gunasekaran P et al (2015) COSMIC: exploring the world's knowledge of somatic mutations in human cancer. Nucleic Acids Res 43:D805–D811

Part V

The Experimental Models

Chapter 19

Time-Lapse Imaging for the Detection of Chromosomal Abnormalities in Primate Preimplantation Embryos

Brittany L. Daughtry and Shawn L. Chavez

Abstract

The use of time-lapse microscopic imaging has proven to be a powerful tool for the study of mitotic divisions and other cellular processes across diverse species and cell types. Although time-lapse monitoring (TLM) of human preimplantation development was first introduced to the in vitro fertilization (IVF) community several decades ago, it was not until relatively recently that TLM systems were commercialized for clinical embryology purposes. Traditionally, human IVF embryos are assessed by successful progression and morphology under a stereomicroscope at distinct time points prior to selection for transfer. Due to the high frequency of aneuploidy, embryos may also be biopsied at the cleavage or blastocyst stage for preimplantation genetic screening (PGS) of whole and/or partial chromosomal abnormalities. However, embryo biopsy is invasive and can hinder subsequent development, and there are additional concerns over chromosomal mosaicism and resolution with PGS. Moreover, embryos are typically outside of the incubator in suboptimal culture conditions for extended periods of time during these procedures. With TLM systems, embryos remain in the stable microenvironment of an incubator and are simultaneously imaged for noninvasive embryo evaluation using a fraction of the light exposure as compared to a stereomicroscope. Each image is then compiled into a time-lapse movie, the information from which can be extrapolated to correlate morphological, spatial, and temporal parameters with embryo quality and copy number status. Here, we describe the various TLM systems available for clinical and/or research use in detail and provide step-by-step instructions on how the measurement of specific timing intervals and certain morphological criteria can be implemented into IVF protocols to enhance embryo assessment and avoid the selection of aneuploid embryos. We also discuss the biological significance of processes unique to mitotically dividing embryos and the likelihood that complex chromosomal events such as chromothripsis occur during preimplantation development in humans and other mammals, particularly nonhuman primates.

Key words Aneuploidy, Preimplantation, Cellular fragmentation, Embryo, Human, Imaging, Micronuclei, Mitosis, Multipolar, Rhesus, Time-lapse

1 Introduction

It has now been well established by comprehensive chromosome screening (CCS) of all 24 chromosomes that human preimplantation embryos endure a high incidence of aneuploidy following in vitro fertilization (IVF). With the use of high-resolution techniques, estimates are that between 50 and 80% of cleavage-stage

Franck Pellestor (ed.), *Chromothripsis: Methods and Protocols*, Methods in Molecular Biology, vol. 1769, https://doi.org/10.1007/978-1-4939-7780-2_19,

human embryos contain one or more chromosomally abnormal cells [1–5]. More recent reports indicate that a similar proportion of human blastocysts capable of implanting upon uterine transfer are also aneuploid to some degree [6–8]. As a consequence, embryos that exceed the threshold for aneuploidy tolerance will either arrest prior to implantation or later result in spontaneous miscarriage if allowed to proceed in development. Although aneuploidy increases in oocytes with maternal age [9, 10], postfertilization mitotic errors are known to occur at an equal or greater frequency than meiotic chromosomal mis-segregation and irrespective of age or fertility status [1–3, 5, 11, 12]. There are a multitude of ways by which aneuploidy can arise as a cell is undergoing mitosis, including defective microtubule attachment to the kinetochore region of the centromere, abnormal centrosome number, compromised spindle assembly checkpoints, and premature loss or prolonged persistence of chromosome cohesion [13]. Regardless of the mechanism(s), these defects will result in mitotic nondisjunction, chromosomal breakage, and/or chromosome lagging during anaphase if not resolved. A distinct process called endoreduplication can also occur, whereby the nuclear genome is replicated in the absence of cell division, leading to loss of heterozygosity if haploid cells are duplicated, or polyploidy [14]. While much of what we know about chromosomal mis-segregation events was largely determined from cancer studies, all of these phenomena have also been identified in human IVF embryos and contribute to the high degree of mosaicism observed among cells during preimplantation development [3, 15].

Once anaphase is complete and homologous chromosomes or sister chromatids have migrated to opposite poles of the meiotic or mitotic spindle, respectively, lagging chromosomes that failed to properly connect to the spindle apparatus will be excluded from the nucleus upon cell division and reformation of the nuclear membrane. Despite separation from the primary nucleus, these so-called laggards will still become encapsulated in nuclear membrane to form a micronucleus (MN) in the resultant daughter cells [5, 16]. Studies with cancer cells and transformed cell lines have shown that the chromosomes within MN are fragile and highly susceptible to double-stranded breaks, complex rearrangements, and segmental losses through a process called chromothripsis that is discussed in more detail in other chapters [17–20]. Recent evidence suggests that when chromosome-containing micronuclei rejoin the primary nucleus of daughter cells after nuclear envelope rupture during mitosis, chromothripsis can occur. Consequently, even if total chromosome copy number is maintained in a cell, the extensive genomic rearrangements and segmental deletions that occurred in the chromosomes of micronuclei may be integrated into the genome and propagated [21]. Although chromosomal events specific to chromothripsis were postulated to occur in preimplantation embryos [22, 23], it has not yet been adequately demonstrated. Moreover, embryonic blastomeres are inherently different from immortalized cells in culture, and little is known about the potential fate and detrimental effects of micronuclei formation during preimplantation development. One study has suggested that micronuclei in mouse embryos do not rejoin with the primary nucleus and are instead propagated unilaterally through multiple cell divisions [16]. In this case, the DNA repair machinery provided by the primary nucleus that is necessary for generating genomic rearrangements in micronuclei would not be available [21]. Our group has previously suggested and recently confirmed that micronuclei can also be eliminated from blastomeres through cellular fragmentation (Daughtry et al., under review) [5, 24], which may represent a mechanism to remove damaged chromosomes or further confound aneuploidy and genotypic complexity.

Under current clinical IVF practice, human preimplantation embryos are traditionally scored for quality at the 1-cell stage and then again on day 2 and/or day 3 between 48 and 72 h after fertilization at the 4- to 8-cell stage. Morphological features, including cell number, degree of cellular fragmentation, presence of multinuclei, and blastomere symmetry in size, are typically used to evaluate embryo developmental potential at the cleavage stage [25, 26]. If embryos are allowed to proceed to the blastocyst stage, additional criteria such as the cohesiveness of the inner cell mass (ICM) and trophectoderm (TE) are considered in the decision of which embryo(s) to select for transfer [26, 27]. Although these snapshots in time greatly assist in the identification of the embryos that are likely to arrest, subtle and yet, informative developmental features might be missed from the hours in between designated embryo assessments. To address some of the chromosomal aspects that conventional embryo scoring may fail to detect, embryos are often biopsied at the cleavage stage and/or blastocyst stage for preimplantation genetic screening (PGS). However, embryo biopsy and additional screening measures require removal of the embryos from the physiological conditions that an incubator provides and increased exposure to light from the stereomicroscope. In addition, the biopsy of embryos is considered invasive, and there is some evidence to suggest that it may hinder subsequent preimplantation development [28].

Considering the limitations of evaluating human embryos based on static time points and the risks associated with embryo biopsy for PGS, noninvasive time-lapse imaging techniques have been established to measure the timing of key developmental events and morphological features throughout early embryogenesis. Initially, these studies focused on the safety and efficacy of time-lapse imaging and whether so-called morphometric parameters could be used to predict successful development to the blastocyst stage (reviewed in Daughtry and Chavez [24]).

More recently, some of these parameters were used to aid in the identification and/or selection of human embryos that are chromosomally normal [5, 6, 29–32] and showed similar benefits for prognosis of preimplantation embryos from other mammalian species [33–36]. Multiple TLM systems have been commercialized for clinical embryology purposes, and all available microscopes are equipped with software to aid in embryo selection (Table 1). To our knowledge, however, only the Eeva™ system has incorporated automated cell tracking for predicting which embryo(s) have the greatest chance of reaching the blastocyst stage [37–40]. By incorporating TLM into clinical practice, the selection of the best embryo to transfer is still at the discretion of the Reproductive Endocrinology and Infertility (REI) physician and/or a professionally trained embryologist, but additional and nonsubjective data is made available to assist in the decision [41]. Not only can this information be used to predict embryo developmental success in cases of cleavage-stage transfers, it may also assist in the determination of the best embryo for transfer when multiple blastocysts are obtained from a patient.

Early time-lapse imaging studies used a common procedural start point such as the time of intracytoplasmic sperm injection (ICSI) to measure the time to the first cell division as well as subsequent divisions [42]. However, this can have confounding effects on the measurement of other overlapping cell cycle parameters as recently shown by a study that implemented a biological start point instead [43]. Therefore, it is important to calculate the time interval between each cell division separately rather than the total time from a shared starting point that may not be physiologically relevant. Using this approach (Fig. 1), it was shown that the duration of the first cytokinesis should be 14.3 ± 6.0 min in those zygotes that reach the blastocyst stage [44] and 14.4 ± 4.2 min in embryos that are chromosomally normal at the 4-cell stage [5]. For embryos that are entirely euploid at the 4-cell stage, the time between the completion of the first cytokinesis and the appearance of a cleavage furrow in one of the two daughter cells during the second mitotic division is estimated at 11.8 ± 0.71 h. Since the division of both daughter cells is fairly synchronous, the time between the beginning of the second cell division and the appearance of a cleavage furrow in the other daughter cell during the third mitotic division should be 0.96 ± 0.84 h [5]. Besides the timing between cell divisions, the time interval from the disappearance of the maternal and paternal pronuclei (PN) to the formation of the first cleavage furrow is also predictive of embryo ploidy status and should be in the range of 2.1–2.9 h for euploid embryos [45]. Additional time-lapse embryo studies have identified other imaging parameters beyond the 4-cell stage that are also predictive of blastocyst formation and may be used to differentiate between chromosomally normal and abnormal human embryos. This includes the time to the

Table 1
TLM systems for clinical human embryology

Time-lapse monitoring system	Illumination options	Microscope	Image resolution	Imaging interval	Custom culture dishes	Max # of embryos	Incubator set-up	Software
Eeva™ 2.2.1 (Progyny, USA and Merck Serono, Germany)	Darkfield, Red (625 nm) LED	×10 Olympus objective, 1 focal plane	2560 × 1920 pixels	5 min	Eeva Dish; 12 group culture microwells	48	Incubator installation	Eeva Test; day 3 blastocyst prediction
EmbryoScope/ Embryoscope+ (Vitrolife, Sweden)	Brighfield, Red (635 nm) LED	Leica 20×, 0.40 LWD Hoffman Modulation contrast objective, 3–17 focal planes	1280 × 1024 pixels; 0.3 μm/px	5–20 min	EmbryoSlide; 12 individual culture microwells OR EmbryoSlide+; 2 × 8 individual culture microwells	72 or 96	Stand-alone	EmbryoViewer™; annotation tools, customized modeling, KIDScore D3 Basic for implantation potential of day 3 embryos, remote access
Primo Vision™ EVO (Vitrolife, Sweden)	Brightfield, Green (550 nm) LED	0.26 NA, Hoffmann-like contrast, 3–11 focal planes	2560 × 1920 pixels; 1 μm/px	10–60 min	9 group culture wells	54	Incubator installation	Primo Vision EVO+ Programmable, annotation tools, remote access
Primo Vision™ EVO+ (Vitrolife, Sweden)	Brightfield, Amber (590 nm) LED	25× 0.4 NA, Hoffmanlike contrast, 3–11 focal planes	2560 × 1920 pixels; 0.3 μm/px	5–60 min	16 group culture wells	96	Incubator installation	Primo Vision EVO+ Programmable, annotation tools, remote access
Miri® TL (Esco Medical, Denmark)	Brighfield, (635 nm) LED	Zeiss 20×, 0.35 numerical aperture, 3–7 focal planes	1280 × 1024 pixels; 0.45 μm/px	5 min	CultureCoin; 14 individual culture microwells	84	Stand-alone	Miri® TL Viewer; customizable grading options, ideal timing reference, cleavage navigator, data tracking, programmable

A comprehensive table of each TLM system available to human IVF clinics for embryo assessment with the specifics of their key hardware and software components

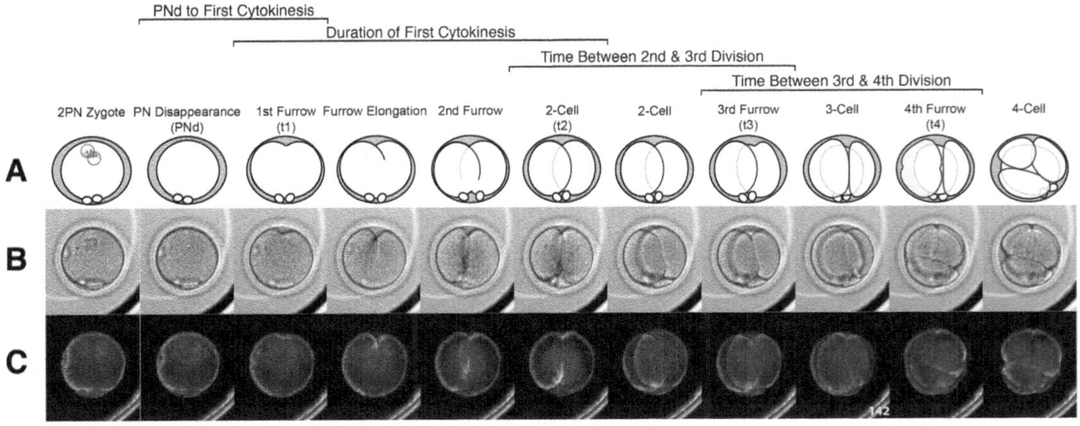

Fig. 1 Cell division parameter timing up to the 4-cell stage. (**a**) Schematic illustration of a rhesus macaque embryo from the zygote to 4-cell stage (below) imaged with the Auxogyn Eeva™ 2.3.5 bimodal TLM system. The large white circles are blastomeres, whereas the small white ovals are polar bodies. Intervals between morphokinetic time points (above) outline cell cycle parameters used to determine if dividing embryos are within the timing windows for normal development. A zygote containing two pronuclei and two polar bodies was imaged using (**b**) dark-field and (**c**) bright-field illumination and undergoes syngamy as represented by pronuclei disappearance (PNd) before cleavage up to the 4-cell stage

appearance of the 5th cell, the initiation of compaction, and the completion of blastulation [29–31].

On occasion polyploid embryos may be observed, which contain an extra copy of all chromosomes if triploid and are visibly apparent by the presence of 3PN at the zygote stage. Gynogenesis, whereby an oocyte is activated by sperm, but embryonic development proceeds without paternal genomic contribution, can also be inferred if 2PN are visible in the zygote with only one polar body present or 1PN and two polar bodies are detected [46]. In either of these cases, errors in centrosome duplication can result, leading to multiple spindle poles and abnormal cytokinesis. Moreover, diploid embryos may also have multipolar divisions if the centrosome deposited by the sperm is defective or if there are insufficient maternal centrosomal components in the MII oocyte [47]. Delineating the completion of cytokinesis and accurate cell counts are often confounded by the incidence of multipolar divisions, cellular fragmentation, and/or unsuccessful attempts at cleavage divisions (Fig. 2). Under certain circumstances, polyploidy is correlated with multipolar divisions that commonly appear as 1- to 3-cell tripolar divisions [48] as shown in Fig. 2a, but higher-order divisions can also occur to produce four, five, or six cells and so on. Approximately 12% of human zygotes undergo multipolar divisions [49] and are less likely to form blastocysts and successfully implant than zygotes that exhibit a bipolar first mitotic division [50]. Regardless of whether an embryo underwent a multipolar division at the zygote stage or during subsequent cell

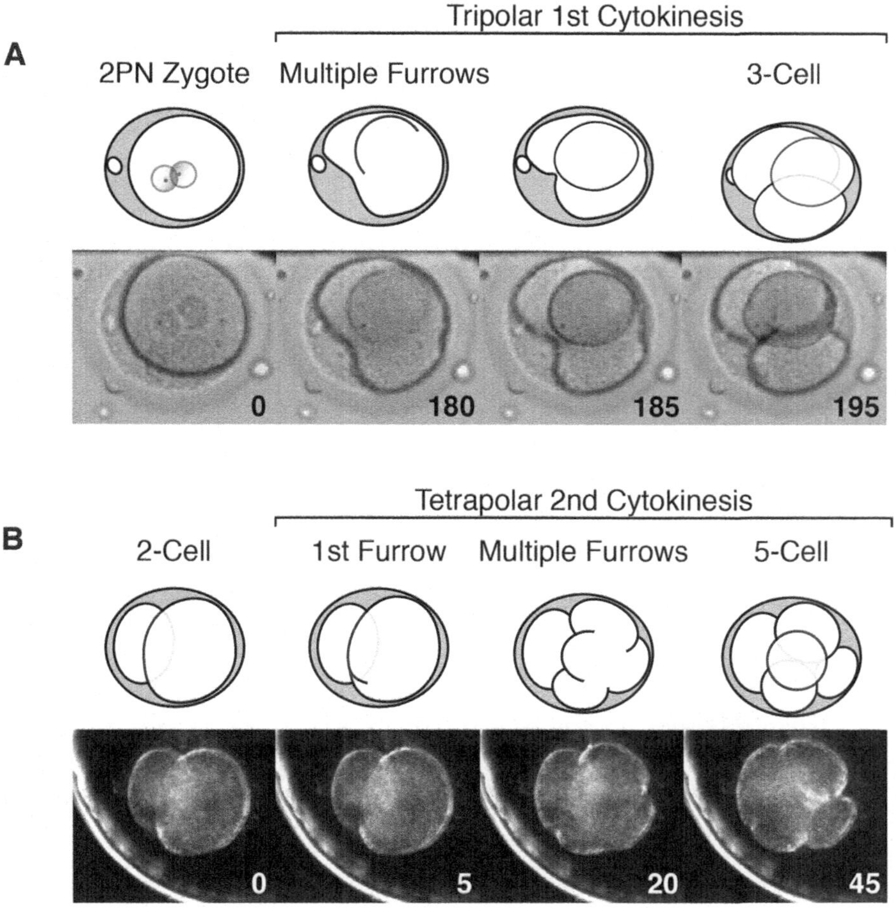

Fig. 2 Multipolar divisions in the 1- and 2-cell embryo. (**a**) Schematic illustration and bright-field images of a human zygote with two pronuclei and one visible polar body that develops multiple cleavage furrows and subsequently divides one to three cells in a tripolar division lasting approximately 15 min. The embryo was imaged with the Auxogyn Eeva™ V2.0 bright-field TLM. (**b**) Schematic illustration and dark-field images of a rhesus macaque 2-cell embryo imaged with the Auxogyn Eeva™ 2.2.1 dark-field TLM that undergoes a tetrapolar second division that lasted approximately 40 min. Note how the daughter cell on the left never divides during this time. (**a, b**) Numbers in the bottom right corner of images represent the duration in minutes from the far-left image ($t = 0$)

divisions, it was recently shown that the chromosomal status of a blastomere is clonally propagated as development proceeds [51]. Moreover, there is evidence to suggest that the allocation of chromosomes to blastomeres and other cell types is randomly distributed during tripolar divisions [48].

Another key advantage of using time-lapse imaging for embryo assessment is the ability to detect abnormal nuclear structures that can reveal chromosome mis-segregation events and ploidy errors. In contrast to normal primary nuclei, wherein a single and relatively large nucleus with intact nuclear envelope is observed (Fig. 3a, b, first panels), there are several different types of nuclear

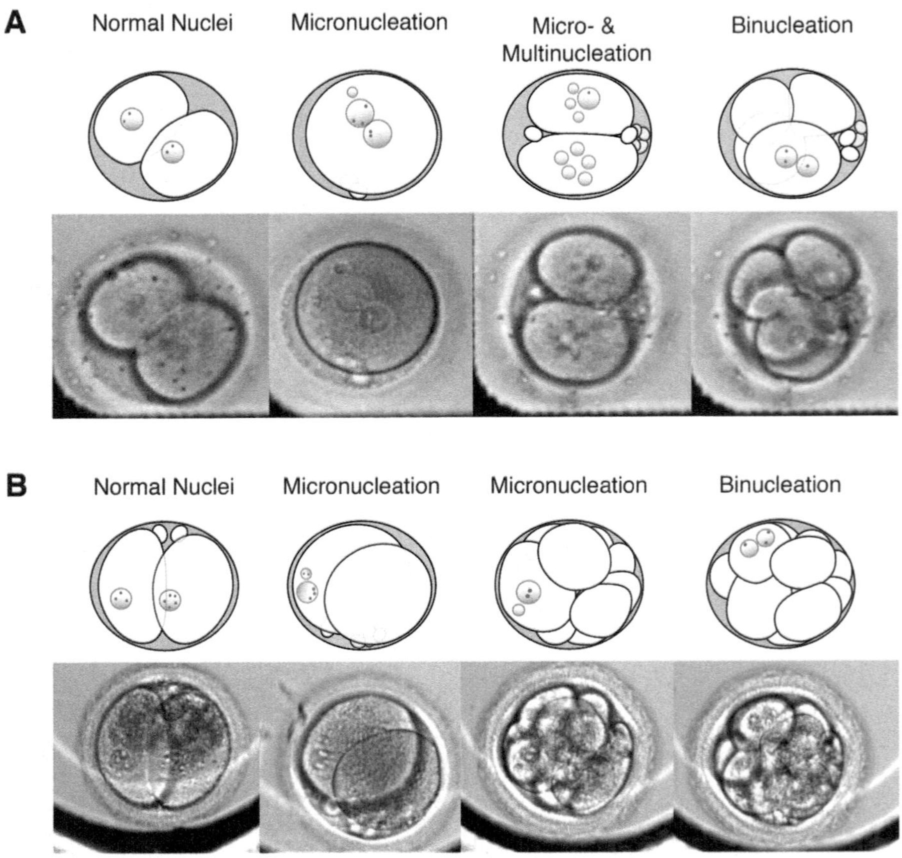

Fig. 3 Nuclear morphologies indicative of chromosomal instability. (**a**) Human and (**b**) rhesus macaque embryos at different stages of development with various types of nuclear morphologies ranging from normal to micronucleation, multinucleation, and binucleation. Cellular fragments are depicted as light yellow ovals. Bright-field images of human embryos were captured with the Auxogyn Eeva™ 2.0 TLM system, while the rhesus macaque embryos were monitored with the Auxogyn Eeva™ 2.3.5 TLM system

defects that could arise in embryos such as multinucleation, micronucleation, and anucleation [52]. As the name implies, micronuclei are smaller versions of primary nuclei and contain only one or a few chromosomes, whereas anucleated blastomeres lack any nuclear structure altogether. While there is still much to be learned regarding the fate of an embryo as a whole, when abnormal nuclear morphologies are observed, some conclusions can be drawn depending on when and to what extent they occur. If micronuclei are present in a zygote (Fig. 3a, second panel), the embryo likely inherited a meiotic error from the oocyte, or chromosome missegregation occurred upon maternal and paternal pronuclear assembly and fusion. This chromosomal error would be propagated in all resultant daughter cells unless selectively eliminated in subsequent mitotic divisions. Aneuploidy at the 1-cell stage is generally less tolerated except in cases when the affected chromosomes are compatible with live birth such as chromosomes 21, X, and Y in humans. Furthermore, the detection of multi- and/or micronuclei in both blastomeres at the 2-cell stage (Fig. 3a, third panel) would also have a poor prognosis compared to only one blastomere (Fig. 3b, second panel) since all subsequent cells would be chromosomally unbalanced [53]. The occurrence of binucleation, a unique type of multinucleation, is also fairly common in the early cleavage stages (Fig. 3a, 4th panel) and indicative of successful karyokinesis, or nuclear division during mitosis, in the absence of cytokinesis [54]. Since EGA does not occur until approximately the 8-cell stage in human embryos [55], abnormal nuclear structures in blastomeres are permissible up to this point. Upon the onset of EGA, however, such nuclear and chromosomal defects may limit the transcription of genes necessary to proceed in development and instead lead to embryo arrest.

An additional morphological indicator of embryo quality is cellular fragmentation, a dynamic process by which cytoplasmic bodies pinch off from the blastomeres of embryos during cytokinesis [56, 57] (Fig. 4a). While relatively common in humans, affecting approximately 50% of human embryos or more, mouse embryos rarely exhibit cellular fragmentation even in suboptimal culture conditions [5, 16, 58, 59]. However, embryos from other mammalian species, including rhesus monkeys, cattle, and horses, also undergo cellular fragmentation to differing degrees [60, 61]. Distinct from the cell death-induced DNA fragmentation that occurs at the morula and blastocyst stages [62–64], cellular fragmentation has been reported in human embryos conceived in vivo [65, 66], indicating that it is not an artifact of in vitro culture. Given that they can vary in size, time of appearance, spatial distribution, and organelle content [67–69], cellular fragments are sometimes difficult to differentiate from small blastomeres. To further complicate matters, cellular fragments may persist as separate units from blastomeres or be reabsorbed by the embryo [70, 71]. High levels of cellular fragmentation are associated with aneuploidy [72, 73] and, depending on timing, can be used to distinguish embryos that underwent a meiotic or mitotic error [5]. Although extensive fragmentation often correlates with low implantation and pregnancy potential [57, 62, 74–76], taken alone it is not a clear indicator of embryo quality since live births have resulted from the transfer of fragmented embryos [56, 57, 77–79]. Rather, we would argue that the stage at which it initially occurs and the degree or pattern of cellular fragmentation, are better predictors of embryo viability.

Despite the relatively high prevalence of multipolar divisions, abnormal nuclear morphologies, and cellular fragmentation in human embryos, many will still reach the blastocyst stage [80, 81], and several mechanisms by which an embryo may overcome aneuploidy have been proposed. This includes increased abundance of DNA repair gene products, a cell growth advantage of diploid cells in mosaic embryos, cell division lagging in aneuploid blastomeres,

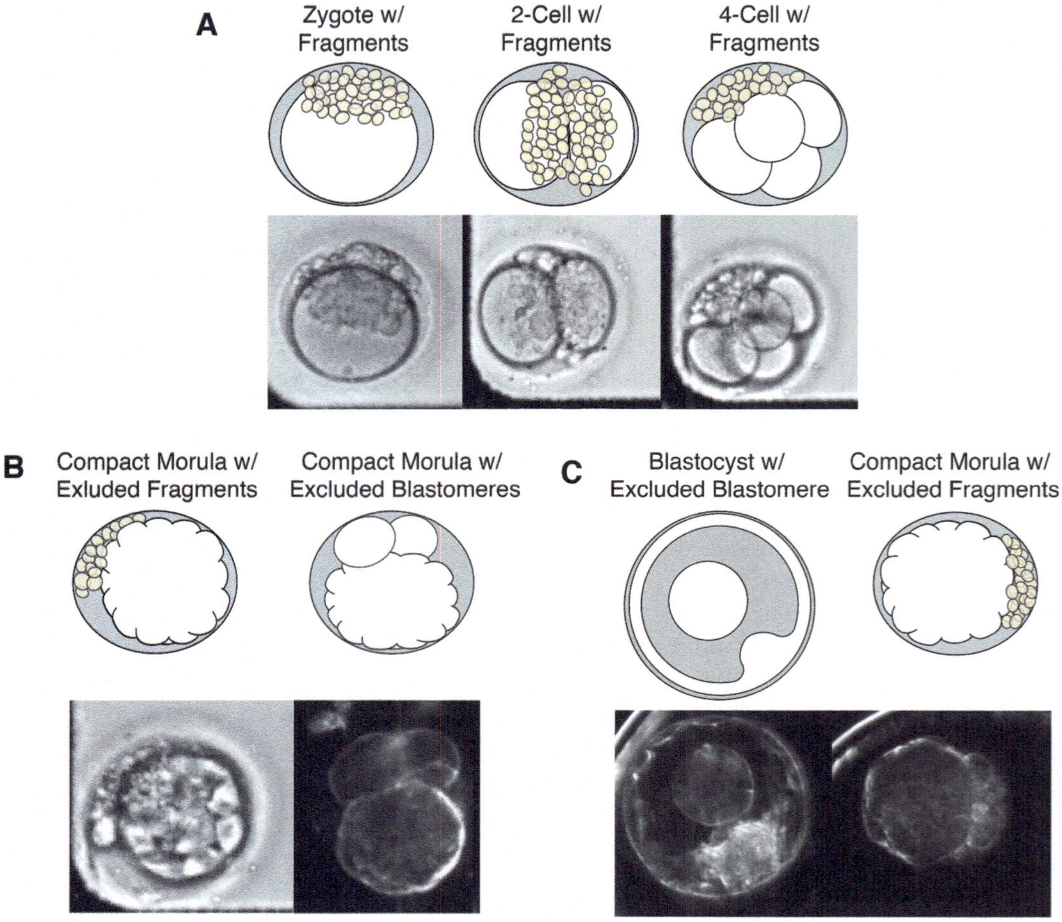

Fig. 4 Cellular fragmentation and blastomere exclusion dynamics. (**a**) Human embryos at different stages of development with a high degree of cellular fragmentation imaged with the Auxogyn Eeva™ 2.0 TLM system. (**b**) Compacted human morulas that have excluded cellular fragments (left) or large, nondividing blastomeres (right). Bright-field images were acquired with the Auxogyn Eeva™ V2.0 and dark-field images were captured with the Auxogyn Eeva™ V2.1 TLM system. (**c**) A rhesus macaque blastocyst (right) with a large, nondividing blastomere and a compact morula (right) with excluded fragments imaged with the Auxogyn Eeva™ 2.2.1 TLM system

extrusion or duplication of an aneuploid chromosome, and preferential allocation of aneuploid cells to the trophectoderm [82]. A more recent suggestion is that an embryo is capable of excluding any unwanted cells and/or cellular fragments from the remaining viable cells during the morula-to-blastocyst transition [Daughtry et al., under review, 24, 51] (Fig. 4b). While the frequency of cellular exclusion in embryos is unknown since it is not a well-documented process, cleavage-stage embryos generally tolerate the loss of one or more cells during blastomere or trophectoderm biopsy for PGS. In addition, we have evidence to suggest that blastomeres, which cease to divide at the early cleavage stages, are excluded from the rest of the embryo at the blastocyst stage. These nondividing blastomeres often exhibit complex or polyploid genotypes and high levels of DNA damage as shown by numerous γ-H2AX foci (Daughtry et al., under review). Thus, it would appear that the extrusion rather than the reincorporation of aneuploid blastomeres as well as cellular fragments may be more beneficial for the genomic integrity of an embryo and represents an additional mechanism to potentially circumvent chromosomal instability.

In the remainder of this chapter, we describe which TLM parameters and morphological criteria should be considered during embryo assessments and provide detailed instructions on how to properly measure mitotic timing and other imaging features for embryo classification and potential ranking.

2 Materials

Various TLM systems are commercially available for the evaluation of preimplantation embryos during culture and, in general, are offered in one of two configurations, either installed into an existing conventional incubator or as a stand-alone dual-purpose TLM incubator unit (Table 1). The materials described below encompass overall considerations for all current TLM systems used to assess human embryos from the zygote to blastocyst stage. If a different animal model or TLM system is used, readers are encouraged to adapt these materials to conditions most suitable for their experiments. Special permission may be required for research using these units with nonhuman animal models, and certain TLM systems may not be available in all countries. In selecting a TLM system, the type of illumination, pixel density of each image, frame rate, and number of focal planes are important factors to consider for detecting informative events in developing embryos.

2.1 Incubator Settings

The incubator should be calibrated at least 24 h prior to the initiation of each embryo culture experiment:

1. 37°C and medical grade 6% CO_2 + 5% O_2 + 89% N_2 (hypoxic) are preferable since these conditions are more physiologically relevant, but 5% CO_2 (atmospheric) may also be used.
2. Sterile water for the humidified chamber.

2.2 Embryo Micromanipulation Station

1. Stereomicroscope.
2. 275–300 μm and 75–100 μm polycarbonate or glass capillary micropipettes.
3. Embryo micropipettor.

2.3 Embryo Culture Components

Imaging dish containing culture media and covered with mineral oil should be equilibrated at culture conditions preferably overnight or at least 4 h prior to embryo culture:

1. LifeGlobal® Media (LifeGlobal® Group) (*see* **Note 1**).
   (a) Stored at 2–8 °C and protected from light for up to 10 weeks from the date of manufacture.
2. LifeGlobal® Protein Supplement (LifeGlobal® Group).
   (a) Stored at −20 °C or −80 °C and protected from light for up to 1 year from the date of manufacture.
3. Mineral oil for tissue culture (Sage™).
4. 60 mm sterile petri dish.
5. 1.5 mL microcentrifuge tubes.
6. Custom TLM embryo imaging culture dish.
7. Denuded human zygotes with verified two pronuclei (2PN) and two polar bodies (PB) (*see* **Notes 2 and 3**).

3 Methods

3.1 Preparation of TLM Embryo Culture Dishes and Zygote Loading into Micro-wells for Image Analysis of Preimplantation Development

1. Combine 9 parts media with 1 part protein in a sterile microcentrifuge tube, and mix by vortexing or pipetting up and down. Slowly add the designated amount of media (~100 μL) to each well of the custom imaging dish. Then slowly add the appropriate amount of mineral oil (~3 mL) until the media is completely covered to prevent desiccation, and equilibrate in the incubator for at least 4 h. Overnight is recommended.
2. The following morning, examine the TLM embryo culture dish under a stereomicroscope to determine if there are air bubbles trapped in any of the wells. If air bubbles are present, use a sterile 75–100 μm polycarbonate or glass capillary attached to a micropipettor to gently remove the bubbles either by aspirating or gently tapping the bubble to dislodge it from the well (*see* **Notes 4 and 5**).
3. Using a 275–300 μm polycarbonate or glass micropipette, gently load each zygote into an individual alphanumerically labeled well.
4. Load the custom TLM embryo dish containing the zygotes onto the time-lapse microscope, and start the experiment according to the manufacturer's instructions. Avoid opening the incubator door until a media change.

3.2 Tracking Embryonic Cleavage Divisions for Predicting Blastocyst Formation and Normal Chromosome Number up to the 4-Cell Stage

Delays in the progression through the cell cycle can occur if an embryo contains one or more blastomeres with an abnormal chromosome number. Therefore, cleavage divisions that are outside the ideal time frame for normal cell division parameters may indicate that the embryo has reduced developmental potential due to aneuploid blastomeres. Morphokinetics of embryo development likely varies across species, culture conditions, and the age of the female donor [83]. Whether age-related or not, sperm quality may also play a factor in embryo cell division timing. Thus, cell cycle parameter measurements should be conducted under controlled conditions and stratified according to age and/or other patient characteristics. The following steps can assist in accurately determining the beginning and end of each cell division as an embryo progresses in development from the zygote to 4-cell stage.

1. Review the entire length of the movie for a given embryo to determine whether it successfully divides, to which developmental stage, and if it exhibits cellular fragmentation or other morphological characteristics that might affect your ability to measure the imaging parameters.
2. Replay the movie to the first cell division, and after the embryo has reached the 2-cell stage, pause and reverse the TLM movie to the first frame where the pronuclei are visible in the zygote if a bright-field system is available (Fig. 1), and record the time or frame number (*see* **Note 6**).
3. If using bright-field imaging, advance the movie to when pronuclei disappearance (PNd) occurs, and record the time or frame number (*see* **Note 7–9**).
4. Regardless of whether a bright-field or dark-field imaging system is being used, record the time or frame number at which a cleavage furrow is first observed in the zygote as *t*1 (Fig. 1; *see* **Note 10**).
5. Visually track the furrow of the first cleavage to the opposite side of the blastomere until two distinct blastomeres separated by cell membranes are observed, and record this time or frame number as *t*2 (Fig. 1; *see* **Note 11**).
6. Advance the movie to the 3-cell stage and then backtrack to when a cleavage furrow is first detected in one of the two daughter cells as the embryo progresses from the 2- to 3-cell stage. Record this time or frame number as *t*3 (Fig. 1; *see* **Note 12**).
7. Repeat this procedure for the other daughter cell once the embryo advances from the 3- to 4-cell stage, and record this time or frame number as *t*4 (Fig. 1; *see* **Note 13**).

8. If the data is collected as frame numbers, use the following formulas to calculate the time between cell divisions in hours.
   (a) Time between pronuclei disappearance and first cytokinesis: [(PNd − *t*1) × (frame interval in minutes)/60)].
   (b) Duration of first cytokinesis: [(*t*2 − *t*1) × (frame interval in minutes)/60)].
   (c) Time between 2- and 3-cell: [(*t*3 − *t*2) × (frame interval in minutes)/60)].
   (d) Time between 3- and 4-cell: [(*t*4 − *t*3) × (frame interval in minutes)/60)].
9. Compare the time intervals between the first three mitotic divisions of each embryo to the ideal timing parameters described above for chromosomally normal 4-cell embryos. Significant deviations from these intervals may indicate abnormal chromosome number in one or more blastomeres (*see* **Note 14**).

3.3 Inferring Chaotic Aneuploidy, Parthenogenesis, and Polyploidy in Early Embryonic Development from Multipolar Divisions

A multipolar division at the zygote stage is often detrimental for an embryo and is expected to produce different types of aneuploidy in all subsequent blastomeres. In contrast, a multipolar division at the 2-cell stage and beyond could result in a mosaic embryo containing a mixture of chromosomally normal and abnormal cells if at least one of the two original daughter cells were euploid. Below, we provide step-by-step instructions for detecting multipolar divisions distinguishable by TLM in embryos as they undergo early cleavage.

1. Re-review the first part of the movie up to the ~4-cell stage, and count the number of blastomeres in the embryo to determine if the cell number correlates with the number of times cleavage furrows and successful mitotic divisions were observed. Multipolar divisions are often recognizable as having more than two cleavage furrows in a blastomere at one time to produce more cells than would be expected from a bipolar division (*see* **Note 15**).
2. For a multipolar division at the 1-cell stage, stop the movie after pronuclei disappearance, and slowly advance each frame until a cleavage furrow begins to form and record the time or frame number as *t*1(Fig. 2a).
3. Advance the video to the end of the first cytokinesis and determine if the blastomere number is greater than two cells to indicate that a multipolar division had occurred (*see* **Note 16**).
4. Record the time or frame number for the completion of the first cytokinesis as *t*2 and the beginning of the second cell division as *t*3 (Fig. 2a). In cases of tripolar divisions, *t*2 and *t*3 may be the same or similar values so that the time between the first and second mitosis is zero or close to zero.

5. For a multipolar division at the 2-cell stage, record the time or frame number when a cleavage furrow is detected to signal the start of the second mitosis as *t*3 and the beginning of the third mitosis as *t*4, which might equal zero as explained in **step 4** (Fig. 2b).
6. Continue to advance the movie frame-by-frame to determine if additional multipolar divisions are detected. Keep in mind that the embryo will increase in cell number, but not overall size, and as blastomeres become smaller, it is more difficult to observe multipolar divisions by TLM.

3.4 Identifying and Tracking Abnormal Nuclear Morphologies in Preimplantation Embryos by Bright-Field TLM

Identification of abnormal nuclear morphology at any stage of development should be noted as an indication of chromosomal mis-segregation. Some specific guidelines for detecting multi- and micronucleation in preimplantation embryos using a bright-field TLM system are provided as follows.

1. Start the movie again and manually forward and reverse the imaging frames at different speeds from the zygote to 2-cell stage to acquire a sense of pronuclei and primary nuclei size and location, respectively (*see* **Note 17**).
2. Prior to undergoing mitosis, the nuclear envelope of primary nuclei will disappear and then reappear following cell division. Besides dynamic movement within the blastomere, the primary nuclei may also be recognized as having lighter spots, which are nucleolar structures, against the background of the darker shaded nucleus (*see* **Note 18**).
3. Once the primary nuclei are observed, search the remainder of the cell for other nuclear structures that exhibit movement, nucleoli, and/or nuclear envelope breakdown and reassembly.
4. Zygotes with micronucleation will have two large pronuclei and one or more smaller nuclei. Not only will the micronuclei in blastomeres of cleavage-stage embryos be small in size compared to primary nuclei, but they may also be a similar shade or color and contain nucleoli as primary nuclei do (*see* **Note 19**).
5. A binucleated blastomere will have two nuclei of relatively equal size within one cell. If a blastomere divides, but later re-fuses in a process called reverse cleavage [64, 65], check to see if binucleation is present in the fused cell (*see* **Note 20**).
6. Multinucleated cells will contain multiple smaller nuclei of approximately equal size, which may or may not have nucleoli.
7. As an embryo continues to divide, blastomeres often exhibit intense movements, and this may cause the embryo to move out of focus. Even if the embryo is not in the correct focal plane, nuclear structures should be distinguishable by dark

spots in the blastomere that appear to move with each cell division (*see* **Note 21**).

8. Cellular fragments on the posterior side of a blastomere may produce a shadow that appears similar to nuclear structure. As mentioned above, the nuclear envelope will disappear several frames prior to cell division and can be used to determine whether the shadow in question is a cellular fragment or nucleus.
9. Continue tracking nuclear morphology in all blastomeres of cleavage-stage embryos until cell density obscures detection (*see* **Note 22**).
10. Take special notice if cells with abnormal nuclear morphology continue to divide or arrest and no longer participate in embryo development (discussed below). A lack of cytoplasmic streaming may also signal blastomere arrest.
11. If abnormal nuclear structure(s) are observed in only one or two blastomeres, whereas the remaining cells have normal nuclear morphology (Fig. 3b, third and 4th panel), this indicates that the embryo is likely euploid-aneuploid mosaic with the potential ability to overcome aneuploidy (*see* **Note 23**).

3.5 Evaluating the Incidence of Cellular Fragmentation as an Indication of Chromosomal Mis-segregation During Meiosis and Mitosis

When examined in conjunction with the timing of the first three mitotic divisions as described above, the assessment of cellular fragmentation dynamics may also be used to largely differentiate between euploid and aneuploid embryos at the 4-cell stage [5]. This is because we postulate that the embryo responds to chromosomal mis-segregation and abnormal nuclear morphologies such as micronucleation by fragmenting. We describe the steps for evaluating the significance of cellular fragmentation in embryos, particularly at the early cleavage stages, below.

1. Replay the movie from the beginning again and manually forward and reverse the imaging frames to visualize at which stage cellular fragmentation initially occurred in an embryo. If cellular fragments first appear at the zygote stage, the embryo likely inherited a meiotic error from the oocyte or chromosomal mis-segregation arose as a result of defective pronuclei fusion (Fig. 4a, first panel). In contrast, if cellular fragmentation instead begins at the 2-cell stage or later, then the embryo probably incurred a mitotic error (*see* **Note 24**).
2. Determine the number of cells that underwent cellular fragmentation by comparing the symmetry between blastomeres. If a blastomere appears smaller than neighboring cells, it may have been the source of cellular fragments (Fig. 4a, second panel). Because multipolar divisions may also produce blastomere asymmetry, this should also be evaluated to discern which or if both of these cellular events occurred.

3. In some instances, you may need to determine whether a large fragment is actually a blastomere and visa versa. This decision may be based on the cell size as well as the overall behavior of the embryo, including assessing whether the large fragment or small blastomere remains quiescent or continues to divide.
4. Special consideration should be given to an embryo that exhibits fragment resorption or exclusion, the latter of which is discussed more in the next section. For fragment resorption, it should be determined if the fragment fused back with the same blastomere that it originated from or a neighboring blastomere (*see* **Note 25**).

3.6 Assessing if Cellular Fragment and Blastomere Exclusion Provide a Means for Embryos to Overcome Aneuploidy

The following steps may assist in the determination of whether an aneuploid embryo is attempting to correct chromosomal copy number or ploidy status by fragment or blastomere exclusion as it reaches the blastocyst stage.

1. Restart the movie toward the beginning and manually forward and reverse the imaging frames of an embryo with cellular fragmentation up to the morula stage to determine whether the fragments appear to be expelled out into the perivitelline space and separated from the rest of the cells (Fig. 4b, first panel).
2. As the embryo transitions from the early cleavage stages to the compact morula stage, take notice if there are any large blastomeres that stop dividing and do not appear to participate in compaction (Fig. 4b, second panel).
3. Upon the initiation of blastulation and blastocyst formation, the exclusion of nondividing blastomeres (Fig. 4c, left panel) and cellular fragments (Fig. 4c, right panel) may be easier to visualize, but more difficult to isolate (*see* **Note 26**).

4 Notes

1. Alternative medias are often used in human IVF clinical practice such as Quinn's Advantage Cleavage Medium (SPS; Sage™) supplemented with 10% Quinn's Advantage Serum Protein Substitute (SPS; Sage™). Use medias and mineral oils that are consistent with current laboratory protocols in TLM experiments before switching to a new method that has not been validated.
2. Verification of 2PN presence should be checked approximately 16–18 h after conventional IVF or ICSI. If 2PN are not detected, but two polar bodies are clearly visible, the TLM experiment may proceed. The pronuclei may have already undergone syngamy and therefore were not observed. To

detect 2PN at an earlier time point, users may transfer oocytes to the TLM system immediately after fertilization and culture overnight in IVF media and with incubator settings that are compatible with fertilization.

3. Zygotes should be denuded of all cumulus cells and sperm. Residual cumulus cells and sperm can remain alive or active for several hours to days and may visually obstruct nuclear structures, cell division dynamics, or rotate the embryo in and out of the focal plane.
4. When dislodging air bubbles from micro-wells, take precautionary measures not to touch the bottom or sides of each well as this can deposit oil into the embryo culture dish. For TLM systems with autofocus, oil droplets in a well can compromise embryo focus.
5. If the TLM embryo culture dishes consistently trap air bubbles, add media slowly to the side of the well to avoid trapping air. While not required, rinsing the TLM embryo culture dish with media prior to adding the media drop can decrease the occurrence of bubbles.
6. The first cytokinesis for a chromosomally normal (euploid) human zygote is typically observed approximately 24–27 h after fertilization.
7. Once the pronuclei undergo syngamy, discerning uniparental or triploidy errors can become indistinguishable from a diploid zygote, especially if the first polar body undergoes cytokinesis or cellular fragmentation occurs at the zygote stage to obscure polar body number.
8. Zygote cleavage prior to syngamy and pronuclei disappearance is a poor prognosis of embryo quality and subsequent development. However, cellular fragmentation prior to syngamy may be tolerated.
9. Pronuclei are generally only visible in zygotes with the use of bright-field TLM systems.
10. The first cleavage furrow will appear after the zygote horizontally elongates and the edges of the cell pinch inward. In dark-field TLM movies, the cleavage furrow will be apparent as a bright spot at the edge of the cell prior to invagination. For bright-field TLM movies, shortly after syngamy when the 2PN disappear in the zygote, the first cleavage furrow should emerge. If a cleavage furrow is not observed prior to the first cell division, record the start of cytokinesis as when the embryo significantly elongated. In some instances, a cleavage furrow may be observed, but the embryo does not actually divide until several frames later. As long as the embryo continues to exhibit significant movement during this time period, then the image frame where the first cleavage furrow was detected should be

recorded. If instead the embryo becomes quiescent and ceases movement for an extended period of time (greater than approximately 2 h), then the image frame where the next cleavage furrow is observed should be used.

11. The end of the first cytokinesis may be subtle or unclear due to blastomeric movement or if a transverse division occurs that is parallel to the surface of the dish. Examine the space in between the two blastomeres for a sudden brightening in the dark-field images, which should signal the completion of the first division. Given this short time frame, the ideal TLM system should take an image every 5 min to capture the full duration of the first cleavage.
12. In the case where a cleavage furrow is detected, but the blastomere never actually divides, it should be recorded as a nonevent.
13. There may be an instance where one of two daughter cells at the 2-cell stage never actually divides and the other daughter cell divides twice to produce a 4-cell embryo. This should be considered a nonevent and is not a measurement for *t*4.
14. It is recommended to repeat all of the above at least once to confirm your measurements and compare the consensus between three independent observers.
15. If there are more than two cells after the zygote cleaves and the duration of the cytokinesis was between 10 min and approximately 1 h, then a multipolar division took place. This is based on findings that the time interval from the 2- to 3-cell stage averages 12 h long, and thus, the presence of more than two cells before this time indicates a multipolar first division. Because multipolar divisions can occur relatively quickly, the ideal time in between imaging frames should be as short as possible, but not longer than 1 h especially if it occurs in the transition from two to three cells since this is nearly synchronous with the 3- to 4-cell stage.
16. Particularly with asymmetrical multipolar cell divisions, a small blastomere may appear to be a cellular fragment or visa versa. Track this blastomere though embryo development to determine if a nucleus appears or it goes onto divide since an anuclear cellular fragment will not divide.
17. Similar to pronuclei disappearance, visualizing nuclear structures require that a bright-field TLM system be used since dark-field imaging focuses more on the cell membranes of blastomeres [37].
18. Frame rate, pixel density, and the number of focal planes affect the quality of bright-field movies and thus, will affect your ability to visualize nuclear structures. Nuclei can also move around within the cell due to cytoplasmic streaming.

If the TLM settings are adjustable, reduce the frame rate to less than 5 min, and maximize pixel density and/or the number of focal planes.

19. The presence of MN with or without DNA damage is indicative of anaphase lagging. Therefore, it is important to determine when MN first appeared since damaged chromosomes at the zygote or 2-cell stage could have devastating consequences for the embryo, whereas MN formation at later stages of development might be tolerated if other surrounding cells are euploid and lack MN.
20. A binucleated blastomere may also originate from the fusion of mononucleated cells, and in both instances, this could produce a tetraploid cell [85]. It is also possible for a binucleated cell to be diploid [86], whereby a haploid genome underwent endoreduplication [87]. These mechanisms are possible due to the apparent lack of cell cycle control in human embryos prior to embryonic genome activation (EGA) [88–90].
21. The embryos shown in Fig. 3a (panels 1, 3, and 4) were out of focus, but nuclear structures were detectable by actively moving dark objects within the cell.
22. Analogous to multipolar divisions, the detection of multi- and micronucleation becomes increasingly difficult past the 4-cell stage as nuclei become smaller and multiple cells crowd the field of view.
23. Recently, several groups have demonstrated that mosaic embryos are capable of producing healthy offspring in both the mouse and humans by an unknown selective mechanism [8, 91]. This is especially true if micronuclei in human embryos undergo perpetual unilateral inheritance as has been shown to occur in mice [16].
24. In general, the later in preimplantation development that cellular fragmentation occurs, the better embryo prognosis will be since fewer cells will likely be impacted.
25. If a cellular fragment containing mis-segregated chromosomes fuses with the blastomere from which it arose, euploidy could potentially be restored in an embryo following breakdown of the nuclear envelope. However, fragment resorption may also have no impact on embryo ploidy status, or it may even increase genotypic complexity in embryos upon fusion with a different blastomere.
26. In our experience, excluded blastomeres often become trapped in the blastocoel cavity and are difficult to remove. Therefore, it is recommended that these cells be removed during the morula-to-blastocyst transition if chromosomal or nuclear analysis of these excluded cells is desired.

References

1. Vanneste E, Voet T, Le Caignec C et al (2009) Chromosome instability is common in human cleavage-stage embryos. Nat Med 15(5):577–583. https://doi.org/10.1038/nm.1924
2. Johnson DS, Cinnioglu C, Ross R et al (2010) Comprehensive analysis of karyotypic mosaicism between trophectoderm and inner cell mass. Mol Hum Reprod 16(12):944–949. https://doi.org/10.1093/molehr/gaq062
3. Chow JF, Yeung WS, Lau EY et al (2014) Array comparative genomic hybridization analyses of all blastomeres of a cohort of embryos from young IVF patients revealed significant contribution of mitotic errors to embryo mosaicism at the cleavage stage. Reprod Biol Endocrinol 12:105. https://doi.org/10.1186/1477-7827-12-105
4. Huang J, Yan L, Fan W et al (2014) Validation of multiple annealing and looping-based amplification cycle sequencing for 24-chromosome aneuploidy screening of cleavage-stage embryos. Fertil Steril 102(6):1685–1691. https://doi.org/10.1016/j.fertnstert.2014.08.015
5. Chavez SL, Loewke KE, Han J et al (2012) Dynamic blastomere behaviour reflects human embryo ploidy by the four-cell stage. Nat Commun 3:1251. https://doi.org/10.1038/ncomms2249
6. Minasi MG, Colasante A, Riccio T et al (2016) Correlation between aneuploidy, standard morphology evaluation and morphokinetic development in 1730 biopsied blastocysts: a consecutive case series study. Hum Reprod 31(10):2245–2254. https://doi.org/10.1093/humrep/dew183
7. Gleicher N, Vidali A, Braverman J et al (2016) Accuracy of preimplantation genetic screening (PGS) is compromised by degree of mosaicism of human embryos. Reprod Biol Endocrinol 14(1):54. https://doi.org/10.1186/s12958-016-0193-6
8. Greco E, Minasi MG, Fiorentino F (2015) Healthy babies after intrauterine transfer of mosaic aneuploid blastocysts. N Engl J Med 373(21):2089–2090. https://doi.org/10.1056/NEJMc1500421
9. Nagaoka SI, Hodges CA, Albertini DF et al (2011) Oocyte-specific differences in cell-cycle control create an innate susceptibility to meiotic errors. Curr Biol 21(8):651–657. https://doi.org/10.1016/j.cub.2011.03.003
10. Wang S, Hassold T, Hunt P et al (2017) Inefficient crossover maturation underlies elevated aneuploidy in human female meiosis. Cell 168(6):977–989e917. https://doi.org/10.1016/j.cell.2017.02.002
11. McCoy RC, Demko Z, Ryan A et al (2015) Common variants spanning PLK4 are associated with mitotic-origin aneuploidy in human embryos. Science 348(6231):235–238. https://doi.org/10.1126/science.aaa3337
12. Vanneste E, Voet T, Melotte C et al (2009) What next for preimplantation genetic screening? High mitotic chromosome instability rate provides the biological basis for the low success rate. Hum Reprod 24(11):2679–2682. https://doi.org/10.1093/humrep/dep266
13. Gordon DJ, Resio B, Pellman D (2012) Causes and consequences of aneuploidy in cancer. Nat Rev Genet 13(3):189–203. https://doi.org/10.1038/nrg3123
14. Lee HO, Davidson JM, Duronio RJ (2009) Endoreplication: polyploidy with purpose. Genes Dev 23(21):2461–2477. https://doi.org/10.1101/gad.1829209
15. Coonen E, Derhaag JG, Dumoulin JC et al (2004) Anaphase lagging mainly explains chromosomal mosaicism in human preimplantation embryos. Hum Reprod 19(2):316–324
16. Vazquez-Diez C, Yamagata K, Trivedi S et al (2016) Micronucleus formation causes perpetual unilateral chromosome inheritance in mouse embryos. Proc Natl Acad Sci U S A 113(3):626–631. https://doi.org/10.1073/pnas.1517628112
17. Janssen A, van der Burg M, Szuhai K et al (2011) Chromosome segregation errors as a cause of DNA damage and structural chromosome aberrations. Science 333(6051):1895–1898. https://doi.org/10.1126/science.1210214
18. Liu P, Erez A, Nagamani SC et al (2011) Chromosome catastrophes involve replication mechanisms generating complex genomic rearrangements. Cell 146(6):889–903. https://doi.org/10.1016/j.cell.2011.07.042
19. Crasta K, Ganem NJ, Dagher R et al (2012) DNA breaks and chromosome pulverization from errors in mitosis. Nature 482(7383):53–58. https://doi.org/10.1038/nature10802
20. Hatch EM, Fischer AH, Deerinck TJ et al (2013) Catastrophic nuclear envelope collapse in cancer cell micronuclei. Cell 154(1):47–60. https://doi.org/10.1016/j.cell.2013.06.007
21. Zhang CZ, Spektor A, Cornils H et al (2015) Chromothripsis from DNA damage in micronuclei. Nature 522(7555):179–184. https://doi.org/10.1038/nature14493

22. Pellestor F (2014) Chromothripsis: how does such a catastrophic event impact human reproduction? Hum Reprod 29(3):388–393. https://doi.org/10.1093/humrep/deu003
23. Pellestor F, Gatinois V, Puechberty J et al (2014) Chromothripsis: potential origin in gametogenesis and preimplantation cell divisions. A review. Fertil Steril 102(6):1785–1796. https://doi.org/10.1016/j.fertnstert.2014.09.006
24. Daughtry BL, Chavez SL (2016) Chromosomal instability in mammalian pre-implantation embryos: potential causes, detection methods, and clinical consequences. Cell Tissue Res 363(1):201–225. https://doi.org/10.1007/s00441-015-2305-6
25. Racowsky C, Vernon M, Mayer J et al (2010) Standardization of grading embryo morphology. Fertil Steril 94(3):1152–1153. https://doi.org/10.1016/j.fertnstert.2010.05.042
26. Luke B, Brown MB, Stern JE et al (2014) Using the Society for Assisted Reproductive Technology Clinic Outcome System morphological measures to predict live birth after assisted reproductive technology. Fertil Steril 102(5):1338–1344. https://doi.org/10.1016/j.fertnstert.2014.07.1242
27. Gardner DK, Lane M, Stevens J et al (2000) Blastocyst score affects implantation and pregnancy outcome: towards a single blastocyst transfer. Fertil Steril 73(6):1155–1158
28. Scott RT Jr, Upham KM, Forman EJ et al (2013) Cleavage-stage biopsy significantly impairs human embryonic implantation potential while blastocyst biopsy does not: a randomized and paired clinical trial. Fertil Steril 100(3):624–630. https://doi.org/10.1016/j.fertnstert.2013.04.039
29. Campbell A, Fishel S, Bowman N et al (2013) Modelling a risk classification of aneuploidy in human embryos using non-invasive morphokinetics. Reprod Biomed Online 26(5):477–485. https://doi.org/10.1016/j.rbmo.2013.02.006
30. Basile N, Nogales Mdel C, Bronet F et al (2014) Increasing the probability of selecting chromosomally normal embryos by time-lapse morphokinetics analysis. Fertil Steril 101(3):699–704. https://doi.org/10.1016/j.fertnstert.2013.12.005
31. Yang Z, Zhang J, Salem SA et al (2014) Selection of competent blastocysts for transfer by combining time-lapse monitoring and array CGH testing for patients undergoing preimplantation genetic screening: a prospective study with sibling oocytes. BMC Med Genet 7:38. https://doi.org/10.1186/1755-8794-7-38
32. Chawla M, Fakih M, Shunnar A et al (2015) Morphokinetic analysis of cleavage stage embryos and its relationship to aneuploidy in a retrospective time-lapse imaging study. J Assist Reprod Genet 32(1):69–75. https://doi.org/10.1007/s10815-014-0372-3
33. Sugimura S, Akai T, Somfai T et al (2010) Time-lapse cinematography-compatible polystyrene-based microwell culture system: a novel tool for tracking the development of individual bovine embryos. Biol Reprod 83(6):970–978. https://doi.org/10.1095/biolreprod.110.085522
34. Sugimura S, Akai T, Imai K (2017) Selection of viable in vitro-fertilized bovine embryos using time-lapse monitoring in microwell culture dishes. J Reprod Dev 63(4):353–357. https://doi.org/10.1262/jrd.2017-041
35. Weinerman R, Feng R, Ord TS et al (2016) Morphokinetic evaluation of embryo development in a mouse model: functional and molecular correlates. Biol Reprod 94(4):84. https://doi.org/10.1095/biolreprod.115.134080
36. Burruel V, Klooster K, Barker CM et al (2014) Abnormal early cleavage events predict early embryo demise: sperm oxidative stress and early abnormal cleavage. Sci Rep 4:6598. https://doi.org/10.1038/srep06598
37. Wang Y, Moussavi F, Lorenzen P (2013) Automated embryo stage classification in time-lapse microscopy video of early human embryo development. Med Image Comput Comput Assist Interv 16(Pt 2):460–467
38. Conaghan J, Chen AA, Willman SP et al (2013) Improving embryo selection using a computer-automated time-lapse image analysis test plus day 3 morphology: results from a prospective multicenter trial. Fertil Steril 100(2):412–419e415. https://doi.org/10.1016/j.fertnstert.2013.04.021
39. VerMilyea MD, Tan L, Anthony JT et al (2014) Computer-automated time-lapse analysis results correlate with embryo implantation and clinical pregnancy: a blinded, multi-centre study. Reprod Biomed Online 29(6):729–736. https://doi.org/10.1016/j.rbmo.2014.09.005
40. Kieslinger DC, De Gheselle S, Lambalk CB et al (2016) Embryo selection using time-l apse analysis (Early Embryo Viability Assessment) in conjunction with standard morphology: a prospective two-center pilot study. Hum Reprod 31(11):2450–2457. https://doi.org/10.1093/humrep/dew207
41. Diamond MP, Suraj V, Behnke EJ et al (2015) Using the Eeva Test adjunctively to traditional day 3 morphology is informative for con-

sistent embryo assessment within a panel of embryologists with diverse experience. J Assist Reprod Genet 32(1):61–68. https://doi.org/10.1007/s10815-014-0366-1
42. Meseguer M, Herrero J, Tejera A et al (2011) The use of morphokinetics as a predictor of embryo implantation. Hum Reprod 26(10):2658–2671. https://doi.org/10.1093/humrep/der256
43. Liu Y, Chapple V, Roberts P et al (2014) Time-lapse videography of human oocytes following intracytoplasmic sperm injection: events up to the first cleavage division. Reprod Biol 14(4):249–256. https://doi.org/10.1016/j.repbio.2014.08.003
44. Wong CC, Loewke KE, Bossert NL et al (2010) Non-invasive imaging of human embryos before embryonic genome activation predicts development to the blastocyst stage. Nat Biotechnol 28(10):1115–1121. https://doi.org/10.1038/nbt.1686
45. Vera-Rodriguez M, Chavez SL, Rubio C et al (2015) Prediction model for aneuploidy in early human embryo development revealed by single-cell analysis. Nat Commun 6:7601. https://doi.org/10.1038/ncomms8601
46. Palermo G, Munne S, Cohen J (1994) The human zygote inherits its mitotic potential from the male gamete. Hum Reprod 9(7):1220–1225
47. Schatten H, Sun QY (2011) New insights into the role of centrosomes in mammalian fertilization and implications for ART. Reproduction 142(6):793–801. https://doi.org/10.1530/REP-11-0261
48. Kalatova B, Jesenska R, Hlinka D et al (2015) Tripolar mitosis in human cells and embryos: occurrence, pathophysiology and medical implications. Acta Histochem 117(1):111–125. https://doi.org/10.1016/j.acthis.2014.11.009
49. Chamayou S, Patrizio P, Storaci G et al (2013) The use of morphokinetic parameters to select all embryos with full capacity to implant. J Assist Reprod Genet 30(5):703–710. https://doi.org/10.1007/s10815-013-9992-2
50. Hlinka D, Kalatova B, Uhrinova I et al (2012) Time-lapse cleavage rating predicts human embryo viability. Physiol Res 61(5):513–525
51. Ottolini CS, Kitchen J, Xanthopoulou L et al (2017) Tripolar mitosis and partitioning of the genome arrests human preimplantation development in vitro. Sci Rep 7(1):9744. https://doi.org/10.1038/s41598-017-09693-1
52. Aguilar J, Rubio I, Munoz E et al (2016) Study of nucleation status in the second cell cycle of human embryo and its impact on implantation rate. Fertil Steril 106(2):291–299e292. https://doi.org/10.1016/j.fertnstert.2016.03.036
53. Chavez SL, McElroy SL, Bossert NL et al (2014) Comparison of epigenetic mediator expression and function in mouse and human embryonic blastomeres. Hum Mol Genet 23(18):4970–4984. https://doi.org/10.1093/hmg/ddu212
54. Iwata K, Mio Y (2016) Observation of human embryonic behavior in vitro by high-resolution time-lapse cinematography. Reprod Med Biol 15:145–154
55. Braude P, Bolton V, Moore S (1988) Human gene expression first occurs between the four- and eight-cell stages of preimplantation development. Nature 332(6163):459–461. https://doi.org/10.1038/332459a0
56. Antczak M, Van Blerkom J (1999) Temporal and spatial aspects of fragmentation in early human embryos: possible effects on developmental competence and association with the differential elimination of regulatory proteins from polarized domains. Hum Reprod 14(2):429–447
57. Alikani M, Cohen J, Tomkin G et al (1999) Human embryo fragmentation in vitro and its implications for pregnancy and implantation. Fertil Steril 71(5):836–842
58. Dozortsev D, Ermilov A, El-Mowafi DM et al (1998) The impact of cellular fragmentation induced experimentally at different stages of mouse preimplantation development. Hum Reprod 13(5):1307–1311
59. Winston NJ, Johnson MH (1992) Can the mouse embryo provide a good model for the study of abnormal cellular development seen in human embryos? Hum Reprod 7(9):1291–1296
60. Enders AC, Hendrickx AG, Binkerd PE (1982) Abnormal development of blastocysts and blastomeres in the rhesus monkey. Biol Reprod 26(2):353–366
61. Tremoleda JL, Stout TA, Lagutina I et al (2003) Effects of in vitro production on horse embryo morphology, cytoskeletal characteristics, and blastocyst capsule formation. Biol Reprod 69(6):1895–1906. https://doi.org/10.1095/biolreprod.103.018515
62. Hardy K (1999) Apoptosis in the human embryo. Rev Reprod 4(3):125–134
63. Hardy K, Spanos S, Becker D et al (2001) From cell death to embryo arrest: mathematical models of human preimplantation embryo development. Proc Natl Acad Sci U S A 98(4):1655–1660. https://doi.org/10.1073/pnas.98.4.1655

64. Xu J, Cheung T, Chan ST et al (2001) The incidence of cytoplasmic fragmentation in mouse embryos in vitro is not affected by inhibition of caspase activity. Fertil Steril 75(5):986–991
65. Buster JE, Bustillo M, Rodi IA et al (1985) Biologic and morphologic development of donated human ova recovered by nonsurgical uterine lavage. Am J Obstet Gynecol 153(2):211–217
66. Pereda J, Croxatto HB (1978) Ultrastructure of a seven-cell human embryo. Biol Reprod 18(3):481–489
67. Van Blerkom J, Davis P, Alexander S (2001) A microscopic and biochemical study of fragmentation phenotypes in stage-appropriate human embryos. Hum Reprod 16(4):719–729
68. Yang HW, Hwang KJ, Kwon HC et al (1998) Detection of reactive oxygen species (ROS) and apoptosis in human fragmented embryos. Hum Reprod 13(4):998–1002
69. Jurisicova A, Varmuza S, Casper RF (1996) Programmed cell death and human embryo fragmentation. Mol Hum Reprod 2(2):93–98
70. Hardarson T, Lofman C, Coull G et al (2002) Internalization of cellular fragments in a human embryo: time-lapse recordings. Reprod Biomed Online 5(1):36–38
71. Lemmen JG, Agerholm I, Ziebe S (2008) Kinetic markers of human embryo quality using time-lapse recordings of IVF/ICSI-fertilized oocytes. Reprod Biomed Online 17(3):385–391
72. Munne S (2006) Chromosome abnormalities and their relationship to morphology and development of human embryos. Reprod Biomed Online 12(2):234–253
73. Pellestor F, Girardet A, Andreo B et al (1994) Relationship between morphology and chromosomal constitution in human preimplantation embryo. Mol Reprod Dev 39(2):141–146. https://doi.org/10.1002/mrd.1080390204
74. Giorgetti C, Terriou P, Auquier P et al (1995) Embryo score to predict implantation after in-vitro fertilization: based on 957 single embryo transfers. Hum Reprod 10(9):2427–2431
75. Ebner T, Yaman C, Moser M et al (2001) Embryo fragmentation in vitro and its impact on treatment and pregnancy outcome. Fertil Steril 76(2):281–285
76. Ziebe S, Petersen K, Lindenberg S et al (1997) Embryo morphology or cleavage stage: how to select the best embryos for transfer after in-vitro fertilization. Hum Reprod 12(7):1545–1549
77. Edwards RG, Fishel SB, Cohen J et al (1984) Factors influencing the success of in vitro fertilization for alleviating human infertility. J In Vitro Fert Embryo Transf 1(1):3–23
78. Pelinck MJ, Hoek A, Simons AH et al (2010) Embryo quality and impact of specific embryo characteristics on ongoing implantation in unselected embryos derived from modified natural cycle in vitro fertilization. Fertil Steril 94(2):527–534. https://doi.org/10.1016/j.fertnstert.2009.03.076
79. Hoover L, Baker A, Check JH et al (1995) Evaluation of a new embryo-grading system to predict pregnancy rates following in vitro fertilization. Gynecol Obstet Investig 40(3):151–157
80. Desai N, Ploskonka S, Goodman LR et al (2014) Analysis of embryo morphokinetics, multinucleation and cleavage anomalies using continuous time-lapse monitoring in blastocyst transfer cycles. Reprod Biol Endocrinol 12:54. https://doi.org/10.1186/1477-7827-12-54
81. Lagalla C, Tarozzi N, Sciajno R et al (2017) Embryos with morphokinetic abnormalities may develop into euploid blastocysts. Reprod Biomed Online 34(2):137–146. https://doi.org/10.1016/j.rbmo.2016.11.008
82. Bazrgar M, Gourabi H, Valojerdi MR et al (2013) Self-correction of chromosomal abnormalities in human preimplantation embryos and embryonic stem cells. Stem Cells Dev 22(17):2449–2456. https://doi.org/10.1089/scd.2013.0053
83. Kirkegaard K, Sundvall L, Erlandsen M et al (2016) Timing of human preimplantation embryonic development is confounded by embryo origin. Hum Reprod 31(2):324–331. https://doi.org/10.1093/humrep/dev296
84. Liu Y, Chapple V, Roberts P et al (2014) Prevalence, consequence, and significance of reverse cleavage by human embryos viewed with the use of the Embryoscope time-lapse video system. Fertil Steril 102(5):1295–1300e1292. https://doi.org/10.1016/j.fertnstert.2014.07.1235
85. Balakier H, Cabaca O, Bouman D et al (2000) Spontaneous blastomere fusion after freezing and thawing of early human embryos leads to polyploidy and chromosomal mosaicism. Hum Reprod 15(11):2404–2410
86. Xanthopoulou L, Delhanty JD, Mania A et al (2011) The nature and origin of binucleate cells in human preimplantation embryos: relevance to placental mesenchymal dysplasia. Reprod

Biomed Online 22(4):362–370. https://doi.org/10.1016/j.rbmo.2011.01.001

87. Barlow P, Owen DA, Graham C (1972) DNA synthesis in the preimplantation mouse embryo. J Embryol Exp Morphol 27(2):431–445
88. Harrison RH, Kuo HC, Scriven PN et al (2000) Lack of cell cycle checkpoints in human cleavage stage embryos revealed by a clonal pattern of chromosomal mosaicism analysed by sequential multicolour FISH. Zygote 8(3):217–224
89. Los FJ, Van Opstal D, van den Berg C (2004) The development of cytogenetically normal, abnormal and mosaic embryos: a theoretical model. Hum Reprod Update 10(1):79–94
90. Mantikou E, Wong KM, Repping S et al (2012) Molecular origin of mitotic aneuploidies in preimplantation embryos. Biochim Biophys Acta 1822(12):1921–1930. https://doi.org/10.1016/j.bbadis.2012.06.013
91. Bolton H, Graham SJ, Van der Aa N et al (2016) Mouse model of chromosome mosaicism reveals lineage-specific depletion of aneuploid cells and normal developmental potential. Nat Commun 7:11165. https://doi.org/10.1038/ncomms11165

Check for updates

Chapter 20

Correlative Live Imaging and Immunofluorescence for Analysis of Chromosome Segregation in Mouse Preimplantation Embryos

Cayetana Vázquez-Diez and Greg FitzHarris

Abstract

Chromothripsis is a phenomenon observed in cancer cells, wherein a single or few chromosome(s) exhibit vast genomic rearrangements. Recent studies elucidated a striking series of events in which defective segregation of chromosomes causes their incorporation into micronuclei, where they are subject to extensive DNA damage prior to re-joining the main mass of chromosomes in a subsequent cell cycle, which provide an appealing mechanism for the etiology of chromothripsis. Micronuclei are well known to be common in human preimplantation embryos. We recently showed that, unlike in cancer cells, in mouse preimplantation embryos the micronuclei are maintained during multiple cell generations and apparently fail to re-join the main set of chromosomes. This unexpected finding could safeguard the early embryonic genome from chromothripsis. Here, we describe an approach that combines live and immunofluorescence imaging methods that was pivotal in that study to reveal the lack of a functional kinetochore in chromosomes from mouse embryo micronuclei.

Key words Mitosis, Live cell imaging, Chromosome segregation, Preimplantation development, Immunofluorescence, Micronuclei

1 Introduction

A recently elucidated feature of cancer in both primary tumors and immortalized cell lines are clustered genomic rearrangements restricted to one or few chromosomes, a phenomenon termed chromothripsis [1–3]. While conventionally it is considered that mutations and translocations are progressively acquired over time, the nature of rearrangements observed in chromothripsis, restricted to a defined genomic region and showing limited changes in gene copy number, suggests that these emerge concurrently over a short period of time, possibly in a single catastrophic event [2, 4]. Such rapid and complex genomic rearrangements have the potential to simultaneously activate oncogenes while disabling

Franck Pellestor (ed.), *Chromothripsis: Methods and Protocols*, Methods in Molecular Biology, vol. 1769,
https://doi.org/10.1007/978-1-4939-7780-2_20,

checkpoints, hence posing an important novel pathway for cellular transformation.

Micronuclei (MN) are a hallmark of chromosomal instability and cancer and may arise through a variety of different mechanisms [5]. A common cause of MN in human cancer cells are lagging anaphase chromosomes [6, 7], leading to the encapsulation of single chromosomes into a MN at the time of nuclear envelope (NE) formation. Recently it has been shown that during the subsequent cell cycle, MN-enclosed chromosomes acquire DNA double-strand breaks during S/G2 phases, resulting from atypical nuclear membrane function [8–11]. MN-chromosomes are then reincorporated into daughter cells during the subsequent mitosis [8, 11]. Elegant experiments showed that MN-free daughter cells of MN-containing cells harbor extensive genomic rearrangements usually confined to a single chromosome as revealed by a combined live imaging and single cell whole genome sequencing approach [11]. Hence, MN have emerged as the likely structural intermediate between chromosome mis-segregation and the emergence of chromothripsis.

Chromosomal abnormalities, such as mosaic aneuploidy and MN, have been observed at high incidences in human preimplantation embryos (up to 70% of embryos have mosaic aneuploidy) [12, 13]. We recently found that, in cleavage-stage mouse embryos, MN-enclosed chromosomes are rarely reincorporated into the genome as they have shown to in cancer cells. Instead, MN are repeatedly inherited by daughter cells during subsequent divisions of preimplantation development as a result of faulty kinetochores [14]. Central to this discovery was a series of experiments in which we combined live imaging and fixed cell immunofluorescence in the same embryo to show that lagging chromosomes destined for MN possess an apparently normal kinetochore but that the ability to subsequently assemble a functional kinetochore is lost while the chromosome is in the MN, and kinetochores are not observed on the MN-chromosome in subsequent mitoses.

Live imaging is an essential tool for the analysis of chromosome segregation dynamics during cell division, providing single cell and even single chromosome tracking of mitotic events over time [15]. However, live cell imaging usually requires overexpression of fluorescent fusion proteins to look at subcellular localization of proteins of interest during cell division. In some cases, fusion protein overexpression may have biological effects causing artifacts, or the addition of the reporter, e.g., GFP, may cause the protein to mislocalize or loose function. To circumvent these limitations, some somatic cell studies have observed chromosome segregation dynamics using live imaging followed by fixation of single cells at defined moments of mitosis to perform correlative immunofluorescence and analyze localization of endogenous proteins [16–18]. Here, we describe such an approach, applied to preimplantation mouse embryos, to investigate the origin and fate of MN-enclosed chromosomes during mitosis.

2 Materials

2.1 Production of Synthetic cRNA

1. H2B:RFP plasmid DNA in pRN3 vector backbone (*see* **Note 1**).
2. Restriction enzyme for plasmid linearization (*see* **Note 2**).
3. Agarose.
4. 1× TAE buffer: Tris-acetate 0.04 M, EDTA 0.01 M.
5. DNA loading dye (NEB) or equivalent.
6. SYBR Safe DNA Gel Stain (Thermo Fisher Scientific).
7. DNA 1 kb ladder (NEB).
8. Gel electrophoresis tank and power supply.
9. UV gel illuminator to visualize DNA.
10. Sterile scalpels.
11. DNA Gel Extraction Kit (QIAGEN) or equivalent.
12. mMessage mMachine® Kit (Thermo Fisher Scientific, SP6/T3/T7) (*see* **Note 3**).
13. Poly(A) Tailing Kit (Thermo Fisher Scientific).
14. RNeasy Cleanup Kit (QIAGEN) or equivalent.
15. Molecular biology-grade ethanol.
16. RNase AWAY Reagent (Invitrogen) or equivalent.

2.2 Embryo Collection, Culture, and Microinjection

1. Male and female CD-1 mice of approximately 6–12 weeks of age (Charles River Laboratories) (*see* **Note 4**).
2. Pregnant mare's serum gonadotropin (PMSG) (*see* **Note 5**).
3. Human chorionic gonadotropin (hCG) (*see* **Note 5**).
4. Phosphate-buffered saline (PBS) tablets.
5. 27-gauge needles.
6. M2 medium (Sigma-Aldrich) (*see* **Note 6**).
7. KSOM medium (EMD/Millipore) (*see* **Note 6**).
8. Mineral oil.
9. 0.22 μm syringe filter.
10. 10 mL syringe to filter media.
11. Digital dry bath at 37 °C.
12. Petri dishes (BD Falcon, 35 × 10 mm).
13. Pasteur pipette (soda-lime glass 9 in.) (*see* **Note 7**).
14. Mouth-controlled aspiration tube (*see* **Note 7**).
15. Cell culture incubator at 37 °C, 5% CO_2.
16. Ice.
17. Centrifuge (Eppendorf centrifuge 5430) or equivalent.

18. Oocyte and embryo microinjection station, with components and setup as previously described [19]. Inverted light microscope (Leica DMIL-LED), fitted micromanipulators and injectors (Narishige), a Pneumatic PicoPump (World Precision Instruments), and an intracellular amplifier electrometer (Harvard Apparatus). The apparatus is assembled and placed on an anti-vibration table (*see* **Note 8**).
19. Holding pipettes (*see* **Note 9**).
20. Vertical micropipette puller for the injection pipette (Sutter Instrument Company).
21. Glass capillaries with filament for cRNA injection (Harvard Apparatus, 1.5 mm OD × 0.86 mm ID × 150 mm L).

2.3 Confocal Live Imaging

1. Leica SP8 confocal fitted with HyD detectors or equivalent live cell imaging scopes with highly sensitive emitted light detection and/or reduced illumination such as spinning disk or structured illumination microscopes may also be used for this purpose.
2. Temperature-controlled microscope-stage incubator with 5% CO_2 supply.
3. 50 mm uncoated glass bottom culture dishes (MatTek Corporation).
4. KSOM medium (EMD/Millipore) (*see* **Note 6**).
5. Mineral oil.

2.4 Embryo Fixation, Immunofluorescence, and Confocal Imaging

1. The following working solutions should be prepared. The solutions should be filtered and are stable for several weeks at 4 °C:
   (a) 4% paraformaldehyde (PFA) in PBS for fixing.
   (b) 0.25% Triton X-100 in PBS for permeabilization.
   (c) 1% bovine serum albumin (BSA) in PBS for washing.
   (d) 3% BSA in PBS for blocking.
2. 96-well plates with round bottom.
3. 35 mm glass bottom culture dishes (MatTek Corporation) (*see* **Note 10**).
4. Mouse anti-α-tubulin primary antibody (Sigma-Aldrich).
5. Human CREST or ACA primary antibody (*see* **Note 11**).
6. Alexa 405 anti-mouse IgG secondary antibody (Thermo Fisher scientific).
7. Alexa 488 anti-human IgG secondary antibody (Thermo Fisher scientific).
8. Alexa 647 phalloidin (Thermo Fisher Scientific).
9. A confocal microscope equipped with three different colored lasers, a 63× oil immersion lens, and a computer-controlled Z-motor allowing to take Z-stacks.

3 Methods

The workflow for this experimental approach follows four defined stages, which are described herein: production of cRNA for H2B:RFP, embryo collection and microinjection, live imaging of embryo mitosis, fixation and reimaging by immunofluorescence.

3.1 Production of cRNA

Here, we briefly describe our standard approach for producing cRNAs. We use QIAGEN kits for DNA and RNA purification and Ambion kits for IVT and Poly(A) tailing. These steps are performed according to manufacturer's instructions.

1. **Linearization and purification of plasmid DNA:** Incubate 3–5 μg of H2B:RFP plasmid DNA with the appropriate restriction enzyme according to manufacturer's instructions (*see* **Notes 1, 2,** and **12**). Mix reactions with appropriate volumes of loading dye, and load on a 1% agarose gel supplemented with SYBR-DNA dye in an electrophoresis tank. Run gel at 100 V for 40 min. Visualize DNA using a UV gel illuminator. Using a clean scalpel, excise the gel fragment containing linearized DNA. Purify linearized DNA following the QIAGEN Gel Extraction Kit protocol (*see* **Note 13**). We recommend using a NanoDrop to measure concentration and purity of DNA.
2. **IVT of linearized plasmid DNA:** Use the mMessage mMachine® in vitro transcription kits (Ambion) to produce capped RNA from the linearized template plasmid. Reaction is performed according to manufacturer's instructions (*see* **Note 14**).
3. **Poly(A) Tailing:** To increase cRNA stability and enhance translation efficiency, following IVT, perform a 3′ polyadenylation reaction, using the Poly(A) Tailing Kit (Ambion) (*see* **Note 15**).
4. **RNA purification:** Purify the resulting capped and polyadenylated H2B:RFP cRNA using the RNeasy Mini Kit Cleanup Protocol (QIAGEN), and elute in 30 μL of RNase-free water. We recommend using a NanoDrop to measure the concentration and purity. Aliquot H2B:RFP cRNA in 1 μL batches in RNAase-free tubes, and store at −80 °C.

3.2 Two-Cell Embryo Collection, Culture, and Microinjection

3.2.1 Two-Cell Embryo Collection

1. ~96 h before collection of two-cell embryos, stimulate 2–3-month-old CD-1 female mice with an intraperitoneal injection of 5 IU of PMSG. 40 h before collection of two-cell embryos, inject the same mice with 5 IU of hCG (*see* **Note 16**), and transfer immediately into cages with a male mouse for mating (1 female:1 male).
2. To collect two-cell embryos, sacrifice the female mice using isoflurane anesthesia followed by cervical dislocation, and dissect out the reproductive tract (*see* **Note 17**). For more detailed description, *see* [20].

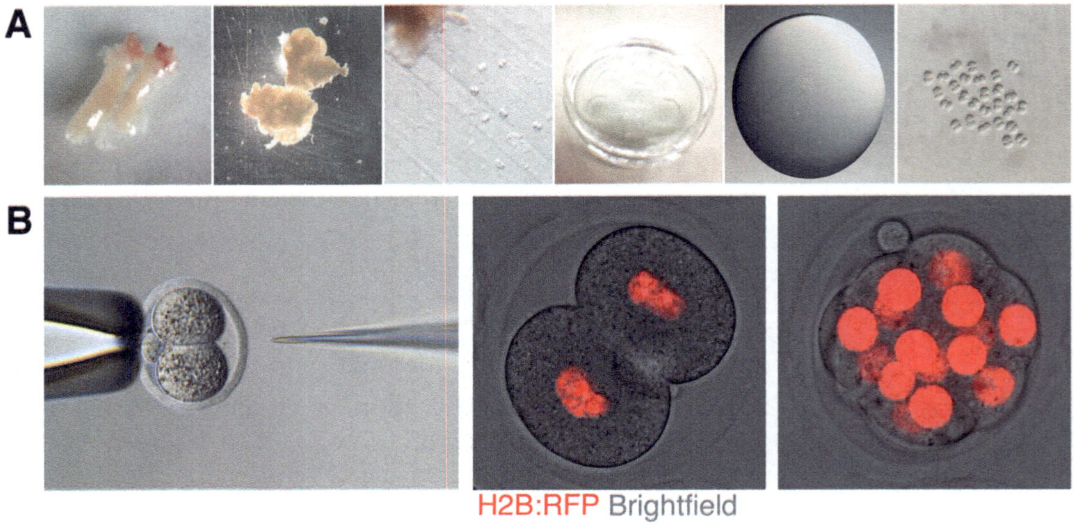

Fig. 1 Embryo collection, culture, and microinjection. (**a**) From left to right: Mouse female reproductive tracts, isolated oviducts, torn oviducts with embryos released, 35 mm culture dish with 40 μL drops of KSOM and mineral oil, KSOM drop containing embryos for culture, magnified image of embryos within a media drop. (**b**) From left to right: Microinjection of two-cell embryos, with holding and microinjection pipettes. Two-cell embryo expressing H2B:RFP. Morula-stage embryo expressing H2B:RFP

3. Using a dissection microscope: In a culture dish with ~2 mL of M2 media, use the forceps and needles to isolate the oviducts, leaving the ovaries, uterine horns, and other tissues in this dish (Fig. 1a) (*see* **Note 18**).
4. Transfer the oviducts into a fresh dish with ~2 mL M2 media. Carefully tear open the oviducts using forceps and a needle or two needles, so that the embryos are released into the media (*see* **Note 19**).
5. At higher magnification, find and collect two-cell embryos using a pulled glass Pasteur of approximately 80–100 μm in diameter, mounted on a mouth-controlled aspiration tube (*see* **Note 7**), and transfer into a dish containing ~40 μL drops of M2 covered by mineral oil.
6. Wash the embryos from residual tissues from dissection by consecutively transferring them into fresh M2 drops. This process is usually repeated three times or until no more tissue debris is found in the media drop containing the embryos.
7. For embryo culture, transfer the embryos into a dish containing ~40 μL drops of pre-equilibrated KSOM. Wash away any residual M2 medium by transferring embryos through multiple KSOM drops. Place the dish in an incubator with 5% CO_2 at 37 °C (Fig. 1a) (*see* **Note 20**).

3.2.2 Embryo Microinjection

Our standard microinjection components, setup, and procedure have been described in extensive details elsewhere [19]; therefore, here we will briefly describe the protocol and variations required for two-cell embryo microinjections (Fig. 1b):

1. Set up the microinjection station: Prepare the microinjection chamber using 400–800 μL M2 media covered with mineral oil on an upturned lid of a 35 mm tissue culture dish, and place on the microscope stage. Prepare and place holding pipette in the field of view just above the bottom surface of the microinjection chamber. Pull fresh microinjection pipettes using a vertical pipette puller.
2. Pipette embryos into the M2 media in the microinjection chamber (*see* **Note 21**).
3. Keep cRNA on ice, and dilute to appropriate concentration as necessary. Centrifuge briefly before opening. Using a microloader tip, backload ~1 μL of cRNA into a microinjection pipette, and mount it on the injection probe [19] (*see* **Note 22**).
4. Focus on the embryos, and lower holding pipette until it becomes leveled with the embryos; vary the holding pressure such that the embryo is stably held, and then lift and raise the holding pipette and embryo 100–200 μm above the chamber surface (*see* **Note 23**).
5. Using micromanipulators, lower the microinjection pipette into the M2 media such that the tip of the pipette is in the same plane as the embryos (*see* **Note 24**).
6. Insert the tip of the injection pipette into the cytoplasm of one of the blastomeres through the zona pellucida, and apply a short pulse of negative capacitance with the intracellular electrometer, and microinject the H2B:RFP cRNA by applying a timed pressure using the PicoPump. Successful microinjection should be evidenced by cytoplasmic displacement. Then, remove the pipette smoothly. Repeat procedure for the second blastomere (*see* **Note 25**).
7. Once the settings have been adjusted for the desired injection size, use the same settings to microinject the main group of embryos to be used for the experiment, to ensure consistent microinjection volume and comparable cRNA translation levels between blastomeres and embryos (*see* **Note 26**).
8. Following microinjection, wash the embryos through three drops of pre-equilibrated KSOM media using a mouth-controlled pipette. Culture embryos in KSOM in an incubator with 5% CO_2 at 37 °C for 2–3 h, and check fluorescent protein expression using an inverted epifluorescence microscope

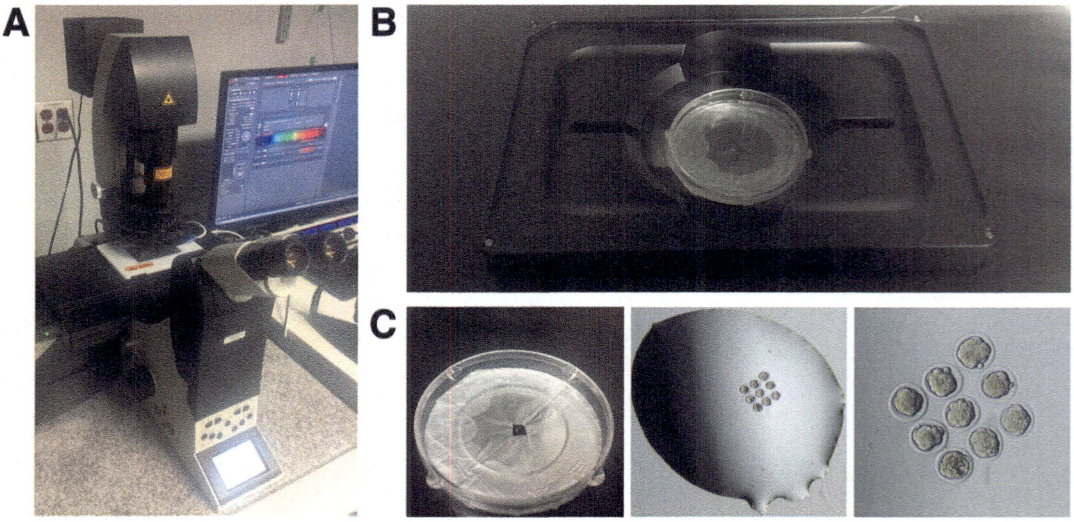

Fig. 2 Live imaging confocal and top-stage incubator setup. (**a**) Confocal microscope fitted with top-stage incubator chamber. (**b**) 50 mm glass bottom culture dish and aluminum foil bottom for live cell imaging in top-stage incubator. (**c**) From left to right: 50 mm glass bottom culture dish with a 5 μL drop of KSOM covered with mineral oil, embryos positioned in 5 μL drop of KSOM, magnified image of embryos in culture dish

equipped with fluorescence optics. Return embryos to the incubator and culture until the desired developmental stage is reached (*see* **Note 20**).

3.3 Live Cell Imaging Setup for Mouse Embryos

We aim to observe the dynamics of chromosome segregation in embryos using confocal live cell imaging conditions as non-damaging as practically possible. To this end, we use a temperature-controlled incubator with precise 5% CO_2 concentration supply mounted on the microscope stage, and use imaging protocols designed to minimize light exposure to embryos. The setup of the live cell imaging chamber is described below and depicted in Fig. 2.

1. Fit the top-stage incubator on the microscope stage, turn on the temperature controller device, and set temperature at 37 °C (*see* **Note 27**). Connect the lid of the incubator to a 5% CO_2 supply (*see* **Note 28**).
2. Place a 50 mm glass bottom culture dish on the stage incubator, pipette a 5 μL drop of KSOM in the center of the dish, and cover with warm mineral oil pre-equilibrated with 5% CO_2 (*see* **Notes 29** and **30**).
3. At a low magnification (5×, 10×), focus on the KSOM drop, and pipette H2B:RFP cRNA microinjected embryos into the middle of the drop. Positioning the embryos closely together will allow to image multiple embryos within the same frame (*see* **Note 31**).

4. Embryos are imaged using a 0.75 NA 20× lens and 552 nm laser at 0.1–0.5% power, collecting 560–700 nm emission with a HyD detector and spectral detection system. Select an optical section of 2 μm, a Z-stack size of ~60 μm, and step size of 2 μm. Acquire images bidirectionally at 400 Hz speed in 1024 × 1024 pixel format at an 8 bit depth; two-scan averaging may be used to further increase image clarity. A bright-field image is simultaneously acquired by means of a PMT_{trans} detector. To capture the dynamic events of chromosome segregation in H2B:RFP-expressing embryos, we perform time-lapse imaging using a time interval of 3–5 min (*see* **Notes 32** and **33**).
5. Adjust imaging settings such that minimal 552 nm laser power is used to achieve a clear unsaturated image and start three-dimensional (Z-stack) time-lapse imaging.
6. To investigate the fate of MN-chromosomes in preimplantation embryos, identify embryos with blastomeres containing MN and monitor cells entering mitosis; this is first evident by condensation of chromatin into chromosomes during nuclear envelope breakdown (NEBD). Mitotic divisions at this stage have a duration of 40–60 min, so we recommend to systematically and periodically (every 10 min) examine the Z-slices for each embryo for NEDB.
7. Upon observation of a MN-containing cell entering mitosis, closely monitor progression into mitosis until most chromosomes are aligned at the metaphase plate; in most cases, the MN are distinctly separate from the metaphase plate.
8. At this time, stop the image acquisition, switch back to a lower magnification lens, and using a mouth-controlled pipette individually collect the embryo of interest for immediate fixation.
9. For later identification and relation in correlative immunofluorescence, for each selected embryo, it is useful to draw a map of the layout of the embryos, and mark the relative position selected embryo. It is also recommended to annotate the live imaging file name, last time-point and Z-slice in which the MN-containing metaphase cell is located (*see* **Note 34**).

3.4 Embryo Fixation and Immunofluorescence

Prior to initiation of live imaging, prepare a round-bottom 96-well plate for immunofluorescence staining, arranging solutions in vertical columns and individual selected embryos in horizontal rows. Place 50 μL of solution in each well, covering each 50 μL-filled well with a drop of mineral oil, except wells containing 0.25% Triton X-100. Stain embryos with CREST and tubulin antibodies to label kinetochores and microtubules, respectively. Alexa phalloidin staining will enable visualization of filamentous actin at the cell cortex, hence delineating individual blastomeres within the embryo.

1. Using a bent Pasteur pipette mounted on the mouth-controlled aspiration tube, transfer the embryo into the first column containing 4% PFA in PBS. Incubate for 30 min at room temperature. Subsequent steps described below entail moving embryos to successive wells in the 96-well plate using the mouth-controlled pipette.
2. Permeabilize embryos with 0.25% Triton X-100 in PBS for 10 min at room temperature.
3. Wash with 3% BSA in PBS three times.
4. Block in 3% BSA in PBS for 60 min at 37 °C or overnight at 4 °C.
5. Incubate with CREST primary antibody for 60 min at 37 °C (*see* **Note 35**).
6. Wash with 1% BSA in PBS three times to remove residual primary antibody.
7. Incubate with secondary antibody (anti-human IgG Alexa) for 60 min at 37 °C. From this step onward, wrap the plate with aluminum foil to protect it from light exposure (*see* **Note 36**).
8. Wash with 1% BSA in PBS three times to remove residual secondary antibody.
9. Incubate with mouse anti-α-tubulin primary antibody for 60 min at 37 °C (*see* **Note 35**).
10. Wash with 1% BSA in PBS three times to remove residual primary antibody.
11. Incubate with secondary antibody (anti-mouse IgG Alexa) for 60 min at 37 °C (*see* **Note 36**).
12. Wash with 1% BSA in PBS three times to remove residual secondary antibody.
13. Incubate with Alexa-conjugated phalloidin for 60 min at 37 °C (*see* **Note 36**).
14. Wash with 1% BSA in PBS three times to remove residual Alexa phalloidin staining.
15. DNA-staining dyes such as DAPI/Hoechst are not required since chromosomes are fluorescently labeled with H2B:RFP.

3.5 Correlative Immunofluorescence Confocal Imaging

This section describes our standard method for confocal immunofluorescence imaging to obtain high-resolution three-dimensional images of individual embryos and individual mitotic cells (*see* **Note 37**), hence allowing to analyze the localization of kinetochore proteins and interaction with microtubules of MN-chromosomes during mitosis (Fig. 3):

1. For confocal immunofluorescence imaging, place a 5 μL drop of 1% BSA in PBS in a 35 mm glass bottom culture dish, and cover it with mineral oil.

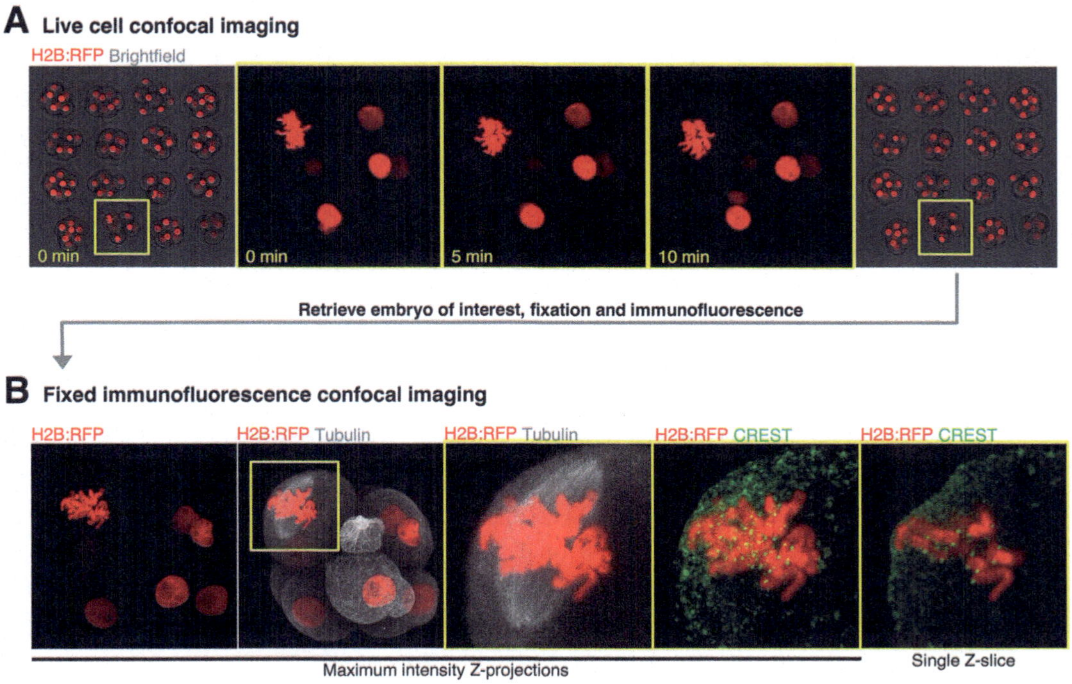

Fig. 3 Correlative live and immunofluorescence imaging. (**a**) Still frames from time-lapse live cell confocal imaging of H2B:RFP-expressing embryos. Yellow boxes indicate the embryo of interest. (**b**) Immunofluorescence imaging of the embryo of interest in (**a**) selected by live imaging. Chromosomes/DNA appear in red (H2B:RFP), microtubules appear in grayscale (anti-α-tubulin), and kinetochores appear in green (CREST). Yellow boxes indicate cell of interest

2. Transfer the embryo into the imaging dish using the mouth-controlled pipette.
3. Using a confocal microscope, locate the embryo using a lower magnification (5×, 10× lens). After having applied a small amount of immersion oil on the objective lens, switch to a 63× (1.40 NA) oil immersion lens, and perform immunofluorescence confocal imaging (*see* **Note 37**).
4. Manually rotate and reposition the embryo by means of a mouth-controlled pulled glass pipette to place the embryo in a similar orientation to that in the initial live imaging experiment (*see* **Note 34**).
5. Increase the zoom further to make the cell of interest fit the image frame. Adjust the optical section to 1 μm and excitation laser power for the different channels as necessary. Use the Alexa phalloidin signal marking the cell cortex to set a Z-stack front end to the back end of the cell of interest, with a step size of 1 μm.

4 Notes

1. The H2B:RFP plasmid we routinely use is a gift from Alex McDougall (Observatoire Océanologique de Villefranche-sur-Mer, Villefranche Sur Mer, France) [14]. However, equivalent alternatives are available from Addgene: pCS-H2B-mRFP1 Plasmid #53745, H2B-mCherry Plasmid #20972 (https://www.addgene.org/53745/, https://www.addgene.org/20972/).
2. The restriction enzyme required depends on the DNA plasmid to be used as a template. Choice of an appropriate restriction enzyme for DNA plasmid linearization requires a unique restriction site that must lie downstream of the promoter and fusion protein.
3. mMessage mMachine® Kit used must match the promoter present on template plasmid DNA: T3, T7, or SP6.
4. We typically use CD-1 mice. However, different mouse strains may be used. Animals should be acquired and stored in accordance with local animal welfare regulations.
5. Hormones are prepared in sterilized filtered PBS on ice (we use a dilution of 25 international units (IU) per mL). 1 mL aliquots are stored at −20 °C. Hormones are injected into mice immediately after thawing.
6. We normally use commercially provided M2 and KSOM media, but these solutions can also be made in-house and stored frozen at −20 °C, supplementing with the appropriate amount of BSA after thawing [21, 22].
7. Pasteur pipettes need to be pulled to an appropriate width using an alcohol lamp or Bunsen burner. Pipettes with an internal diameter of ~80–100 μm are recommended for embryo collection. Pasteur pipettes are attached to a mouth-controlled aspiration tube to allow embryos to be transferred between drops. Moving embryos with a mouth-controlled Pasteur pipette takes several weeks' practice, but once mastered allows very precise and controlled movement. As an alternative to mouth pipetting, manual pipettes may also be used (Cook Medical).
8. Further details on components and setup of our microinjection system can be found in [19]. Precise micromanipulation setup may vary. The amplifier and negative capacitance pulse is essential when performing injections into metaphase-stage eggs and embryos, which are extremely sensitive to injection. An anti-vibration assembly is essential, such as commercial vibration isolation tables, though a marble slab mounted on tennis or squash balls can also suffice.
9. Holding pipettes can be made in-house using a pipette puller and microforge. Alternatively, we recommend investigators

with limited micromanipulation experience to use commercially available premade holding pipettes that may be purchased from Hunter Scientific (http://www.hunterscientific.com).

10. We normally use petri dishes fitted with No. 0 cover glass (0.085–0.13 mm) for immunofluorescence microscopy.
11. Our ACA antibody is derived from CREST patients' human sera and obtained from Marvin J Fritzler from University of Calgary. However, CREST antibodies are also commercially available.
12. We typically assemble control and restriction enzyme digestion reactions with final volumes of 10 μL and 50 μL, respectively. To verify complete digestion, run 5 μL of each reaction on a 1% agarose gel; a single band is observed if plasmid is completely linearized. Use a DNA ladder to check the linearized fragment is the correct size.
13. Linearized DNA template may be purified by alternative methods such as DNA cleanup and DNA precipitation.
14. Before starting IVT and following steps, ensure that all surfaces and equipment to be used have been cleaned with 70% ethanol and then wiped with RNase decontamination solution reagent. Use certified RNase-free or autoclaved filter tips. It is also recommended to change gloves before each step.
15. Prior to the Poly(A) tailing step, incubate the IVT reaction mix with Turbo DNase to remove the DNA template.
16. We recommend optimizing hormone doses for the chosen mouse strain and age.
17. The method of euthanasia must be performed in accordance with local animal welfare regulations.
18. M2 media must be pre-warmed to 37 °C on a dry bath. During the dissection process, dishes should be kept on the dry bath when not being used on the dissection microscope.
19. Alternatively, embryos can be retrieved by oviduct flushing [20].
20. In our hands, CD-1 two-cell embryos will be 4 cells ~60 h post-hCG and mating (post-hCG + m) and 8 to 16 cells ~72 h post-hCG + m, morula ~96 h post-hCG + m, and blastocysts ~120 h post-hCG + m.
21. To wash residual KSOM medium, transfer embryos through three drops of M2 before pipetting into the M2 media in the microinjection chamber.
22. We typically inject H2B:RFP cRNA at a concentration of 400–800 ng/μL (microinjection pipette concentration); however, these may be adjusted for different cRNA batches to achieve efficient expression. Optimal cRNA concentrations and injection size can be tested by microinjection as described in [19] (*see* **Note 25**).

23. Orienting two-cell embryos such that both cells are in the same focus plane is advised to facilitate successful microinjection of both blastomeres.
24. Upon contact of the microinjection tip with the M2 media, the negative capacitance circuit should be closed and displayed on the intracellular electrometer as described in [19] (*see* **Note 8**).
25. Since microinjection volume may vary with pipette tip diameter, use a small group of embryos to adjust the pressure and time to achieve a modest, but visible cytoplasmic displacement. Using a small cohort of test embryos, inject separate groups with different injection sizes, by adjusting the ejection pressure and injection time as described in [19]. Confirm injection size is not detrimental to embryo health by culturing two-cell embryos in KSOM for ~72 h, and compare blastocyst rates in non-injected vs. water-injected embryos. Cell death directly attributable to the microinjection technique should not exceed 15% for standard microinjections.
26. We recommend applying gated pressure pulses before and after every microinjection to prevent backflow of media into the pipette and ensure consistent content of microinjection.
27. We advise to test and calibrate the temperature of the top-stage incubator required to maintain the drop of media at 37 °C by using a thermocouple.
28. We use centrally supplied CO_2 which is equilibrated at 5% using a proportional gas mixer. CO_2 gas cylinders may also be purchased, and concentration is adjusted using a pressure regulator. In any case, we recommend the use of a CO_2 meter to confirm CO_2 concentration is 5% before starting an experiment.
29. For optimal heat transfer throughout the culture dish, we place a piece of aluminum foil on the bottom of the dish, a small hole is cut out, and the 5 μL drop of KSOM is placed right above it to allow light transmission and imaging of embryos. A sparing amount of Vaseline on the dish circumference may be used to better affix the aluminum foil to the dish.
30. We recommend the live cell imaging chamber to be set up at least 30 min before starting the experiment, to allow sufficient time for temperature and pH equilibration. This will also prevent focus shifts during live cell imaging should there be any expansion of the live cell imaging chamber as the temperature equilibrates.
31. Before beginning experiments, it is important to verify that the live cell imaging conditions are suitable for embryo development. This can be done by setting up the chamber and placing a small group of 10–15 two-cell embryos on the microscope

stage (without any imaging) for 2 or 3 days, and compare developmental rates with embryos placed in a similar dish in the cell culture incubator.

32. Our confocal settings have been optimized to reduce embryo light exposure while still yielding high-resolution images. For this reason, the RFP fluorophore is most suitable, since shorter wavelengths may cause phototoxicity and longer wavelengths may cause sample heating. HyD detectors as opposed to conventional PMT detectors are preferable due to their superior sensitivity, hence requiring reduced amounts of laser power and light exposure to achieve a high contrast image with minimal noise.

33. To validate that the imaging is not causing damage to the embryos, we recommend performing a live imaging run for 8–12 h at the earlier stages, two-cell up to eight-cell, and then placing embryos back in the culture incubator. Compare developmental rates to blastocyst with control embryos that were left in the incubator and not imaged. The settings, and in particular the 552 nm laser power, might have to be slightly modified on an experiment-to-experiment basis to obtain bright but unsaturated images.

34. In the case that more than one cell is in mitosis in a given embryo, it is advised to take notes of other features of the embryo that may later help identify the dividing MN-containing blastomere, for example, the other mitotic cell being in a different stage of mitosis, prometaphase/anaphase. Positional information such as proximity to polar body, metaphase plate orientation, and inner/outer location of blastomere within the embryo may also be valuable. The brightfield image will aid in discerning which blastomere the MN belongs. When performing confocal immunofluorescence imaging, the H2B:RFP signal may be used to positively identify the cell of interest within the embryo. Manipulating embryo orientation, comparison to the last time-point of the live cell imaging Z-stack of that same embryo, and the use of previous annotations of embryo features may be required to facilitate identification of the MN-containing cell at metaphase.

35. Optimal fixation method, antibody dilution, incubation times, and temperatures should have been previously and empirically determined on similar developmental stage embryos. Alternative fixing methods that may be used include [1] using 2% PFA in PBS fixation, [2] a brief (5 s) pre-permeabilization in 0.5% Triton X-100 in PHEM buffer (60 mM pipes, 25 mM HEPES, 10 mM EGTA, 2 mM $MgCl_2$, pH 6.9) prior to fixation in 4% PFA in PHEM followed by normal permeabilization, or [3] simultaneously fixing and permeabilizing in 4% PFA + 0.25% Triton X-100 in PHEM buffer.

36. Use different Alexa fluorophores to distinguish the microtubules from CREST and phalloidin. After performing the immunofluorescence staining, embryos may be stored in 1% BSA in PBS solution at 4 °C, protected from light. In our experience, Alexa phalloidin staining provides best results when applied immediately prior to confocal imaging.
37. For whole embryo immunofluorescence imaging, we typically acquire 1024 × 1024 format images at a speed of 400 Hz with bidirectional scanning and an optical section of 1.5 μm. Different channels are sequentially acquired per frame with two-scan averaging. Excitation laser power is adjusted as required and emission collection selected based on emission spectra of each fluorophore, preferentially using a HyD over PMT detectors, since they provide a more defined image with a better signal-to-noise ratio. Even if using sequential scanning, we set the non-overlapping collection spectra to prevent crosstalk between acquired channels. We recommend having the live cell imaging Z-stack for the last time-point of the embryo to be analyzed, open, and easily available on the software simultaneously. The digital zoom is adjusted such that the whole embryo fits the image frame. The Z-stack is set to span the whole embryo, starting from the side that is closest to the glass coverslip, using a step size of 1.5 μm. It is recommended to manually reposition the embryo if necessary, such that spindles are parallel to the coverslip to obtain a high-quality image and facilitate analysis of CREST and microtubule localization to mitotic MN-chromosomes.

Acknowledgments

Work in GF's lab is supported by Fondation JL Lévesque, NSERC, and CIHR.

We thank Gaudeline Rémillard-Labrosse and Angus Macaulay for critical reading and discussion of this manuscript.

References

1. Kloosterman WP, Hoogstraat M, Paling O et al (2011) Chromothripsis is a common mechanism driving genomic rearrangements in primary and metastatic colorectal cancer. Genome Biol 12:R103. https://doi.org/10.1186/gb-2011-12-10-r103
2. Stephens PJ, Greenman CD, Fu B et al (2011) Massive genomic rearrangement acquired in a single catastrophic event during cancer development. Cell 144:27–40
3. Adey A, Burton JN, Kitzman JO et al (2013) The haplotype-resolved genome and epigenome of the aneuploid HeLa cancer cell line. Nature 500:207–211
4. Forment JV, Kaidi A, Jackson SP (2012) Chromothripsis and cancer: causes and consequences of chromosome shattering. Nat Rev Cancer 12:663–670
5. Storchova Z, Kloosterman WP (2016) The genomic characteristics and cellular origin of chromothripsis. Curr Opin Cell Biol 40:106–113
6. Thompson SL, Compton DA (2011) Chromosome missegregation in human cells

arises through specific types of kinetochore–microtubule attachment errors. Proc Natl Acad Sci 108:17974–17978

7. Cimini D, Howell B, Maddox P et al (2001) Merotelic kinetochore orientation is a major mechanism of aneuploidy in mitotic mammalian tissue cells. J Cell Biol 153:517–528
8. Crasta K, Ganem NJ, Dagher R et al (2012) DNA breaks and chromosome pulverization from errors in mitosis. Nature 482:53–58
9. Hatch EM, Fischer AH, Deerinck T et al (2013) Catastrophic nuclear envelope collapse in cancer cell micronuclei. Cell 154:47–60
10. Ly P, Teitz LS, Kim DH et al (2017) Selective Y centromere inactivation triggers chromosome shattering in micronuclei and repair by non-homologous end joining. Nat Cell Biol 19:68–75
11. Zhang CZ, Spektor A, Cornils H et al (2015) Chromothripsis from DNA damage in micronuclei. Nature 522:179–184
12. Fragouli E, Alfarawati S, Spath K et al (2013) The origin and impact of embryonic aneuploidy. Hum Genet 132:1001–1013
13. Kort DH, Chia G, Treff NR et al (2016) Human embryos commonly form abnormal nuclei during development: a mechanism of DNA damage, embryonic aneuploidy, and developmental arrest. Hum Reprod 31:312–323
14. Vázquez-Diez C, Yamagata K, Trivedi S et al (2016) Micronucleus formation causes perpetual unilateral chromosome inheritance in mouse embryos. Proc Natl Acad Sci U S A 113:626–631
15. Rieder CL, Khodjakov A (2003) Mitosis through the microscope: advances in seeing inside live dividing cells. Science 300:91–96
16. Hinchcliffe EH, Day CA, Karanjeet KB et al (2016) Chromosome missegregation during anaphase triggers p53 cell cycle arrest through histone H3.3 Ser31 phosphorylation. Nat Cell Biol 18:668–675
17. Hornick JE, Mader CC, Tribble EK et al (2011) Amphiastral mitotic spindle assembly in vertebrate cells lacking centrosomes. Curr Biol 21:598–605
18. Gregan J, Polakova S, Zhang L et al (2011) Merotelic kinetochore attachment: causes and effects. Trends Cell Biol 21:374–381
19. Nakagawa S, FitzHarris G (2016) Quantitative microinjection of morpholino antisense oligonucleotides into mouse oocytes to examine gene function in meiosis-I. Methods Mol Biol 1457:217–230
20. Nagy A (2003) Manipulating the mouse embryo: a laboratory manual, 3rd edn. Cold Spring Harbor, New York
21. Erbach GT, Lawitts JA, Papaioannou VE et al (1994) Differential growth of the mouse preimplantation embryo in chemically defined media. Biol Reprod 50:1027–1033
22. Gardner DK, Leese HJ (1986) Non-invasive measurement of nutrient uptake by single cultured pre-implantation mouse embryos. Hum Reprod 1:25–27

Chapter 21

Experimental Induction of Genome Chaos

Christine J. Ye, Guo Liu, and Henry H. Heng

Abstract

Genome chaos, or karyotype chaos, represents a powerful survival strategy for somatic cells under high levels of stress/selection. Since the genome context, not the gene content, encodes the genomic blueprint of the cell, stress-induced rapid and massive reorganization of genome topology functions as a very important mechanism for genome (karyotype) evolution. In recent years, the phenomenon of genome chaos has been confirmed by various sequencing efforts, and many different terms have been coined to describe different subtypes of the chaotic genome including "chromothripsis," "chromoplexy," and "structural mutations." To advance this exciting field, we need an effective experimental system to induce and characterize the karyotype reorganization process. In this chapter, an experimental protocol to induce chaotic genomes is described, following a brief discussion of the mechanism and implication of genome chaos in cancer evolution.

Key words Chromosomal or karyotype coding, Chromosome instability (CIN), Chromothripsis, Genome chaos, Genome theory, Macro-cellular evolution, Punctuated cancer evolution

1 Introduction

Genome chaos, chromosome chaos, or karyotype chaos refers to the phenomenon of rapid and massive reorganization of genome topology under high levels of cellular stress [1]. In the early 2000s, these structures were initially described by Heng's group in both in vitro drug treatment model and in vitro immortalization model [2–6]. These drastically altered chromosomes were clearly associated with the pattern of cancer evolution as they were frequently detectable within the punctuated phase of somatic cell evolution [3, 4]. Detailed analyses demonstrated that these genome-level alterations can be classified into complex structural and numerical changes. While translocations often involve multiple chromosomes, the total number of chromosomes in a given individual cell can reach 500–700 [7].

Even though these striking karyotypes have been observed from cancer samples, which are often clinically associated with aggressive disease progression and poor prognosis, the mechanism

Franck Pellestor (ed.), *Chromothripsis: Methods and Protocols*, Methods in Molecular Biology, vol. 1769,
https://doi.org/10.1007/978-1-4939-7780-2_21, © Springer Science+Business Media, LLC, part of Springer Nature 2018

and the common functional consequences are less clear. In addition to the fact that it has been extremely difficult to study these seemingly stochastic structures even by cytogeneticists (most molecular researchers are not familiar with them at all), chaotic genomes have been ignored by many. One common argument is that these drastically altered chromosomes will surely lead to cell death as massive genomic alteration has no evolutionary advantages, so there is no reason to study their biological meaning. Another viewpoint is that these structures represent in vitro artifacts, as it is hard to imagine these drastic genetic changes can be tolerated by in vivo systems. These viewpoints have made it a challenge for us to publish chaotic genome data in high-profile journals, despite our efforts to point out that chaotic genomes have even been frequently observed in primary tumors and in cultured lymphocytes from normal individuals after radiation exposure. Furthermore, some chaotic genomes do survive and can generate new stable genomes during evolution. Given that chaotic genomes are closely associated with all major transitions of cancer evolution (the macro-cellular evolutionary phase), including immortalization, transformation, metastasis, and drug resistance, we continued the efforts to characterize them and push the concept of chaotic genome-mediated macro-cellular evolution.

Recently, the cancer genome sequencing project has generated large amounts of data, which inevitably confirmed the importance of studying genome chaos [8–10]. Various chaotic genomes were detected within nearly all types of cancers. In some cancer types such as prostate cancer, chaotic genomes were detected in a majority of cases. Interestingly, these fragmented and stitched chromosomes were given many different names by different investigators including "chromothripsis," "chromoplexy," "chromoanagenesis," "chromoanasynthesis," "chromosome catastrophes," and "structural mutations" [11–21]. Based on their descriptions, chromothripsis refers to the chaotic genome mainly involving local reorganization (within a single chromosome), while chromoplexy refers to a more whole genome reorganization involving many individual chromosomes. We prefer using "genome chaos," "karyotype chaos," or "chromosomal chaos" to refer to these structures due to their broad coverage (from local to global reorganization, from structural to numerical changes) and for the simplicity of terminology [1, 3, 22, 23].

In the past few years, chromothripsis studies have become a hot topic, evidenced by a large number of high-profile publications [8, 11–21, 24, 25]. Large-scale genome reorganization supports the new mechanism of cancer formation which contradicts the current doctrine of cancer formation: accumulation of gene mutations in a stepwise fashion. However, most of the analyses have been based on sequencing data from only one stage of the tumor profile without comparing the dynamics of the entire evolutionary process of genome chaos, ignoring the important information from karyotype analyses. Limited mechanistic studies mainly focused on specific gene mutations, pathways, and specific cellular mechanisms, such as how the p53 mutation and micronucleus formation contribute to genome chaos. For example, it was elegantly illustrated that chromothripsis may directly originate from DNA damage in micronuclei, linking the restriction of chromothriptic rearrangements to a single chromosome. Alternatively, chromothripsis was linked to telomere crisis [24, 25]. From a genome evolutionary perspective, however, there should be large numbers of specific molecular mechanisms that can contribute to genome chaos, and only a small portion of them belong to chromothripsis [1, 3, 4]. Additional studies have now linked many individual factors, such as hyperploidy and radiation, to chromothripsis [26, 27], and the list of the contributing factors should be increasing, as predicted by the linkage between diverse cellular stresses and chromosome fragmentation observed in cell death [28–30], and more generally, by the evolutionary mechanism of cancer and other diseases [1, 31–33].

Based on the prediction of the genome theory of cancer evolution [1, 34–37], stress-mediated genome chaos represents a powerful tool for the survival and triumph of cancer cells. It is thus essential to study the entire process of genome chaos, from its induction to its consequences in evolution, which requires a user-friendly in vitro model. Our initial models of immortalization (human fibroblast cells) and transformation (mouse ovarian surface epithelial cells) are time consuming, and the cell cultures were hard to control when these cells were within the phase of punctuated evolution. Since the evolutionary phase transition (from stepwise to punctuated phase) can be effectively achieved by drug treatment, we then developed a simple protocol to induce genome chaos to study its mechanism and biological consequences [38].

This protocol has offered us an experimental system with which to generate and monitor the entire process of genome chaos and to study various factors that can trigger genome chaos, as well as a platform to classify and characterize different types of chaotic genomes. Further synthesis has led to our current understanding of genome chaos. Genome chaos represents a survival strategy for the genome system when under crisis. By reshuffling the genome, newly formed karyotypes display new potential for survival, as the karyotype defines the order of genes/regulating elements along and between chromosomes, and similar gene content can result in different system inheritance simply by changing the genomic topology. According to our proposed genome theory, the karyotype functions as a key genomic information organizer to guide the interactive and emergent relationship among genes [1, 35–38]. Thus, karyotype-mediated evolution can explain the important portion of missing inherit-

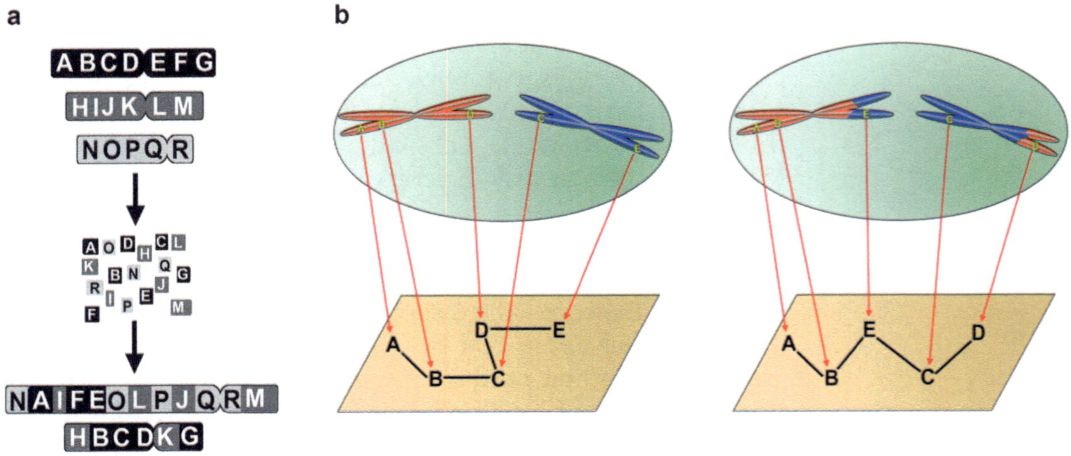

Fig. 1 (**a**) Schematic model of how genome chaos changes the coding system by altering the order of gene/regulatory elements along and among chromosomes (reproduced from [4]). Three normal chromosomes are shown at the top. Each letter within the chromosomes represents a distinct region (genes or regulatory elements). Following exposure to sufficient levels of various stressors, the genome is partially fragmented (middle panel). Chromosomal regions are recombined and rejoined, resulting in the drastically altered chromosomes illustrated at the bottom of the figure. The newly formed chimeric chromosomes can be a mixture of various chromosomal origins or restricted local regions or from a single chromosome. The process of genome chaos represents an evolutionary process of genome shattering that reorganizes the new genomic systems. (**b**) A diagram demonstrating how changing the genome topology alters the structure of genetic networks. For simplicity, only two chromosomes are drawn within the nucleus representing the genome. Genes/regulatory elements are designated A, B, C, D, and E within the chromosomes. When a chromosomal translocation occurs, the genome topology is altered, affecting the physical relationship between chromatin domains, which changes the overall interactive genetic network structure. While DNA codes the individual genes or parts, the interactive relationship among all genes and noncoding sequences are achieved by the 3D genome topology which serves as the platform or matrix where the self-organization principle works to form the genetic network [5, 35]. Diagram is reproduced from [35]

ability [39, 40]. Our current mechanistic understanding of genome chaos and its impact on creating new genome systems are illustrated in Fig. 1a, b. There are two main points: (1) genome reshuffling (as a result of genome chaos) changes the order of genes along and among chromosomes [4], and (2) the genome topology (reflected by the order of genes and regulatory elements) defines the interactive gene network [5, 35, 41].

Recently, inter-patient and intra-tumor heterogeneity has become a key challenge for cancer research [1, 10]. The mechanistic understanding of how genome chaos contributes to cancer heterogeneity represents a research priority. Increased studies have also started to pay more attention to these complex karyotypes, and many of them belong to chaotic genomes [42–45]. With the establishment of an experimental platform to induce genome chaos, as introduced in this protocol, various subtypes of genome chaos can be compared, and mechanisms of each type can be directly examined. Moreover, many challenging questions can be systematically addressed, including if an individual gene's function can be altered when the karyotype reshuffles, how chaotic genomes can be used to explain the contribution of individual driver genes or pathways, and how karyotype coding defines the new genome system by providing the emergent transcriptome. Finally, chaotic genomes represent important types of non-clonal chromosome aberrations or NCCAs [46–49]. With the new realizations that the karyotype encodes the genomic blueprint, and that cells with NCCAs contribute to cellular emergent properties, the interest in profiling NCCAs will play an important role in understanding and monitoring various human diseases.

2 Materials

Apart from standard cell culture and cytogenetic equipment, including culture hood, CO_2 incubator, clinical centrifuge, water bath, slide warmer, phase microscope, fluorescent microscope, and imaging system, standard supplies (e.g., pipette, culture flask, microscope slide, cover slip, etc.), the following specialized reagents and equipment are required.

2.1 Cell Culture and Drug Treatment

1. Cell lines.
2. DMEM Media (Gibco)—Dulbecco Modified Eagle Medium.
3. FBS (fetal bovine serum).
4. Antibiotics (penicillin/streptomycin/fungizone) (Hyclone).
5. Trypsin solution (1×) (Sigma-Aldrich).
6. PBS-azide solution pH 7.4: 1 mM ethylene glycol tetraacetic acid (EGTA) (38 mg); 0.01% sodium azide (NaN_3) (1 mL); 150 mM NaCl (877 mg); 10 mM sodium phosphate ($NaPO_4$) (10 mL of 100 mM, pH 7.4); distilled water to 100 mL.
7. Chemicals/drugs: Doxorubicin (Fisher) or Mitomycin C (Fisher).
8. Culture flasks (75 cm).

2.2 Chromosome Preparation

1. Fixative *solution* (ratio of methanol to acetic acid 3:1).
2. Hypotonic solution (0.4% KCl).
3. PBS-azide solution.
4. 15 mL centrifuge tubes.

2.3 SKY: Spectral Karyotyping

All SKY reagents (SKY probes and detection kit) can be purchased from Applied Spectral Imaging.

1. DAPI (Sigma): 0.2 mg mL^{-1} of stock solution in H_2O. Store in the dark at 4 °C.

2. DAPI/Antifade: 0.2 μg mL^{-1}.
3. Prepare BSA solution using prewarmed (37 °C) PBS. Vortex well and leave at RT until dissolved.
4. 20× SSC (saline sodium citrate) stock solution: 3 M NaCl, 300 mM Na citrate.
5. Cy5 staining reagent: 5 μL of anti-Digoxin (Sigma, D8156); 5 μL of Cy5 Strep Avidin (1 mg mL^{-1}, Amersham); 1 mL of 4 × SSC.
6. Cy5.5 staining reagent: 5 μL of Cy5.5 sheep anti-mouse (1 mg mL^{-1}) (Rockland); 1 mL of 4×SSC. For long-term storage of the diluted Cy5 and Cy5.5 needed, add 1% of BSA fraction V (Roche) to the 4×SSC solution.
7. SKYPaints™ (painting probes in hybridization buffer can be purchased from Applied Spectral Imaging).
8. 20×SSC (saline sodium citrate) stock solution: 3 M NaCl, 300 mM Na citrate. Prepare 1×SSC, 2×SSC, 4×SSC in distilled water.
9. Denaturation solution: 70% deionized formamide (Sigma) in 2×SSC.
10. Washing solution I: 50% formamide in 2×SSC.
11. Washing solution II: 1×SSC.
12. Washing solution III: 0.1% TWEEN® 20 in 4×SSC.

3 Methods

Carry out all procedures at room temperature unless otherwise specified.

3.1 Cell Culture

1. Human or mouse cells (*see* **Note 1**) are cultured at 37 °C in a humidified incubator with 5% CO_2 in DMEM media with 10% FBS and penicillin/streptomycin/fungizone.
2. When cells reach 50–70% confluence, they are ready for the induction of genome chaos (*see* **Note 2**).

3.2 Inducing Genome Chaos and Collecting Cells

1. Treat cultured cells with doxorubicin at 0.5–2.0 μg/mL or MMC at 5–20 μg/mL for 2 h (*see* **Note 3**).
2. Briefly wash off the drug using PBS solution (gently rinsing once) (*see* **Note 4**).
3. Re-culture the treated cells with fresh medium at 37 °C (*see* **Note 5**).
4. Harvest cells at different time points to collect cells that display chromosome fragmentation (12–24 h after re-culture), chaotic genome (24–72 h after re-culture), or stable genomes after selection (3–4 weeks after re-culture) (*see* **Note 6**).

5. Shake off (to enrich mitotic cells) or use trypsin solution to remove cells from the flask. The collected sample can be used for further analyses, including isolating DNA, RNA, and protein (*see* **Note 7**). In our protocol, we describe the cytogenetic characterization (*see* Subheading 3.4).

3.3 Time Course to Profile the Dynamic Pattern of Genome Chaos (Optional)

When performing the time course experiments, drug-treated cells will be continuously harvested throughout a 2–8-week period following re-culture. Multiple flasks of cultured cells are therefore needed for the treatment and sample collection (*see* **Note 8**).

1. Culture cells in 5–10 flasks (75 cm).
2. When cells reach 50–70% confluence, add the drug and treat cells for 2 h.
3. Briefly wash off the drug using PBS solution (gently rinse once).
4. Re-culture the treated cells with fresh medium at 37 °C.
5. Harvest each flask at different time points (to cover the entire process, from 12 h following the re-culture, up to a few weeks) (*see* **Note 9**).

3.4 Cytogenetic Preparation

Routine cytogenetic preparation protocol will be followed [50–53].

1. Collect cells using brief trypsin treatment (or simply shake off mitotic cells for those loosely attached cells), and transfer the cell suspension into a 15 mL tube.
2. Spin down the cells (8 min at 200 g), and discard the supernatant using a transfer pipette (0.3 mL of supernatant is left to avoid loss of cellular material).
3. For hypotonic treatment, resuspend cell pellet in 5 mL 0.4% KCl and incubate at 37 °C for 10 min.
4. Slowly add 0.5 mL of fixative solution and mix well for prefixation.
5. Immediately repeat **step 2**.
6. Resuspend the pellet in 5 mL of fixative for 20 min.
7. Repeat **step 5**.
8. Resuspend the pellet in 5 mL of fixative and repeat **step 5**.
9. Repeat **step 8** twice.
10. The pellet is finally resuspended in 0.3–1 mL of fixative (depending on the density of the cells).
11. Drop 1–2 drops of the suspension onto a clean and humid slide. Quickly air-dry and bake the slides using slide warmer (65 °C).
12. Slides are subjected to Giemsa staining or spectral karyotyping (*see* **Note 10**).

3.5 Spectral Karyotyping

Spectral karyotyping can be performed according to published methods [53–55] (*see* **Note 11**).

3.5.1 Chromosome Denaturation

1. Heat 40 mL of denaturation solution to 72 °C (±2 °C) in a glass coplin jar. Place slides in the solution for 60–90 s.
2. Immediately place slides in cold 70%, 80%, 100% ethanol, 2 min each, and air-dry.

3.5.2 Probe Denaturation and Hybridization

1. Briefly centrifuge the content of the Spectral Karyotyping Reagent (vial #1 supplied by ASI).
2. Mix the contents of the vial well, including the red precipitation, by pipetting up and down several times. Transfer 25 μL to an Eppendorf tube, and denature the probe by incubating at 80 °C in a water bath for 7 min.
3. Transfer the tube into a 37 °C water bath for 30–60 min.
4. Apply 20 μL of denatured probes to the denatured chromosome slide, and place a 24 × 24 mm coverslip over the probe mix. Seal the edges with rubber cement.
5. Transfer the slides to a humidified chamber at 37 °C for 24–36 h.

3.5.3 Detection

1. Remove the slides from the hybridization chamber and carefully remove the rubber cement.
2. Put slides into a coplin jar of washing solution I.
3. Wash the slides twice in washing solution II (1×SSC) at 45 °C for 5 min each.
4. Dip the slides in washing solution III (4×SSC/0.1% TWEEN® 20) at 45 °C for 2 min each.
5. Tilt the slides and allow the fluid to drain. Add 80 μL of Cy5 staining reagent. Place a plastic cover slip (24 × 60 mm) over the top, and incubate at 37 °C for 40 min.
6. Wash slides three times in washing solution III at 45 °C for 2 min each.
7. Apply 80 μL of Cy5.5 staining reagent and cover with a plastic cover slip. Incubate at 37 °C for 40 min.
8. Repeat **step 6**.
9. Tilt slides and allow fluid to drain. Apply 20 μL of Antifade DAPI reagent; place a cover glass (24 × 60 mm) carefully over the top and remove air bubbles.

3.5.4 Image Analysis

Mitotic figures with good quality of hybridization signals will be captured using a CCD camera. Following image acquisition, chromosomes will be analyzed with Applied Spectral Imaging software. A total of 50–100 mitotic figures are needed.

3.6 Score Types and Frequencies of NCCAs

Drug induction can generate chaotic genomes represented by a large number of NCCAs and CCAs, including translocated chromosomes, chromosome fragmentations (C-Frags), defective mitotic figures (DMFs), sticky chromosomes, free chromatin, aneuploidy/polyploidy, and clusters of micro-nuclei, among many others [1, 2, 47].

While it would be ideal to quantitatively compare the contribution of each type to the overall genome-level instability, at present, we suggest that the total number of these major subtypes can be used to measure genome instability, which can be linked to disease status (*see* **Note 12**).

The method of scoring NCCAs in general has been proposed (*see* **Note 13**). Investigators can modify or establish their own scoring system accordingly to tailor to their purpose of research.

3.7 Study the Mechanisms of Genome Chaos

The mechanistic studies include searching for genomic and environmental factors that trigger genome chaos, the process of rapid and massive genome reorganization, the types of chaotic genomes and their related mechanisms and clinical consequences, the patterns of the evolutionary selection process to pick winning genomes among massive cell death, and how cells pass along genetic information during such a dynamic process. Specifically, there is an involvement of mitotic cell death (indicated by massive chromosome fragmentations) followed by nonhomologous end joining (NHEJ)-mediated DNA repair, as well as somatic cell evolution. Our previous studies have suggested that genome chaos exemplifies a survival strategy when cells are under high levels of stress [39], as this goal can only be achieved by creating a new system coupled with changed genomic coding (e.g., it is not sufficient enough simply through gene mutations).

The experimental system introduced here can be utilized to study many aspects of genome reorganization and to validate many assumptions, such as the idea that "there is a single event that is responsible for forming a specific chaotic genome," and the relative low rate of genome chaos detected in some tumor types. It turns out that chaotic genomes can constantly evolve, and a specific configuration of genome reorganization detected at the late clinical stage often involves multiple events; most chaotic genomes will not be detectable as end products during drastic cellular evolution. The reason more chromothripsis can be detected by sequencing is because the limited local rearrangement might have a better chance for the selection process to lead to a clonal population. In contrast, many highly altered genomes are eliminated during the process and for some survivors are not detectable by sequencing due to the status of NCCAs. If single-cell sequencing is performed, especially when the entire process of punctuated cancer evolution is monitored, the chaotic genomes will be detected in much higher frequencies.

Clearly, in addition to continuously studying the mechanism of genome chaos in cancer evolution (*see* **Note 14**), future research needs to search for the genomic basis of genome reorganization under stress, as well as its clinical implications (*see* **Note 15**).

4 Notes

1. Cell lines with different degrees of CIN can be used to induce genome chaos. It is relatively difficult to induce high frequencies of genome chaos using some normal cell lines, possibly due to the normal checkpoints and stable genomes. According to our experience, it is easiest to induce genome chaos in cell populations with high degrees of CIN.
2. Keeping cells in optimal growth conditions is an important factor when inducing chaotic genomes.
3. The concentration of drugs can be tested. According to our experience, a large number of drugs can effectively induce genome chaos, including drugs that interfere with microtubule formation. Interestingly, the morphology of chaotic genomes, and the time course of genome chaos, may vary depending on the drugs and cell line used. For example, there are morphological differences between Dox- and MMC-induced chaotic genomes. Since chromosome fragmentations can be induced by many different types of high level stresses [1, 30], different experimental conditions can also be tested without the use of chemo-drug treatment.
4. One needs to be very careful when handling treated cells, which are easily shaken off. If there are too many detached cells, one can collect all the cells, wash them in a 15 mL centrifuge tube, and then collect them following a 7 min spin using a clinical centrifuge (1000×). Be careful to avoid bacterial contamination, especially for experiments involving longer-term culture.
5. A key for this step is to maintain the optimal culture conditions by washing off dead cells during re-culture. Every 12 h, gently changing the culture medium is recommended. Massive amounts of dead cells will interfere with population recovery.
6. At least three flasks of cells are needed to cover the three stages. If investigators only want to collect the chaotic genome, one should test the time window within 24–72 h, as different treatments may lead to a shift in the peak time. Thus, we recommend performing a time course experiment to determine the peak time of a given cell line and its specific treatment.
7. For cytogenetic analyses, each treated sample needs to immediately go through a chromosomal preparation process. When isolating DNA and RNA, the collected cells can be

immediately frozen down and stored in −70 °C freezers (with special buffers if needed). All samples can be later used for DNA/RNA isolation. Pay more attention to the specific requirement of the storage conditions, so as to avoid DNA/RNA degradation.

8. To cover the entire process of genome chaos, a time course experiment is necessary (even though it is very time consuming). The process of genome chaos can be dissected into three main steps: induction of cell death (indicated by massive chromosome fragmentation), genome reorganization for survival (emergence of massively translocated chromosomes or a high number of total chromosomes), and the evolutionary selection for stable genome (the selection process among changing genomes, which may last from 3 weeks to up to a few years under culture conditions). For the purpose of harvesting chromosome fragmentations (C-Frag), the time window of 12–48 h is recommended. To collect chaotic genomes, 24–72 h are needed. To monitor the whole genome stabilization process, one to a few weeks will be necessary. When the stable clonal population is achieved, indicated by the shared clonal karyotypes and low levels of NCCAs, the cellular population is stabilized.
9. To monitor the process in more detail, additional cultured cells need to be used. Ideally, each collecting point should use two flasks. According to our experience, within the first few days, more flasks are needed. After 10 days, fewer flasks are required. Of course, the number of flasks needed entirely depends on the drug and cell line used, as well as the experimental design.
10. Checking Giemsa-stained slides: the appearance of chromosomes with strange morphologies, including chromosomes that are much longer than normal ones, is an indication that chaotic genomes have been produced. Often, chaotic genomes can be identified using a phase microscope, even without Giemsa staining.
11. SKY and multicolor FISH are the preferred methods when characterizing chaotic genomes. Single-cell monitoring is key when profiling these structures, most of which are non-clonal and can only be characterized by single-cell sequencing and single-cell transcriptome analysis. Therefore, cytogenetics has its unique advantage to monitor both the individual cell and the population, the architecture of subclones, and the evolution of the cellular population, since cytogenetic methods can monitor a large number of cells in one slide. For the same reason, traditional Giemsa staining is also recommended to score large numbers of chaotic genomes. Furthermore, profiling large numbers of individual chaotic genomes by single-cell sequencing could be very costly when the cell number becomes large. Following cytogenetic analyses with SKY analyses, addi-

tional sequencing profiling can provide further molecular characterization of the chaotic genomes, especially when profiling these clonal chaotic genomes.

12. While CIN can be measured based on different levels of genomic variants, a better way to score CIN is mainly based on karyotype variations (both structural and numerical variations). As karyotypes encode/define genomic systems, karyotypic changes often display more impact compared to changing individual genes [1]. Measuring CIN by scoring the copy number variations (CNV) might be less optimal, even though cells with unstable karyotypes can promote CNV. On the other hand, CNV could indirectly promote genome-level changes to a certain degree. Future research should be done to directly compare the contributions from CNV and chaotic genomes.
13. According to the current cytogenetic standard, 20 mitotic figures are needed to score clonal chromosomal aberrations. We have suggested that 50–100 randomly selected mitotic figures be used to score NCCAs in conventional cytogenetic preparations to monitor the overall genome instability present. Following the induction of chaotic genomes, however, there are large numbers of altered mitotic figures and nuclear structures, representing many different types of NCCAs [47], so scoring 100–200 mitotic figures will be needed. We often use a combination: we first score 50 mitotic figures using SKY analysis (to study the detailed translocations including the documentation of how many events occur within one chromosome, and total events for a given cell, as well as the distribution of outliers) and then score 200–500 normal and abnormal mitotic structures on Giemsa-stained slides to classify the portion of different subtypes of chaotic genomes. We score each sample in a double-blind fashion with the help of at least two different individuals.
14. In addition to studying the specific mechanisms of each chaotic genome subtype (e.g., the cluster of micro-nuclei can generate cells with stem cell markers and DMFs can generate aneuploidy [47]), the common evolutionary mechanism of genome reorganization needs to be further illustrated. We have illustrated that during the establishment of immortalization and drug resistance, no fixed pathways increase, but the overall transcriptome dynamics drastically does. This suggests that it is the evolutionary potential that matters the most. Further comparative studies are needed to focus on other major transitions of cancer evolution, including cellular transformation, metastasis, and immunoresponse. We anticipate that the new system formation (by changing karyotype coding) will be the common

mechanism of genome chaos which can unify the diverse individual molecular mechanisms. This explains why evolution can succeed despite massive cell death coupled with genome chaos. While unstable, these chaotic genomes can rapidly evolve until stable clones emerge and thrive.

15. In conclusion, the discovery and further characterization of the chaotic genome is not only useful in understanding cancer evolution beyond gene mutation but can also illustrate a general mechanism for genome evolution. The universal phenomenon detected from earlier embryo development to cancer formation highlights the importance of genome reorganization; the constant dynamics (of cell division and refusion, of chromosomal reorganization using similar DNA materials) represent the mechanism of searching for the minimal workable genome during somatic evolution. When considering the function of genome constraint, such as the main function of sexual reproduction, somatic genome chaos becomes an important window to make sense of organismal macroevolution [1, 56, 57]. Further studies are needed to understand fuzzy inheritance, genome chaos-mediated drug resistance, and how to apply genome constraints to slow down cancer without direct killing strategies. Equally importantly, new methods need to be developed to score chaotic genomes, as current SKY requires mitotic figures, which are hard to obtain in some clinical cases. Single-cell genomic DNA sequencing has the needed resolution, but it is challenging to validate the data without a good reference genome.

    Another important aspect of studying genome chaos is its implication for understanding normal physiological adaptations and disease conditions beyond cancer [1, 33, 34, 56, 58, 59]. Interesting reports have linked rapid genome reorganization to earlier development events, specific diseases, and possibly even the correction of preexisting disease conditions [60–64]. More exciting discoveries will certainly be reported in the future.

Acknowledgments

This manuscript is part of our series of publications on the subject of "the mechanisms of cancer and organismal evolution." This work was also partially supported by the start-up fund for Christine J. Ye from the Department of Internal Medicine, Division of Hematology/Oncology, University of Michigan. Thanks to Julie Heng for editing.

References

1. Heng HH (2015) Debating cancer: the paradox in cancer research. World Scientific, Singapore
2. Heng HH, Stevens JB, Liu G et al (2004) Imaging genome abnormalities in cancer research. Cell Chromosome 3:1
3. Heng HH, Stevens JB, Liu G et al (2006) Stochastic cancer progression driven by non-clonal chromosome aberrations. J Cell Physiol 208:461–472
4. Heng HH, Stevens JB, Bremer SW et al (2011) Evolutionary mechanisms and diversity in cancer. Adv Cancer Res 112:217–253
5. Heng HH, Liu G, Stevens JB et al (2011) Decoding the genome beyond sequencing: the new phase of genomic research. Genomics 98(4):242–252
6. Heng HH, Bremer SW, Stevens JB et al (2013) Chromosomal instability (CIN): what it is and why it is crucial to cancer evolution. Cancer Metastasis Rev 32(3–4):325–340
7. Heng HH, Liu G, Stevens JB et al (2013) Karyotype heterogeneity and unclassified chromosomal abnormalities. Cytogenet Genome Res 139(3):144–157
8. Stephens PJ, Greenman CD, Fu B et al (2011) Massive genomic rearrangement acquired in a single catastrophic event during cancer development. Cell 144(1):27–40
9. Horne SD, Heng HH (2014) Genome chaos, chromothripsis and cancer evolution. J Cancer Stud Ther 1:1–6
10. Heng HH (2017) Cancer genomic landscape. In: Ujvari B, Roche B, Thomas F (eds) Ecology and evolution of cancer. Academic Press, Elsevier, pp 69–86
11. Baca SC, Prandi D, Lawrence MS et al (2013) Punctuated evolution of prostate cancer genomes. Cell 153(3):666–677
12. Crasta K, Ganem NJ, Dagher R et al (2012) DNA breaks and chromosome pulverization from errors in mitosis. Nature 482(7383):53–58
13. Forment JV, Kaidi A, Jackson SP (2012) Chromothripsis and cancer: causes and consequences of chromosome shattering. Nat Rev Cancer 12(10):663–670
14. Holland AJ, Cleveland DW (2012) Chromoanagenesis and cancer: mechanisms and consequences of localized, complex chromosomal rearrangements. Nat Med 18(11):1630–1638
15. Inaki K, Liu ET (2012) Structural mutations in cancer: mechanistic and functional insights. Trends Genet 28(11):550–559
16. Jones MJ, Jallepalli PV (2012) Chromothripsis: chromosomes in crisis. Dev Cell 23(5):908–917
17. Liu P, Erez A, Nagamani SC et al (2011) Chromosome catastrophes involve replication mechanisms generating complex genomic rearrangements. Cell 146(6):889–903
18. Malhotra A, Lindberg M, Faust GG et al (2013) Breakpoint profiling of 64 cancer genomes reveals numerous complex rearrangements spawned by homology-independent mechanisms. Genome Res 23:762–776
19. Righolt C, Mai S (2012) Shattered and stitched chromosomes-chromothripsis and chromoanasynthesis-manifestations of a new chromosome crisis? Genes Chromosomes Cancer 51(11):975–981
20. Setlur SR, Lee C (2012) Tumor archaeology reveals that mutations love company. Cell 149(9):959–961
21. Tubio JM, Estivill X (2011) Cancer: when catastrophe strikes a cell. Nature 470(7335):476–477
22. Heng HH (2007) Karyotypic chaos, a form of non-clonal chromosome aberrations, plays a key role for cancer progression and drug resistance. FASEB: Nuclear Structure and Cancer. Vermont Academy, Saxtons River, VT
23. Duesberg P (2007) Chromosomal chaos and cancer. Sci Am 296(5):52–59
24. Zhang CZ, Spektor A, Cornils H et al (2015) Chromothripsis from DNA damage in micronuclei. Nature 522(7555):179–184
25. Maciejowski J, Li Y, Bosco N, Campbell PJ, de Lange T (2015) Chromothripsis and Kataegis induced by telomere crisis. Cell 163(7):1641–1654
26. Morishita M, Muramatsu T, Suto Y et al (2016) Chromothripsis-like chromosomal rearrangements induced by ionizing radiation using proton microbeam irradiation system. Oncotarget 7(9):10182–10192
27. Mardin BR, Drainas AP, Waszak SM et al (2015) A cell-based model system links chromothripsis with hyperploidy. Mol Syst Biol 11(9):828
28. Stevens JB, Liu G, Bremer SW et al (2007) Mitotic cell death by chromosome fragmentation. Cancer Res 67(16):7686–7694
29. Stevens JB, Abdallah BY, Liu G et al (2011) Diverse system stresses: common mechanisms of chromosome fragmentation. Cell Death Dis 2:e178
30. Stevens JB, Abdallah BY, Liu G et al (2013) Heterogeneity of cell death. Cytogenet Genome Res 139(3):164–173
31. Ye CJ, Stevens JB, Liu G et al (2009) Genome based cell population heterogeneity promotes

tumorigenicity: the evolutionary mechanism of cancer. J Cell Physiol 219(2):288–300
32. Horne SD, Chowdhury SK, Heng HH (2014) Stress, genomic adaptation, and the evolutionary trade-off. Front Genet 5:92
33. Heng HH, Regan S, Ye CJ (2016) Genotype, environment, and evolutionary mechanism of diseases. Environ Dis 1:14–23
34. Heng HH (2007) Cancer genome sequencing: the challenges ahead. BioEssays 29(8):783–794
35. Heng HH (2009) The genome-centric concept: resynthesis of evolutionary theory. BioEssays 31(5):512–525
36. Horne SD, Ye CJ, Heng HH (2015) Chromosomal instability (CIN) in cancer. eLS:1–9
37. Horne SD, Pollick SA, Heng HH (2015) Evolutionary mechanism unifies the hallmarks of cancer. Int J Cancer 136(9):2012–2021
38. Liu G, Stevens JB, Horne SD et al (2014) Genome chaos: survival strategy during crisis. Cell Cycle 13(4):528–537
39. Manolio TA, Collins FS, Cox NJ et al (2009) Finding the missing heritability of complex diseases. Nature 461(7265):747–753
40. Heng HH (2011) Missing heritability and stochastic genome alterations. Nat Rev Genet 11(11):813
41. Stevens JB, Liu G, Abdallah BY et al (2014) Unstable genomes elevate transcriptome dynamics. Int J Cancer 134(9):2074–2087
42. Stepanenko AA, Kavsan VM (2012) Evolutionary karyotypic theory of cancer *versus* conventional cancer gene mutation theory. Biopolym Cell 28(4):267–280
43. Duesberg P, McCormack A (2013) Immortality of cancers: a consequence of inherent karyotypic variations and selections for autonomy. Cell Cycle 12(5):783–802
44. Poot M, Haaf T (2015) Mechanisms of origin, phenotypic effects and diagnostic implications of complex chromosome rearrangements. Mol Syndromol 6:109–133
45. Li P, Cui C (2016) A broader view of cancer cytogenetics: from nuclear aberrations to cytogenetic abnormalities. J Mol Genet Med 10:e108
46. Poot M (2017) Of simple and complex genome rearrangements, chromothripsis, chromoanasynthesis, and chromosome chaos. Mol Syndromol 8:115–117
47. Heng HH, Regan SM, Liu G et al (2016) Why is it crucial to analyze non clonal chromosomal aberrations or NCCAs? Mol Cytogenet 9:15
48. Rangel N, Forero-Castro M, Rondón-Lagos M (2017) New insights in the cytogenetic practice: karyotypic chaos, non-clonal chromosomal alterations and chromosomal instability in human cancer and therapy response. Genes (Basel) 8(6):E155
49. Heng HH, Horne SD, Chaudhry S et al (2018) A postgenomic perspective on molecular cytogenetics. Current Genomics (in press)
50. Heng HH, Squire J, Tsui LC (1992) High-resolution mapping of mammalian genes by in situ hybridization to free chromatin. Proc Natl Acad Sci U S A 89(20):9509–9513
51. Heng HH, Tsui LC, Moens PB (1994) Organization of heterologous DNA inserts on the mouse meiotic chromosome core. Chromosoma 103(6):401–407
52. Ye CJ, Heng HH (2017) High resolution fiber-fluorescence in situ hybridization. Methods Mol Biol 1541:151–166
53. Ye CJ, Liu G, Heng HH (2016) Simultaneous fluorescence immunostaining and FISH. In: Liehr T (ed) Fluorescence in situ hybridization (FISH). Springer, New York, pp 301–325
54. Heng HH, Ye CJ, Yang F et al (2003) Analysis of marker or complex chromosomal rearrangements present in pre- and post-natal karyotypes utilizing a combination of G-banding, spectral karyotyping and fluorescence in situ hybridization. Clin Genet 63(5):358–367
55. Ye CJ, Lu W, Liu G et al (2001) The combination of SKY and specific loci detection with FISH or immunostaining. Cytogenet Cell Genet 93(3–4):195–202
56. Heng HH (2018) Genome chaos: rethinking genomics, evolution and molecular medicine. Academic Press, Elsevier, (in press)
57. Gorelick R, Heng HH (2011) Sex reduces genetic variation: a multidisciplinary review. Evolution 65(4):1088–1098
58. Heng HH, Horne SD, Stevens SB et al (2016) Heterogeneity mediated system complexity: the ultimate challenge for studying common and complex diseases. In: Sturmberg JP (ed) The value of systems and complexity sciences for healthcare. Springer, New York, pp 107–120
59. Heng HH (2017) Heterogeneity-mediated cellular adaptation and its trade-off: searching for the general principles of diseases. J Eval Clin Pract 23(1):233–237
60. Pellestor F, Gatinois V, Puechberty J et al (2014) Chromothripsis: potential origin in gametogenesis and preimplantation cell divisions. A review. Fertil Steril 102(6):1785–1796

61. Furgason JM, Koncar RF, Michelhaugh SK et al (2015) Whole genome sequence analysis links chromothripsis to EGFR, MDM2, MDM4, and CDK4 amplification in glioblastoma. Oncoscience 2(7):618–628. eCollection

62. de Pagter MS, van Roosmalen MJ, Baas AF et al (2015) Chromothripsis in healthy individuals affects multiple protein-coding genes and can result in severe congenital abnormalities in offspring. Am J Hum Genet 96(4):651–656

63. Pellestor F (2014) Chromothripsis: how does such a catastrophic event impact human reproduction? Hum Reprod 29(3):388–393

64. McDermott DH, Gao JL, Liu Q et al (2015) Chromothriptic cure of WHIM syndrome. Cell 160(4):686–699

Check for updates

Chapter 22

Looking for Broken TAD Boundaries and Changes on DNA Interactions: Clinical Guide to 3D Chromatin Change Analysis in Complex Chromosomal Rearrangements and Chromothripsis

Kevin Yauy, Vincent Gatinois, Thomas Guignard, Satish Sati, Jacques Puechberty, Jean Baptiste Gaillard, Anouck Schneider, and Franck Pellestor

Abstract

Apparition of next-generation sequencing (NGS) was a breakthrough on knowledge of genome structure. Bioinformatic tools are a key point to analyze this huge amount of data from NGS and characterize the three-dimensional organization of chromosomes. This chapter describes usage of different browsers to explore publicly available online data and to search for possible 3D chromatin changes involved during complex chromosomal rearrangements as chromothripsis. Their pathogenic impact on clinical phenotype and gene misexpression can also be evaluated with annotated databases.

Key words Topologically associated domains, TAD boundaries, Enhancers, Bioinformatics, 3D genome, Genome browser, Chromothripsis

1 Introduction

The emergence of next-generation sequencing and publication of genome-wide chromatin interaction maps have created a breakthrough on knowledge of genome structure and gene regulation. Noncoding elements conserved in evolution appear to act as regulators. These short sequence regulators are called enhancers. They can bind to transcription factors and are able to regulate expression depending on their spatial pattern [1, 2]. Based on cap analysis of gene expression (CAGE) experiments, a collaboration within the FANTOM5 project produced a freely accessible huge database on specific enhancers and promoters expressed on hundreds of different tissues (human promoter expression atlas and human transcripted enhancer atlas) [3, 4].

Franck Pellestor (ed.), *Chromothripsis: Methods and Protocols*, Methods in Molecular Biology, vol. 1769, https://doi.org/10.1007/978-1-4939-7780-2_22, © Springer Science+Business Media, LLC, part of Springer Nature 2018

Chromatin conformation capture (3C) techniques within the ENCODE project gives to the researchers free access to big data of 3D genome interaction (ChIP-Seq for DNA-protein interaction, CLIP-seq for RNA-protein interaction, 3/4/5/Hi-C Seq for DNA-DNA interaction) [5, 6].

These techniques have made it possible to identify chromatin subdomains, called topologically associating domains (TADs), which share 3D space and separated by specific boundaries. These TADs, which are generally conserved through development and among different cell types, play an essential role in establishing enhancer-promoter interactions by providing a structural scaffold that restricts the expression regulation of genes inside them.

Changes in enhancer-promoter interactions and breaking TAD boundaries have been shown to be pathogenic by causing ectopic promoter interactions and altered gene expression. Thus, TADs emerge as essential structural units whose disruption can lead to pathogenic phenotypes. "TADopathies" are undoubtedly an upcoming new category of human Mendelian disease [7–10].

Knowledge of potential alterations in the enhancer-promoter interactions, as well as the architecture of TADs and the duplication or disruption of their boundaries, are essential data, because they provide new insights on the impact of the 3D organization of chromatin. Depending on their size and position, structural variants may disrupt the chromatin architecture defined by TADs, and thereby they may modify the clinical interpretation of variants found in routine diagnosis by CGH arrays or whole genome sequencing, especially in complex chromosomal rearrangements such as chromothripsis.

These data emphasize the usefulness of considering the 3D architecture of the human genome to analyze but also to predict the effects of structural variations. TAD data for the entire human genome is beginning to be available at increasing resolution. Thus, analysis of chromatin status with ChromHMM [11] provides a concrete idea of the activity of enhancers and promoters in a newly created domain.

In this chapter, we present a pipeline methods based on public online tools to explore these disruptions of 3D chromatin architecture and to predict the pathogenicity of structural variants as well as their possible clinical meanings.

2 Materials

2.1 Promoter and Enhancers

The global FANTOM5 project provides useful tools to access data of human promoterome and enhancerome. A catalogue interface and predefined BED file data are available and accessible through UCSC Genome Browser [12] (http://slidebase.binf.ku.dk/).

2.2 TAD, Hi-C, and Epigenome

We can explore TADs of plenty of different cell types with specific Hi-C browsers. A lot of Hi-C browsers are available online such as the 3D Genome Browser from Penn State [13–15] or in desktop applications such as Juicebox [16]. They give a quick and free access to published 3C experiments. A database integrating 3D genome- and disease-associated chromosome rearrangement data is available [17]. Epigenome browsers such as Washington University Epigenome Browser give access to Hi-C interaction data from Juicebox but also to chromatin states from ChromHMM [11] within the Roadmap Epigenomics Mapping Consortium.

2.3 Clinical Phenotype Database

Explore the genotype-phenotype association is accessible through curated database as OMIM [18] and ClinGen Dosage Sensitivity Map (https://www.ncbi.nlm.nih.gov/projects/dbvar/clingen/) or in physician-feeded database of chromosomal imbalance as DECIPHER [19].

These database and browser websites were tested with Mozilla Firefox (version 52.0.2) and Google Chrome (version 57.0.2987).

3 Methods

3.1 Analysis of Deletion Copy Number Variant (CNV) with Penn State Hi-C Browser

3.1.1 Configure UCSC Genome Browser Session with Chromatin State, Enhancers, and Deletions

1. Add the enhancer track to UCSC Genome Browser:

   Go to http://slidebase.binf.ku.dk/human_enhancers/presets and then choose enhancer track of interest. We advise to use the enhancer transcription start site (tss)-associated bed file (*see* **Note 1**). Then click on UCSC link or download bed and add it to your UCSC session view as a custom track.

2. Manage or load your session in UCSC Genome Browser (https://genome.ucsc.edu/cgi-bin/hgGateway?redirect=manual&source=genome.ucsc.edu) as your configuration preference (*see* **Note 2**).

3. Add the chromatin state track from ENCODE/Broad:

   In UCSC Genome Browser, click on Genome Browser on menu, and then Track Search. Put Broad ChromHMM on search bar, and then check only the Broad ChromHMM track (*see* **Note 3**).

4. Add in a custom track your selection deletion:

   Go to My Data > Custom Tracks > Add Custom Tracks. Then add your CNV position (#Chrom ChromStart ChromStop) (*see* **Note 4**).

5. Save your session:

   Go to My Data > My Sessions. Save your session. Copy the UCSC Genome Browser URL that will be needed in the following steps.

3.1.2 Hands on the 3D Genome Browser from Penn State

1. Configuration to access in-store database:
   Go to http://promoter.bx.psu.edu/hi-c/view.php. You can browse through a multitude of Hi-C-published experiments on different tissues and resolutions (*see* **Note 5**).
2. Submit a new query:
   Choose a window you want to look at to analyze around your deletion (*see* **Note 6**).
3. Upload your UCSC Session:
   Select "Load new external browser session". Click and choose the UCSC Genome Browser and paste the previous UCSC Genome Browser URL.
4. Final results:
   Click on Visualize Interaction to see the heat map of your region of interest (Fig. 1).

3.2 Analysis of Translocation Variant with Washington University Epigenome Browser

3.2.1 Configure and Hands on WashU Epigenome Browser

1. Configuration to access in-store database:
   Go to http://epigenomegateway.wustl.edu/browser. Choose human hg19. Click on public hubs and load HiC interaction from Juicebox, long-range chromatin interaction experiments, Encyclopedia of DNA Elements, and all chromatin state tracks in Roadmap Epigenomics Integrative Analysis Hub.
2. Configuration of display tracks:
   Go on Tracks, write HiC on search bar, and then choose HiC experiments you want to use to analyze TAD (*see* **Note 7**). Add also GENECODE V19 tracks on Tracks and annotation tracks. Add the chromatin state tracks by typing HMM, and select tracks of your interest among almost every human tissue available.
3. Add your breakpoints to WashU Epigenome Browser:
   Create a bed file at a six-column format containing your breakpoints (chrom chromStart chromEnd Name Score Strand). Go to Custom tracks > Add new tracks > Upload them from your computer. Click on Choose File, chose your bed, and click on Setup. Select bed format and add track.
4. Add FANTOM5 enhancer track to WashU Epigenome Browser:
   Select one track on http://slidebase.binf.ku.dk/human_enhancers/presets, and download it. Go to Custom tracks > Add new tracks > Upload them from your computer. Upload your enhancer track, choose bed format, and add tracks (*see* **Note 8**).
5. Check your first region of interest:
   Click on the blue button with genomic position and add window from −2 Mb to your first breakpoint location.

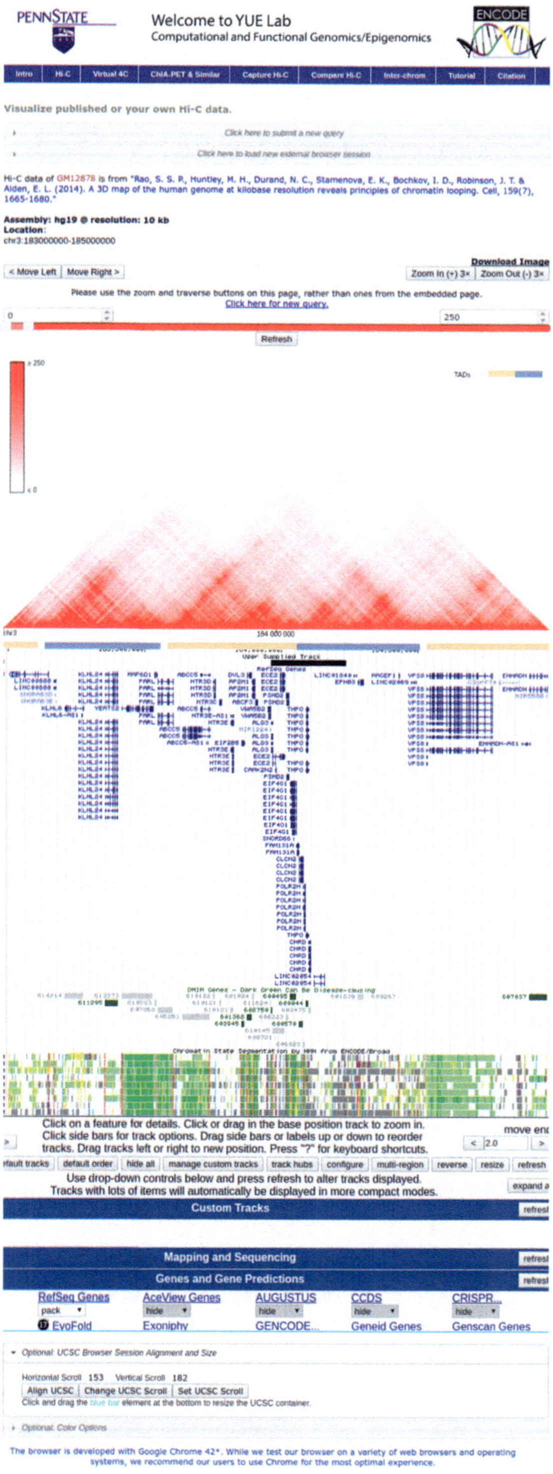

Fig. 1 Overview of Penn State 3D Genome Browser coupled with UCSC Genome Browser with custom session. Analysis of a deletion on chromosome 3 from 183,960,109 to 184,234,764 (represented as black rectangle on first track). See text for details

6. Split screen:

   Click on Apps and split panel (*see* **Note 9**). Choose window from the other breakpoint location to +2 Mb (*see* **Note 10**) (Fig. 2).

3.3 Analysis Around Selected Rearrangement

1. Analysis of conservation of TAD and boundaries:

   You need to perform multiple HiC experiment comparisons to prove conservation between differents cell types before going further on the analysis. You can use the multiple-display HiC maps from WashU Epigenome Browser or make a new query with another cell type on Penn State HiC Browser. Eighty percent of TAD boundaries are conserved across cell type, but 20% of TAD got different boundaries location depending on tissue and species. Broken conserved TAD boundaries are the argument of pathogenicity.

2. Analysis of TAD activity:

   TAD should be size <1 Mb. If you see a big size TAD with only few interactions inside, this area should be silenced or heterochromatin (inactive TAD compartments). The chromatin state prediction from ChromHMM allow us to know if rearrangement changes the enhancer repressor environment that may dysregulate gene expression inside the TAD.

3. Genotype-phenotype associations: OMIM

   OMIM allows you to get curated and updated information for genes and its associated phenotype. We advise to check if haploinsufficiency (nonsense, frameshift, or deletion) mechanisms are known to this gene or triplosensitivity in the Molecular Genetics section of the gene.

   Phenotype analysis between your patient's phenotype and known phenotype on literature may lead to pathogenic hypothesis. These information are available in OMIM gene tracks; dark green genes are related to phenotype in UCSC Genome Browser or at https://www.omim.org.

4. Genotype-phenotype associations: ClinGen and DECIPHER

   ClinGen Dosage Sensitivity Map and DECIPHER are useful tools to get information about known phenotype on the new TAD genomic locations (*see* **Note 11**). ClinGen Dosage Sensitivity Map report curated evaluation of gene based on literature for haploinsufficiency and triplosensitivity pathogenicity. A 0 (benign) to 3 (pathogenic) score is established.

   DECIPHER https://decipher.sanger.ac.uk/ report genotype and phenotype association from patient included by physicians around the world. You get also access to OMIM status, the curated list Development Disorder Genotype-Phenotype Database which included genes associated with development disorder, prediction score as probability of loss of function intolerance (pLI) [20], and haploinsufficiency index [21].

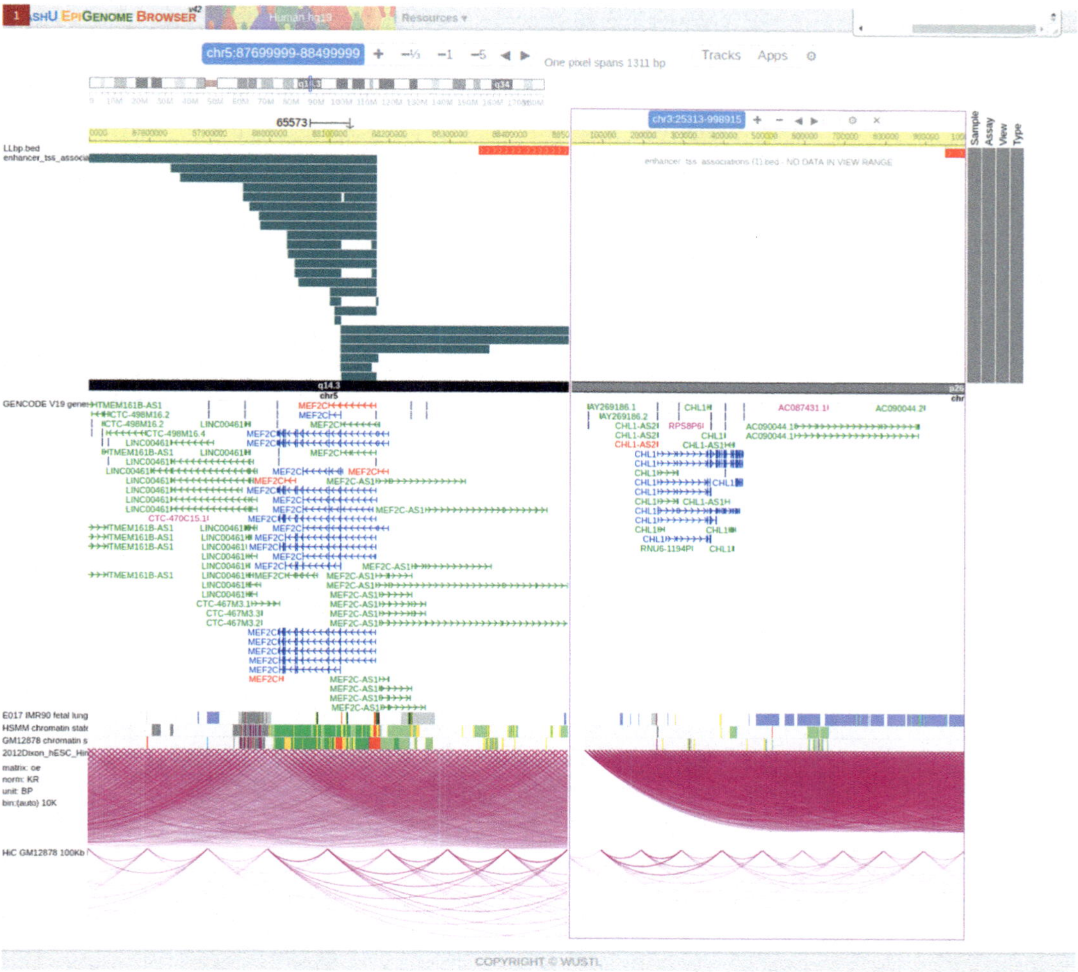

Fig. 2 Overview of split screen of Washington University Epigenome Browser with custom tracks. Analysis of a translocation t(3, 5)(q14.3;3p26.3). See text for details

4 Notes

1. There are numerous enhancer tracks: extensive and specific enhancer tracks depending on cells or on organ/tissues. We use the enhancer tss-associated bed file because it gives you access for each enhancer to the target (gene name and transcript) and the force of the prediction (R).
2. Our configuration advice is enhancer track as pack, RefSeq gene as pack, and OMIM gene as pack to get a quick access to these data. We hide all other tracks.
3. We advise you to choose the pack visibility if you're not comfortable with the color representation of ChromHMM and chromatin state tracks. The dense visibility is more adapted after in routine.

4. If you have numerous CNV to analyze, you should upload a bed track containing all of them in a custom track.
5. We use the GM12878 Hi-C experiments from Rao et al. [22] from the Lieberman lab at resolution 1 kb. It is a high-resolution Hi-C map, and the GM12878 is a classic cell type used in literature.
6. As TAD mean length is 1 Mb, we advise you to select windows −2 Mb/+2 Mb of your genomic breakpoint location. The zoom in and zoom out are too sensitive, so we advise you to change windows by changing the numbers of the start and end position.
7. The advantage of the WashU Epigenome Browser is that it allows to compare multiple HiC experiments. Analysis of multiple heat maps allows to see conservation of TAD across cell type.
8. Add custom track function when bed URL doesn't work in our lab. This is probably because the enhancer bed file is too big. In case it doesn't work for you either, you should download your track and upload it. If it still failed to display track, zip your bed file with bgzip command and index it with tabix; upload these two files on a server and copy the URL to Add Custom Tracks > Add new tracks > bed (as explained http://wiki.wubrowse.org/Simple_bed).
9. WashU Epigenome Browser allows you to split screen, which is very useful in case of translocation or in complex chromosome rearrangement. You can now simulate the newly created TAD and analyze elements inside.
10. If needed, fit windows to see the whole neighboring TADs and their boundaries. WashU Epigenome Browser is very useful because it gives you access to a large set of database easily, but sometimes it freezes, especially if you decide to zoom out too much. We advise not to unfold more than two or three times.
11. You can copy predicted new TAD boundaries' genomic coordinates to do all-in-one check (and not gene by gene) in ClinGen Dosage Sensitivity Map and DECIPHER with the Search by Location bar.

References

1. Lupiáñez DG, Spielmann M, Mundlos S (2016) Breaking TADs: how alterations of chromatin domains result in disease. Trends Genet 32:225–237
2. Bulger M, Groudine M (2011) Functional and mechanistic diversity of distal transcription enhancers. Cell 144:825
3. Andersson R, Gebhard C, Miguel-Escalada I et al (2014) An atlas of active enhancers across human cell types and tissues. Nature 507:455–461
4. FANTOM Consortium, the RIKEN PMI and CLST (DGT), ARR F et al (2014) A promoter-level mammalian expression atlas. Nature 507:462–470

5. Bonev B, Cavalli G (2016) Organization and function of the 3D genome. Nat Rev Genet 17:661–678
6. The ENCODE Project Consortium (2004) The ENCODE (ENCyclopedia Of DNA Elements) Project. Science 306:636–640
7. Matharu N, Ahituv N (2015) Minor loops in major folds: enhancer-promoter looping, chromatin restructuring, and their association with transcriptional regulation and disease. PLoS Genet 11:e1005640
8. Gröschel S, Sanders MA, Hoogenboezem R et al (2014) A single oncogenic enhancer rearrangement causes concomitant EVI1 and GATA2 deregulation in leukemia. Cell 157:369–381
9. Lupiáñez DG, Kraft K, Heinrich V et al (2015) Disruptions of topological chromatin domains cause pathogenic rewiring of gene-enhancer interactions. Cell 161:1012–1025
10. Franke M, Ibrahim DM, Andrey G et al (2016) Formation of new chromatin domains determines pathogenicity of genomic duplications. Nature 538:265–269
11. Ernst J, Kellis M (2012) ChromHMM: automating chromatin-state discovery and characterization. Nat Methods 9:215–216
12. Tyner C, Barber GP, Casper J et al (2017) The UCSC genome browser database: 2017 update. Nucleic Acids Res 45:D626–D634
13. Zhou X, Maricque B, Xie M et al (2011) The human epigenome browser at Washington University. Nat Methods 8:989–990
14. Zhou X, Wang T (2012) Using the Wash U Epigenome Browser to examine genome-wide sequencing data. Curr Protoc Bioinformatics Chapter 10:Unit10.10
15. Wang Y, Zhang B, Zhang L et al (2017) The 3D Genome Browser: a web-based browser for visualizing 3D genome organization and long-range chromatin interactions. https://doi.org/10.1101/112268
16. Durand NC, Robinson JT, Shamim MS et al (2016) Juicebox provides a visualization system for hi-C contact maps with unlimited zoom. Cell Syst 3:99–101
17. Li R, Liu Y, Li T et al (2016) 3Disease Browser: a Web server for integrating 3D genome and disease-associated chromosome rearrangement data. Sci Rep 6:34651. https://doi.org/10.1038/srep34651
18. Amberger JS, Bocchini CA, Schiettecatte F et al (2015) OMIM.org: Online Mendelian Inheritance in Man (OMIM®), an online catalog of human genes and genetic disorders. Nucleic Acids Res 43:D789–D798
19. Firth HV, Richards SM, Bevan AP et al (2009) DECIPHER: database of chromosomal imbalance and phenotype in humans using ensembl resources. Am J Hum Genet 84:524–533
20. Lek M, Karczewski KJ, Minikel EV et al (2016) Analysis of protein-coding genetic variation in 60,706 humans. Nature 536:285–291
21. Huang N, Lee I, Marcotte EM, Hurles ME (2010) Characterising and predicting haploinsufficiency in the human genome. PLoS Genet 6:e1001154
22. Rao SSP, Huntley MH, Durand NC et al (2014) A 3D map of the human genome at kilobase resolution reveals principles of chromatin looping. Cell 159:1665–1680

INDEX

Franck Pellestor (ed.), *Chromothripsis: Methods and Protocols*, Methods in Molecular Biology, vol. 1769,
https://doi.org/10.1007/978-1-4939-7780-2, © Springer Science+Business Media, LLC, part of Springer Nature 2018

D

E

F

G

H

I

J

K

L

M

N

O

P

Q

R

S

T

Printed by Printforce, the Netherlands